EPISTEMIC COMMUNITIES, CONSTRUCTIVISM, AND INTERNATIONAL ENVIRONMENTAL POLITICS

Epistemic Communities, Constructivism, and International Environmental Politics brings together 25 years of publications by Peter M. Haas. The book examines how the world has changed significantly over the last 100 years, discusses the need for new, constructivist scholarship to understand the dynamics of world politics, and highlights the role played by transnational networks of professional experts in global governance. Combining an intellectual history of epistemic communities with theoretical arguments and empirical studies of global environmental conferences, as well as international organizations and comparative studies of international environmental regimes, this book presents a broad picture of social learning on the global scale.

In addition to detailing the changes in the international system since the Industrial Revolution, Haas discusses the technical nature of global environmental threats. Providing a critical reading of discourses about environmental security, this book explores governance efforts to deal with global climate change, international pollution control, stratospheric ozone, and European acid rain. With a new general introduction and the addition of introductory pieces for each section, this collection offers a retrospective overview of the author's work and is essential reading for students and scholars of environmental politics, international relations and global politics.

Peter M. Haas is a Professor of Political Science at the University of Massachusetts Amherst, USA.

Bringing together some of his most important writings during the last quarter of a century, one of the leading scholars of epistemic communities and environmental politics offers a modestly hopeful picture of gradual social learning under conditions of structural change. Scholars interested in some of the best thinking about a vitally important issue in world politics will rely on this collection for many years to come.

Peter J. Katzenstein, *Walter S. Carpenter, Jr. Professor of International Studies, Cornell University, USA*

This volume brings together many of Peter Haas's papers along with a new and lucid introduction tracing the epistemic communities research program that he has pioneered over the last quarter-century.

Robert O. Keohane, *Professor of International Affairs, Princeton University, USA*

Peter Haas has pioneered the study of advocacy groups with claims to consensual knowledge ("epistemic communities") in shaping international cooperation and institution-building. Moreover, he has been among the first to examine international environmental cooperation from a rigorous theoretical perspective. This timely collection of his essays showcases the intellectual development of one of the giants in the field.

Thomas Risse, *Professor of International Politics, Freie Universität Berlin, Germany*

EPISTEMIC COMMUNITIES, CONSTRUCTIVISM, AND INTERNATIONAL ENVIRONMENTAL POLITICS

Peter M. Haas

Routledge
Taylor & Francis Group
LONDON AND NEW YORK

First published 2016
by Routledge
2 Park Square, Milton Park, Abingdon, Oxon OX14 4RN

and by Routledge
711 Third Avenue, New York, NY 10017

Routledge is an imprint of the Taylor & Francis Group, an informa business

British Library Cataloguing in Publication Data
A catalogue record for this book is available from the British Library

Library of Congress Cataloging in Publication Data
A catalog record for this book has been requested

ISBN: 978-1-138-85854-1 (hbk)
ISBN: 978-1-138-85855-8 (pbk)
ISBN: 978-1-315-71790-6 (ebk)

Typeset in Bembo
by Book Now Ltd, London

For Julie and David

CONTENTS

FIGURES

TABLES

PREFACE

This volume presents a collection of previously published works. Most were intended for diverse policy audiences, and published in a wide array of locations. Most share a common empirical focus on global and transboundary environmental risks, and a common analytic focus on the social construction of collective governance responses, focusing in particular on the interplay between organized scientific knowledge and international institutions. The new introduction provides a self-reflective intellectual history of research on epistemic communities.

A common parable in the 1970s was that of the frog and warming water. If the frog was put in hot water it would hop right out. But if it was put in cold water and the water was allowed to gradually boil, the frog would cook and die. Human societies in a new globalized era—what is now called the Anthropocene—are like the frog. If they don't realize what is going on, they are cooked, an analogy which seems less literary as climate change becomes more acute. But the core question is how do societies recognize the problems they face, and how can accurate and timely responses be encouraged? Thus the study of international environmental politics is intimately connected with the study of governance and learning.

Addressing international environmental issues is vitally important for human survival. But we also are learning that environmental issues are tightly coupled to other issues on the international agenda. Effective global governance of the environment is necessary in order to preserve a livable future for the world's population. International environmental issues also provide a critical issue area—a laboratory—for understanding the broader dynamics of other issues on the international agenda as well. Many of the analytic properties of environmental issues (the proliferation of non-state actors, the causal coupling between issues, and the emergence of complexity and uncertainty as systemic conditions) presage shifts in world politics from globalization.

Environmental concerns serve as the leading edge of contemporary world politics in two ways. First, they were the earliest issue to be widely recognized and governed in terms of its complexity and uncertainty. Thus by understanding the politics of international environmental politics we are better positioned to understand and govern other issues as well. Second, because environmental issues are linked to other issues, by starting with the environment decision makers move towards governing causally coupled issues also.

The main argument or big story in this volume is one of gradual accommodation to global structural change via social learning. As societies (governments, their populations, and civil society groups) recognize the changing context of contemporary world politics, as articulated by epistemic communities, governance efforts become more comprehensive in scope. Governance efforts work backwards to address the social causes of environmental degradation. Scientific inputs help inform comprehensive environmental governance and regimes that reflect scientific causal understanding about the behavior of ecosystems. Last, governance efforts work forwards to link the governance of environmental issues with linked social activities. One example of this type of change is Gorbachev's response to the Chernobyl disaster in 1986. Gorbachev realized that Chernobyl wasn't just an instance of the transboundary effects of environmental disasters, but also of the broader failures of centralized planning.

This brief summary of the social mechanisms of large-scale environmental adaptation is an account that relies on hermeneutic changes. Political actors learn about a changed policy environment from experts. Social scientists learn about new political processes by observing policy learning, and through involvement with the principal policy makers and scientists/epistemic community members. Finally, the hermeneutic circle—or more properly the spiral—is followed as new social scientific theories are developed to account for these new modes of governance, and new practices are devised to address material problems.

My epistemology is pragmatic and evolutionary (Haas and Haas 2002). Better understanding of governance occurs through focused work within individual research approaches, which often relies on academic work and joint discussions by academics and policy practitioners. Cumulative knowledge occurs within the academic community through evolutionary epistemology from engaged work between research approaches (not just hypothesis testing, but also reflective considerations about the utility of epistemological and ontological foundations) (Campbell 1993; Adler 2005). Social change may ultimately occur from interplay between academic and policy communities.

The chapters in this volume fall into several of these epistemic categories. Most pieces are intended to build theory about epistemic communities and social learning. Others test alternative hypotheses about patterns of environmental cooperation and governance, primarily by assessing Institutionalist and Constructivist accounts. Others are part of interactive policy dialogues with the practitioner/activist community. Initial drafts of the following chapters were presented to the American Association for the Advancement of Science (Chapter 2); the Nautilus Institute (Chapter 7); the Commission on Global Governance (Chapter 12); the International Maritime

Organization Regional Programme on Building Partnerships in Environmental Management for the Seas of East Asia (Chapter 14), and the Karl Deutsch Memorial Lecture for the WZB (Chapter 18).

The structure of the volume starts with an introduction presenting the intellectual history of epistemic community analysis. Part I provides a historical and ontological overview of the contemporary systemic transformations which justify a Constructivist analysis of epistemic communities in world politics. Part II on regimes and governance patterns elaborates the styles of governance to which epistemic communities contribute, and their dynamics. Part III looks at ways in which international institutions contribute to social learning. Part IV addresses the factors that influence the effectiveness of environmental governance efforts. Part V on science policy looks at design principles for how epistemic communities can better influence international environmental policy making. Finally, Part VI concludes with a look at some of the large-scale social learning dynamics of world politics in which epistemic communities contribute to new organizing principles for world politics and more comprehensive governance arrangements.

These chapters span over 25 years of research and writing. I have a number of co-authors and frequent interlocutors to thank: Emanuel Adler, Steinar Andresen, Nazli Choucri, William C. Clark, Dan Deudney, Sheila Jasanoff, Norichika Kanie, Robert O. Keohane, Marc A. Levy, Ron Mitchell, Craig Murphy, Deborah Poskanzer, John Ruggie, Gus Speth, Casey Stevens, Oran Young, Michael Zürn, and the late Hayward Alker and Ernst B Haas.

Several generations of UMASS Amherst graduate students have contributed to the evolution of my thinking, including Zuhre Aksoy, Kemi George, James Heilman, Inkyoung Kim, Chris Marcoux, and Casey Stevens.

References

Adler, E. (2005). *Communitarian International Relations.* London, Routledge.

Campbell, D. (1993). "Evolutionary Epistemology". *Evolutionary Epistemology, Rationality, and the Sociology of Knowledge.* G. Radnitzky, W. W. Bartlet and K. Popper. Chicago, Open Court Publishing Company.

Haas, P. and E. B. Haas (2002). "Pragmatic Constructivism and the Study of International Institutions." *Millennium* 31(3): 573–602.

ACKNOWLEDGEMENTS

The work in this collection has been published previously in a variety of different forms. I would like to thank the publishers for granting permission to use the following copyright material:

Peter M. Haas, "The Capacity of International Institutions to Manage Bhopal-like Problems"
Copyright © 1994 the University of Pennsylvania Press. This chapter was first published in Sheila Jasanoff (ed.) *Learning from Disaster: Policy Consequences of Bhopal*, Philadelphia: University of Pennsylvania Press, 1994, pp. 225–247. Reprinted with permission of the University of Pennsylvania Press.

Peter M. Haas, "Introduction: Epistemic Communities and International Policy Coordination"
Copyright © 1992 The MIT Press. This chapter was first published in *International Organization*, Vol. 46, No. 1, Winter (1992) pp. 1–37. Reprinted with permission of the MIT Press.

Peter M. Haas, "Constructing Environmental Security from Resource Scarcity"
Copyright © 2002 The MIT Press. This chapter was first published in *Global Environmental Politics*, Vol. 2 No. 1 (2002) pp. 1–19. Reprinted with permission of the MIT Press.

Peter M. Haas, "Social Constructivism and the Evolution of Multilateral Environmental Governance"
Copyright © 1999 Rouledge, the Taylor & Francis Group. This chapter was first published in Aseem Prakash and Jeffrey A. Hart (eds) *Globalization and Governance*, London: Routledge, 1999, pp. 103–133. Reprinted with permission of the Taylor & Francis Group.

Peter M. Haas, "Prospects for Effective Marine Governance in the NW Pacific Region"
Copyright © 2000 Elsevier. This chapter was first published in *Marine Policy*, Vol. 24, No. 4, July (2000) pp. 341–348. Reprinted with permission of Elsevier.

"Letter to the Editor"
Copyright © 2000 Elsevier. First published in *Marine Policy*, Vol. 24, No. 6, December (2000) pp. 499–500. Reprinted with permission of Elsevier.

Peter M. Haas, "Epistemic Communities and International Environmental Law"
Copyright © 2007 Oxford University Press. This chapter was first published as Ch.34: 'Epistemic Communities' by Peter Haas in Daniel Bodansky, Jutta Brunnée and Ellen Hey (eds) *Oxford Handbook of International Environmental Law*, Oxford: Oxford University Press, 2007, pp. 791–806. By permission of Oxford University Press.

Peter M. Haas, "UN Conferences and Constructivist Governance of the Environment"
This chapter was first published in *Global Governance*, Vol. 8, No. 1 (2002) pp. 73–91. Copyright © 2002 by Lynne Rienner Publishers, Inc. Used with permission of the publisher.

Peter M. Haas, "International Institutions and Social Learning in the Management of Global Environmental Risk"
Copyright © 2000 John Wiley & Sons Inc. This chapter was first published in *Policy Studies Journal*, Vol. 28, No. 3 (2000) pp. 558–575. Reprinted with permission of John Wiley & Sons, Inc.

Peter M. Haas, "Addressing the Global Governance Deficit"
Copyright © 2004 The MIT Press. This chapter was first published in *Global Environmental Politics*, Vol. 4, No. 4 (2004) pp. 1–19. Reprinted with permission of the MIT Press.

Peter M. Haas and Ernst B. Haas, "Learning to Learn"
This chapter was first published in *Global Governance*, Vol. 1, No. 3 (1995) pp. 255–284. Copyright © 1995 by Lynne Rienner Publishers, Inc. Used with permission of the publisher.

Peter M. Haas, "Choosing to Comply: Theorizing from International Relations and Comparative Politics"
Copyright © 2000 Oxford University Press. This chapter was first published in Dinah Shelton (ed.) *Commitment and Compliance: The Role of Non-Binding Norms in*

the International Legal System, Oxford: Oxford University Press, 2000, Part 1 section 2. By permission of Oxford University Press.

Peter M. Haas, "Evaluating the Effectiveness of Marine Governance"
Copyright © 2008 PEMSEA. This chapter was first published in Chua Thia-Eng, Gunnar Kullenberg and Danilo Bonga (eds) *Securing the Oceans*, Quezon City, Philippines: Global Environment Facility/United Nations Development Programme/International Maritime Organization Regional Programme on Building Partnerships in Environmental Management for the Seas of East Asia (PEMSEA) and the Nippon Foundation, 2008. Reprinted with permission of PEMSEA.

Peter M. Haas, "When Does Power Listen to Truth? A Constructivist Approach to the Policy Process"
Copyright © 2004 Routledge, the Taylor & Francis Group. This chapter was first published in *Journal of European Public Policy*, Vol. 11, No. 4 (2004) pp. 569–592. Reprinted with permission of the Taylor & Francis Group.

Peter M. Haas and Casey Stevens, "Organized Science, Usable Knowledge, and Multilateral Environmental Governance"
Copyright © 2011 the MIT Press. This chapter was first published in Rolf Lidskog and Goran Sundquist (eds) *Governing the Air: Science–Policy Interactions in International Air Policy Work*, Cambridge, MA: MIT Press, 2011. Reprinted with permission of the MIT Press.

Peter M. Haas, "Promoting Knowledge Based International Governance for Sustainable Development"
Copyright © 2007 Ashgate Publishing. This chapter was first published in Sophie Thoyer and Benoit Martimort-Asso (eds) *Participation for Sustainability in Trade*, Brookfield, VT: Ashgate Publishing, 2007. Reprinted with permission of Ashgate Publishing.

Peter M. Haas, "The Global Spreading of Ideas"
Copyright © 2010 WZB Berlin Social Science Center. This chapter was first published in *WZB-Mittelungen* Vol. 127, March, pp. 40–42. This is the quarterly journal of the WZB Berlin Social Science Center. Reprinted with permission of the WZB Berlin Social Science Center.

Every effort has been made to contact copyright holders for their permission to reprint material in this book. The publishers would be grateful to hear from any copyright holder who is not here acknowledged and will undertake to rectify any errors or omissions in future editions of this book.

1

INTRODUCTION

Reconstructing epistemic communities[1]

Peter M. Haas

The difficulty lies, not in the new ideas, but in escaping from the old ones. What used to be heresy is more and more read as orthodoxy.

John Maynard Keynes 1935, *The General Theory of Employment, Interest and Money*

1 Introduction

The study of epistemic communities (epicoms) forms part of the constructivist turn in International Relations (IR) and Political Science that has flourished since the 1980s. The study of epicoms has since become institutionalized in academic and policy circles. It explains sources of actors' understandings of a complex and uncertain policy environment, and the actor's attendant behaviors or practices under specified conditions.[2]

Epicoms play an important role under conditions of complexity, and the associated systemic uncertainty, in providing information to decision makers that helps them to articulate an understanding of the world and of their own policies and interests. As the world becomes more complex and interdependent (Perrow 1984; Beck 1986; Jervis 1997; Held, McGrew et al. 1999; Giddens 2000), decision makers encounter more novel interconnected issues about which they lack understanding and familiarity. Epicoms provide new causal arguments that enable indivdiual actors to make sense of the world, and contribute to collective (interactive) efforts to deal with shared problems. Epicoms and their ideas contribute to causal and constitutive mechanisms of social change, in particular social learning, which lead to more scientifically informed, and often comprehensive, approaches to policy making and dealing with issues. Analytically, their study provides a way to link the dialectic between micromotives and social facts

(Schelling 1978; Tilly 1984; Ruggie 1998). Ideas priviledge some outcomes by virtue of their causal justification (because they are likely to obtain the desired ends) as well as through constitutive means of creating the policy environment to which their prefered policies are applied, and priviledging the representation of particular expert groups with their own understandings. In terms of methods, the study of epistemic communities seeks to find ways to explain the consequences of understanding by actors, and of the implications of new ideas and policy frames on collective outcomes through process tracing, focused comparisons and large n statistical analysis.

Epistemologically, their study provides a third way between the ominiscient analyst and a relativist world. In terms of social mechanisms they provide a third way between the operation of material forces and the untrammeled imagination (Adler 1997; Pouliot 2007; Adler 2013). Ultimately, the study of epistemic communities and expertise provides an opportunity for reflexive learning that better integrates academic researchers with the policy world they study to yield better IR theory and better broader practices by decision makers who heed the updated advice (Campbell 1993; Haas and Haas 2002).

The study of epistemic communities has become a progressive research program.[3] Its hard core consists of ontological claims about complexity and actors, and epistemic claims about the indeterminacy of actor understandings and interests without a better understanding of how actors perceive and understand the context in which they operate (section 4). A broader auxiliary belt of refinements to the core argument developed during the 1990s and 2000s (section 6). Novel facts have emerged about the reciprocal influence of epistemic communities and international institutions, and the affect of ideas on patterns of governance. Confirmations of the theoretically derived claims have resulted through careful empirical studies (section 8).

The research program was subject to a variety of internal and external influences.

2 Internal intellectual history

My understanding of epistemic commuinities emerged while conducting dissertation field research around the Mediterranean in 1982–1983 (Haas 1990). I began work with a set of hypotheses heavily informed by Lateral Pressure Theory (Choucri and North 1975)—which were built primarily on historical materialist accounts of interests based on the net costs of treating national pollution problems—and expecting that national differences would be resolved through the exercise of power. I quickly discovered that the underlying assumptions were flawed. Through numerous interviews with responsible national policy makers I ascertained that no one really knew the extent or magnitude of their pollution problems, nor did they know the likely domestic interests groups that would be involved. In fact, they asked me if *I* could tell them. I was also lucky to have lengthy, extensive and multiple interviews with United Nations Environment Programme (UNEP) staff that had helped deal with regional environmental pollution problems.[4]

They also shared their personal archival materials. They frankly described their strategy of mobilizing the region's scientific community to address the region's shared environmental problems: what I later dubbed an epistemic community because of the shared features which I identified through elite interviews with many of the members. Epistemic communities were the focus of the two publications from that research, in 1989 and 1990 (Haas 1989; Haas 1990).

3 External intellectual history

The 1970s were rife with significant ontological transformations in IR and politics. There were profound glimmerings of significant changes in the nature of world politics, although many of these were not fully developed until later, or have not yet been fully understood. Key were the proliferation of new actors, groups of actors, and of shifts in the fundamental nature of world politics that made uncertainty the dominant characteristic for analysis. Keohane and Nye (Keohane, Joseph et al. 1971; Keohane and Nye 1977) early identified the role of non-state actors in world politics, as well as the emergence of new issue areas, along with John Burton (Burton 1972). The systemic conditions of uncertainty and complexity emerged, although a full understanding of them only developed later, and the literature on actors and on the structural systemic changes were never adequately unified. World historians have now documented the full ontology and salient historical moments behind constructivist analysis, dating to the Industrial Revolution, with additional break points in the early 20th century and post World War II (Ponting 1992; McNeill 2000; Manning 2003; McNeill and McNeill 2003; Osterhammel and Petersson 2005).

The 1970s and 1980s also saw numerous intellectual themes that laid the partial foundations (along with returns to prior philosophical roots) for contemporary constructivisms. Popular science laid the foundation for recognizing the social construction of an ambiguous world, as well as the social construction of knowledge (Capra 1975; Hofstadter 1979; Ferguson 1980). A confluence of similar insights about the vagaries of perception, and by extension the general social construction of reality, came from numerous locales.[5] Thus, in the face of uncertainty there was no single accepted perspective or understanding. Yet actors' and theorists' understandings were not capricious or structurally uniform: they varied according to social and contextual factors. Consequently a universal and correspondence theory of truth that had provided the general epistemology for IR and political science was brought into question (Kuhn 1970; Feyeraband 1978). But social constructions of reality are not capricious or unique to individuals or circumstances. The challege was to apply such insights within a meaningful framework of understanding about how actors make sense of their world and negotiate political arrangements with one another. With hindsight many of these provocative insights were parts of ongoing disciplinary debates that led to more moderate resolutions—particularly in the area of sociology and philosophy of science—but they were very compelling at the time.

Epicoms analysis provided a delegation model in which decision makers construct their political realities based on the technical advice provided by experts. Intended effects by one set of actors (agents) lead to unintended effects by other actors (principals) with aggregate social benefits through the provision of international public goods from focused collective action. However the state principals act out of mitigating uncertainty and ignorance rather than through a strategic logic.

Philosophers and sociologists of science demonstrated that the world is not directly accessible, implying that our understanding of the world is mediated by pre-existing ideas about the world. While discussions in the 1970s offered the view that understanding lacked any correspondence to the world (Berger and Luckmann 1967; Holzner and Marx 1979; Mulkay 1979; Barnes and Edge 1982), views now have narrowed down to a belief that such understandings are indirectly connected to an independent material reality, with the boundaries resolved through consensus procedures within the scientific community (Gieryn 1999; Latour 2004). Social psychology recognized the role of frames and prior experience in shaping expectations (Boulding 1956; Jervis 1970; Jervis 1976; Stein and Tanter 1980). Decision sciences revealed that actors take cognitive short-cuts when making choices, and thus that purely rational models of political choice are incomplete and often irrelevant to describe the way decisions are actually made (Braybrooke and Lindblom 1963; Sen 1977; Simon 1978; March 1994). Subjective understanding, satisficing, discounting costs in prospect theory, framing and bounded rationality all play roles in making choices. Economics more recently has looked at the psychological factors that shape preference formation and making choices (Nelson 2005; Akerlof and Schiller 2009). The policy sciences literature complemented decision analysis by focusing on the role of distinctive sets of actors responsible for representing the world in which decision makers find themselves and thus the appropriate mode of decision making (Kingdon 1984; Stone 1989). Mannheim and Gramsican analysis also focused on networks of transnational intellectuals involved in educating the populace and instructing the state (Mannheim 1936; Gill 1993). Hayek had even recognized that economic actors needed assistance from experts in articulating the nature of the policy environment and their available options (Hayek 1945).

A diverse range of literatures complemented these constructivist insights. Biology introduced notions of autopoesis (Maturana and Varela 1987). Art history elaborated the ways in which visual understandings could be deliberately shaped by context (Arnheim 1969), reinforcing William James' earlier observation that sensory perception is not objective. Education highlighted gendered differences in understanding and communication (Gilligan 1982; Belenky, McVicker et al. 1986). Historians, anthopologists, sociologists and science scholars documented the social influences over the creation and applicaton of ideas about time (Landes 1998) and statistics (Porter 1986; Hacking 1990).

4 Epistemic communities in IR

"Epistemic communities" is a concept applied by constructivist scholars of political science to focus analytic attention on the process by which states and other political actors formulate their interests and reconcile differences of interest. Epistemic communities are a principal channel by which consensual knowledge about causal understandings is applied to international policy coordination, and by which states may learn through processes of international cooperation.

John Ruggie introduced the term *episteme*, borrowed in turn from Foucault, to describe the overarching perspective through which political relationships are visualized and understood during historical eras (Ruggie 1975). The winter 1992 issue of *International Organization* on Knowledge, Power and International Policy Coordination introduced a focus on *epistemic communities*—the actors responsible for articulating and aggregating knowledge-based understanding in areas of security, environment, and international political economy. It also developed the now-standard four-element definition of an epistemic community. Publication in a high profile and prestigious journal helped legitimate constructivist work for a new generation of graduate students and junior faculty.

The introduction in this issue of *International Organization* (Haas 1992; Chapter 4 this volume) laid out the intellectual foundations for a constructivist research program on the origins and diffusion of ideas in politics. Epistemic communities serve as the originators and transmission mechanisms for causal ideas in politics, and contribute to the social construction of the world. The conclusion (Adler and Haas 1992) provided an evolutionary framework for theory building about epistemic communities and their influence.

Epistemic communities are networks of knowledge-based communities with an authoritative claim to policy-relevant knowledge within their domains of expertise. Their members share knowledge about the causation of social or physical phenomena in an area for which they have a reputation for competence as well as a common set of normative beliefs about what actions will benefit human welfare in such a domain. In particular, they are a group of professionals, often from a number of different disciplines, who share the following characteristics:

1. Shared principled beliefs. Such beliefs provide a value-based rationale for social action by the members of the community.
2. Shared causal beliefs or professional judgment. Such beliefs provide analytic reasons and explanations of behavior, offering causal explanations for the multiple linkages among possible policy actions and desired outcomes.
3. Common notions of validity: intersubjective, internally defined criteria for validating knowledge. These allow community members to differentiate confidently between warranted and unwarranted claims about states of the world, and policies to change those states.
4. A common policy enterprise: a set of practices associated with a central set of problems that have to be tackled, presumably out of a conviction that human welfare will be enhanced as a consequence.

Each characteristic must be present for an epistemic community to exist. In conjunction, they contribute to the broader social cachet and influence of the group. Internal beliefs about causal factors and validity, according to intracommunity standards, provide the glue for collective action among the individuals of the community. The warrants of their claims, according to external social standards, provide the political legitimation for their continued involvement (Haas 2004).

Parallel but independent work on epistemic communities occurred in the sociology of science. Karin Knorr Cetina coined the same term (Knorr Cetina 1991), although these two lines of inquiry proceeded in utter ignorance of one another. Work in IR and comparative politics was already probing the role of ideas and networks of technical experts in formulation national policies (Adler 1987; Sikkink 1991).

By analyzing epistemic communities, constructivist scholars gained leverage in understanding the processes of social construction and collective learning. Epistemic communities are associated with distinctive patterns of social change that involve persuasion and learning. If, as argued by constructivist scholars, contemporary international relations is characterized as a setting of complexity and uncertainty, particularly under contemporary circumstances of complex interdependence, increasing globalization, and the emergence of new technical issues on the international agenda with which traditional decision makers are habitually unfamiliar, then state interests are often unknown or incompletely specified. IR then becomes a matter of applying embedded and institutionalized beliefs about the nature of problems and the appropriate means of collective response rather than the process of resolving rationally formulated state preferences, as is argued by many current theorists of IR.

Epicoms have causal and constitutive effects. The epistemic community, through its institutionalized influence, causes decision makers to pursue choices and practices consistent with the epicom's ideas. In the absence of the epicom the range of collective outcomes would be much wider. Ideas constitute behavior, and social beliefs constitute agency. They also exercise constitutive effects by privileging certain sets of policies and practices, in which they enjoy expertise and primacy as advisors. Deeper functional needs for science and technology in modern industrialized societies also constitute the representative influence and discursive power of science-based epistemic communities.

Epicoms are likely to become involved in policy making after well-publicized shocks or crises. Only at such times are decision makers likely to recognize major anomalies and pursue new policy patterns. During subsequent, less revolutionary periods, these new doctrines, or orthodoxies, assume the status of taken-for-granted assumptions, or dogma, that persist until called into question again by external stimuli. Because of the disjointed-equilibrium nature of policy change, an evolutionary focus on institutional learning and path dependence may provide an appropriate model by which to understand the international recognition and response to global change. Such a research program provides better understanding of factors that influence the introduction of new policy frames, collective understandings, or doctrines, as well as illuminate the mechanisms of lock-in and identify those factors that may influence the degree of irreversibility of national and collective actions.

Epistemic communities are one of the principal actors responsible for aggregating and articulating knowledge in terms of state interests for decision makers, and disseminating those beliefs internationally. In a broader political context, epistemic communities provide one of the major channels by which overarching regime principles, norms, and rules are articulated for the international community and disseminated internationally. While epistemic communities are the primary agents responsible for articulating such principles, norms, and rules, the extent to which they become more deeply diffused and embedded internationally has to do with the political influence of epistemic-community members: their ability to persuade others, their ability to consolidate bureaucratic influence in important institutional venues, and their ability to retain influence over time. State interests and decisions to deploy state power are, thus, identified subject to consensual knowledge.[6]

Epistemic-community members can be identified through cross-checking attendance lists for international negotiations and consultations over time, secondary literature (especially by journalists), and by interviews and related snowball techniques. Recurrent names are eligible candidates. Identifying the consensual knowledge base shared by an epistemic community's members can be accomplished by reading their publications (especially scientific ones with equations, which force precision) and open-ended interviews. The extent to which they feel that they are members of a common community can be ascertained through interviews and by observing snowball techniques, in which prospective members identify others with whom they believe they share beliefs. The combination of self-identified traits and externally confirmed traits (such as the application of consensus approaches to truth) defines the epistemic community. Because epistemic communities are defined by their shared beliefs, the extent of tolerable consensus and contestation is a matter of community belief, and it can be ascertained through the internal norms and practices of the epistemic community in question. The causal mechanisms by which the epistemic community exercises influence—learning—can effectively be pursued by process tracing, focused comparative case studies, counterfactuals, and alternative-hypothesis testing.

More recent methodological advances in network analysis and bibliometric studies offer fruitful techniques for identifying and representing epistemic communities (Roth, Obiedkov et al. 2008), although not for tracking their influence.

5 Reception

The epistemic-community literature is now 25 years old. The epistemic-communities concept initially was favorably received in international relations because it provided a means for focusing on the ideational component of politics, as well as allowing for agency in theorizing about governance and policy making. Studies of the European Union, in particular, analyzed the role of various epistemic communities in shaping EU directives as well as in building a broader sense of European identity. Since the initial response to epistemic communities, their study has become a progressive research program as analysts have developed a variety of clarifications, refined hypotheses, pursued further empirical work, and provided empirical confirmation of the broad predicted social patterns associated with epistemic communities.

Initial critiques of epistemic communities by political scientists called for a clearer theory of the state (Milner 1992) and more attention to the role of domestic politics in mediating the reception of transnational epistemic communities (Risse-Kappen 1994), as well as a clearer metric by which epistemic communities could be recognized and consensus within an epistemic community could be measured.

Critical voices were sounded from the Social Studies of Knowledge literature regarding the degree of political autonomy enjoyed by epistemic communities, and science-based arguments in general, for public policy (Jasanoff 1996; Miller and Edwards 2001). Critics questioned the political consequences of such ideas and also the potential implicit political bias in the research programs pursued by epistemic-community members. The critics feared that epistemic communities might be antidemocratic and antiparticipatory by depoliticizing expertise, and critics also feared the consequences of expert-based advice. Also, because epistemic communities are responsible for simplifying a complex and ambiguous policy environment, critics felt that analysis should be focused on the resolution of contention within the knowledge space of the epistemic community rather than on its consensual beliefs. Relatedly, because epistemic knowledge includes that which epistemic communities agree about as well as that about which they disagree and remain uncertain, critics called for closer attention to the degree to which epistemic knowledge contains large amounts of contestation. Fundamentally, the social construction of knowledge may have biased the type of information which was brought to bear by epicoms, and thus avoid significant social issues worthy of attention (Litfin 1994; Litfin 1995; Walsh 2004).

Many of these points, however, were explicitly acknowledged in the initial formulation of the epistemic-community research program—namely, the political nature of all policy debate. The epistemic-community argument was that, normatively, epistemic communities ultimately provided better advice than other modes of policy advice, because expert advice is likely to be warranted. Unlike other organized interest groups active in politics and policy making, epistemic communities have internal beliefs that make them more likely to provide information that is politically untainted and therefore more likely to "work," in the political sense that this information will be embraced and followed by political authorities concerned about the need to be impartial. Epistemic communities' advice is also more likely to be technically effective, in the sense of obtaining the desired goals while balancing economic and technical trade-offs, than other negotiated approaches to policy making based on political compromise.

Analytically, the epistemic-community approach provided a clear causal pathway by which ideas came to inform political practices. This approach is superior to conventional approaches to the study of politics, which are unable to provide credible explanations for how ideas influence politics, or the conditions under which ideas are likely to be influential.

6 Refinements and evolution of the epicom research program

In response to the initial bout of responses in the literature, a number of refinements and clarifications to the original research program ensued over the following decades, leading to a richer political understanding of the role of causal ideas in world politics. Many of the formal refinements providing for a clearer deductive theory of politics were published in encyclopedias and handbooks (Haas 2001; Haas 2007; Haas and Stevens 2011; Haas 2012; Haas 2014), and thus did not receive widespread attention. There was, however, a significant outpouring of empirical material, discussed later in this chapter.

Refined and clarified assumptions include: states are key juridical actors, and they act willfully; state preferences are not fixed or necessarily given, but are subject to change and require analytic discovery; decision makers are boundedly rational; states are functionally differentiated, and they vary widely according to their state–society relations and the technical capacity of the state to formulate and enforce public policies in technical domains; states operate in a complex policy environment, so information about that environment is highly valued; yet, knowledge is asymmetrically distributed within and among countries; finally, while perceptions are socially constructed, political actors and social scientists are not subject to the same set of influences.

States make choices subject to multiple sources of influence, whose organization varies by issue area. Thus, governance varies by issue area. Consequently, policy networks, organized around specific issues become the appropriate level and unit of analysis, because the array of actors, interests, institutions, and capabilities varies by issue. Even within a particular state the design of policy-making institutions and procedures vary by issue area. Consequently, it is not surprising that international regimes in related areas can differ widely, because the articulation of interests for each regime takes a different form.

Epistemic communities help identify cause-and-effect relationships, elucidate interlinkages among problems, help define the consulting state's or organization's interests, and help to formulate policy. Their aggregate effect depends upon the extent to which their ideas become embedded in influential multilateral institutions more generally: powerful countries and international institutions, and possibly non-governmental organizations (NGOs), which will then deploy their own influence to disseminate the shared ideas enunciated within the epistemic community. Overall, learning will occur in the policy system as new policy-relevant knowledge is identified and applied to a common problem.

Stronger hypotheses about the mechanisms, effects, and variations in epistemic communities' influence were developed. The broader circumstances in which they would exercise an influence—temporally, functionally, and by political institution—were also better understood, following greater immersion in world history and comparative politics literatures.

Historical context

Epistemic communities are probably an artifact of the Modern Era. The 19th century was marked by significant intertwined historical discontinuities. Governments and governance became increasingly rationalized and professional at the national and international levels (Weber 1958a; Weber 1958b; Murphy 1994). Weber writes of "increasing intellectualization and rationalization" (Weber 1958b p. 139). Science became a socially valued and respected enterprise and source of expertise, because it contributed to the technological foundations of wealth and power in an industrialized age, and also a socially valued profession. Capitalism became entrenched as a socially accepted model for organizing economic exchange. Science and the Enlightenment emerged as political antidotes or counterweights to aristocratic, monarchic, and church rule.

Functional or substantive issues and types of countries

Epistemic communities are likely to be formed around substantive issues in which scientific disciplines have been applied to policy-oriented work and in countries with well-established institutional capacities for administration and science and technology. Only governments with such capacities are likely to see the need for the technical skills that epistemic-community members command, and such professionals would be attracted to governmental service only when they believe that their policy enterprise can be advanced.

Timing

Crises or widely publicized shocks are probably necessary precipitants of environmental-regime creation. Crises alone are insufficient to explain how or which collective responses to a perceived joint problem are likely to develop. States feel the need for expert guidance to respond to such crises, as states are often unfamiliar with the political landscape of new globalized issues. State decision makers are typically unaware of the domestic political alignments, or the likely consequences of the putative crisis, nor of the array of interests likely to be affected by the crisis or the array of policies and responses available to mitigate or respond to the problem.

There are several alternatives regarding the likelihood of which ideas will be selected at a moment of crisis. If there is only one new candidate, there is little problem, and the new aspirant will be selected. If there are many, choices may be conditioned by calculations of relative political gain from each idea. Studies of trade policy and the use of ideas in formulating international trade regimes have offered a number of propositions about conditioning factors influencing the selection of competing policy ideas (Odell 1990; Garrett and Weingast 1993; Goldstein 1993). Ideas that provide a potential for building coalitions will be favored over those that do not. Simple ideas that have perceived practical utility will be chosen over more complex approaches. In either case, new ideas will only endure if they are loosely commensurate with deeper seated beliefs and do not endanger strategic political alliances.

Epistemic communities, and the state receptiveness for their ideas, are likely to occur in rich democratic societies, in which scientific capacity and the free flow of information within society are high. More authoritarian countries may also be receptive to epistemic communities' ideas, however, but not as quickly. For instance, in the Mediterranean, the first adopters of epistemic communities and their beliefs were in Israel, Greece, France, and the United Nations Environmental Program (UNEP), while later adopters included Algeria and Egypt. Countries with weak scientific and engineering infrastructure are unlikely to have indigenous epistemic communities, although they may be subject to influence by other countries and international organizations (IOs) acting on behalf of the influence of epistemic communities.

Clearer ideas about the disciplines and issues likely to generate epistemic communities have been provided by the economic historian Paul David (Dasgupta and David 1994; David 1995). He provides some very fruitful and evocative work, pointing to the role of state capacity and the political economy of knowledge. Public acceptance of science—and, consequently, the external authority enjoyed by an epistemic community—is a function of the cost of research and the relevance of that research to the core mission of the sponsoring agency. Thus, in areas in which the cost is relatively small and the relevance is far from the core mission of the sponsor, the independence of researchers and the epistemic community is greater and, thus, likely to be seen as more legitimate. Social authority and political influence is likely to be stronger in those cases where the scientific knowledge is not seen to be politically compromised.

In the environmental sphere these insights seem valid. Similar results were found with regard to physicists involved with nuclear policy (Solingen 1994). Table 1.1 presents examples of degrees of public confidence in various fields of science. Epistemic experts enjoy authoritative status in countries in which their research enjoys social legitimacy away from the direct influence of the state. In countries with highly centralized research funding—such as France and the USSR—science enjoys less independent political and social authority.

TABLE 1.1 Public confidence in epistemic input

	High cost research	*Relatively low cost research*
Research is close to the core mission of the funding body	USA: physics France, USSR: all disciplines USA, USSR, China: space programs	France, USSR: all disciplines USA: seabed exploration and oceanography (Mukerji 1989; Walsh 2004)
Research is far from the core mission of the funding body		USA, Germany: ecology (McIntosh 1985; Worster 1987) Marine biology Ecology-related sciences (Aronova, Baker et al. 2010)

International organizations also play a role in popularizing and disseminating the ideas of epistemic communities as well as the individual influence of epistemic-community members. International organizations that are insulated from the influence of member states and that enjoy bureaucratic resources—such as an adept executive head, adequate budgets, and a professional staff recruited through merit—play a strong role in disseminating epistemic ideas (Haas and Haas 1995; Andresen, Skodin, et al. 2000 ch. 1). They have a subsequent effect on other IO practices, international regimes, and state expectations, interests, and practices. The mechanisms of influence by which IOs deploy epistemic ideas vary by target, however.

Policy styles

Different styles of policy selection occur in different political systems, due to organizational and cultural differences between types of democracies (Katzenstein 1977; Weaver and Rockman 1993; Shepsle and Laver 1994). In presidential systems, with multiple checks and balances, the flow of new ideas to the state will emerge from think tanks, universities, and governmental laboratories; and will be solicited directly by Congressional Committee staff, the judiciary, and executive branch bureaucrats. In parliamentary systems the sources of new ideas are more limited and much more likely to emerge from party think tanks and be deployed through political parties to Ministers and suitably socialized civil servants. At times ideas may go from universities and independent think tanks to mid-level civil servants, but there are fewer universities and independent think tanks in parliamentary systems because of the predominant influence of political parties.

Diffusion and diffusion mechanisms

Over time ideas will be diffused. Whereas they were initially selected by deferring to epistemic communities and recruiting their members by states and international institutions that were seeking new information, the diffusion of ideas occurs by many channels that include direct persuasion, demonstration effects, and inducements and incentives.

Diffusion may occur by institutional learning, as groups apply evolving consensual knowledge to manage institutional practices. Learning occurs directly, through interpersonal persuasion, communication, exchange, and reflection, which leads to the recognition or appreciation of new causal models and shared values. Leaders are socialized to accept new views and to empower their expositors.

One other diffusion mechanism is administrative recruitment, as epistemic-community members or their confidantes replace officials informed with alternative perspectives. Diffusion may also occur by other mechanisms, as actors alter their behavior subject to the influence wielded by institutions that embody the ideas (those that selected an epistemic community and its ideas earlier).

International regimes informed by institutionalized ideas promulgated by strong states and by international institutions will have the effect of disseminating the ideas to other parties in which epistemic communities were not present, and through their socialization and exposure to new ideas presented during regime negotiations.

Diffusion may also occur directly as a consequence of incentives and inducements to adopt those ideas coming from states, international institutions, and international regimes. These incentives and inducements typically take the form of technology assistance and financial transfers. These will only promote diffusion to states with weak capacity, for whom the provision of such resources would be a powerful incentive to change policy. Demonstration effects by multilateral financial institutions will induce borrowing states to adopt the policy being demonstrated. In open democratic systems transnational networks of IOs, NGOs and epistemic communities may reach out to domestic society to create upward pressures on states to adhere to the new collective projects.

States may draw lessons from one another's experiences, and thus borrow policies and ideas from countries that they are trying to emulate (Rose 1993). This process of diffusion is not one of learning though; it is one of mimesis or borrowing because the lesson drawers are not adopting the policies based on the persuasive power of reason by an advocate of the policy. Rather they are imitating the actions of someone who commands respect and appears successful.

These diffusion mechanisms will operate along different time frames. While some effects may be felt immediately through new recruitment decisions, and persuasion, broader shifts in public opinion and societal effects may occur on the order of decades. As these processes overlap, the general diffusion of an epistemic community's ideas will accelerate.

Multiple diffusion mechanisms are possible through various international channels at different political scale, involving inducements, coercion, learning and persuasion, and market responses to price changes. Different mechanisms are likely to operate on different types of countries. Epistemic communities remain a necessary component for learning within a broader pattern of policy diffusion. For instance, in stratospheric ozone protection, the regulatory scheme arose from atmospheric chemists, but the new technologies to satisfy those regulations emerged through the interplay of multinational corporations (MNCs), such as DuPont, ATOCHEM, and ICI; international institutions, such as the Montreal Ozone Protocol's technical advisory panels and the Montreal Ozone Fund, and NGOs, most notably Greenpeace. The appropriate factors by which international institutions may induce other actors to accept the epistemic ideas will vary according to the political and economic conditions in the target country. For instance, advanced industrialized societies are more responsive to public-education campaigns, whereas poor, developing countries respond to material incentives from capacity building.

Persistence

Once in place, ideas are likely to persist once they acquire a taken-for-granted element and as states and their diplomats become socialized to institutionalized regimes and practices generated by the application of new ideas. Eventually ideas get converted to domestic laws, and thus become enforced out of habit and become the basis for policy enforcement by the state. It also becomes politically costly to reverse such practices as new interest groups mobilize around them after recognizing that material gains are possible from the application of the new ideas. Ultimately, new markets may form through the application of these ideas, and thus generate economic incentives for their continued adoption once economic actors associate their profits with the application of such ideas.

Effects

Treaties developed with a strong contribution from epistemic communities may enter into force more rapidly than those developed through other political mechanisms, because the social influence of epistemic-community members will accelerate ratification in domestic legislative venues. National practices are likely to converge around the proposed policies from the epistemic community and the other actors that are deploying influence on behalf of those ideas.

The key analytic point, though, is that without the involvement of epistemic communities to set the agenda and impart new ideas, the political dynamics of collective action are likely to be far more conventional and lack the potential for reflective learning by policy makers.

7 Diffusion of epistemic communities' literature

The epistemic communities' research program has been receiving increased attention since the original 1992 *International Organization* appeared. Citations have spread well beyond the initial source in American political science. In 2014 Google Scholar (Google Scholar) listed 18,900 citations to "epistemic communities," and 12,200 to the *IO* introduction. More sensibly, the Web of Science website (Web of Science) provides 925 references to the introduction of the original 1992 *IO* publication (accessed 2 December 2013).

Figures 1.1 and 1.2 show the growth over time of references to epistemic communities in ISI Web of Knowledge tracked journals. The number of citations to epistemic communities has grown fairly consistently from the initial publication until 2010, with the first notable decline occurring in 2013.

Figures 1.3 and 1.4 show the distribution of citations by nationality of journal. Sixty-four percent of the citations were in English-language journals from the USA, United Kingdom, and Canada. A significant number appeared in journals from other countries as well, indicating a significant international diffusion of the concept.

Figure 1.5 presents citations by discipline. Disciplinary interest has been quite varied, with Political Science accounting for 20% of the citations, followed by International Relations with 16%. It has spread quite broadly beyond the core of political science and IR, though, as the majority of citations are from the social sciences writ large, and also Planning & Development and Geography.

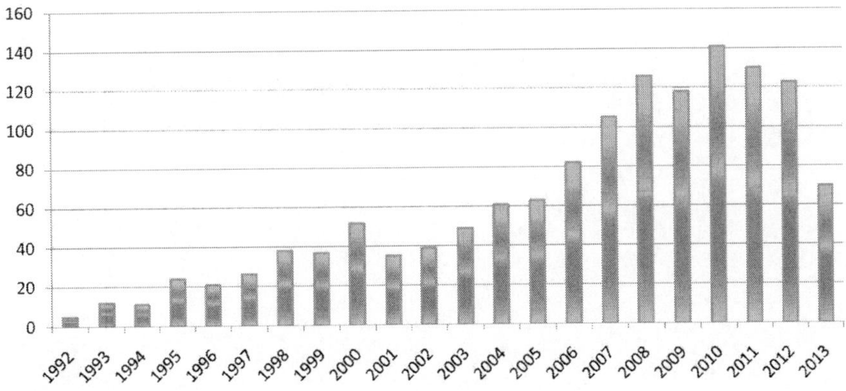

FIGURE 1.1 Number of citations by year, 1992–2013

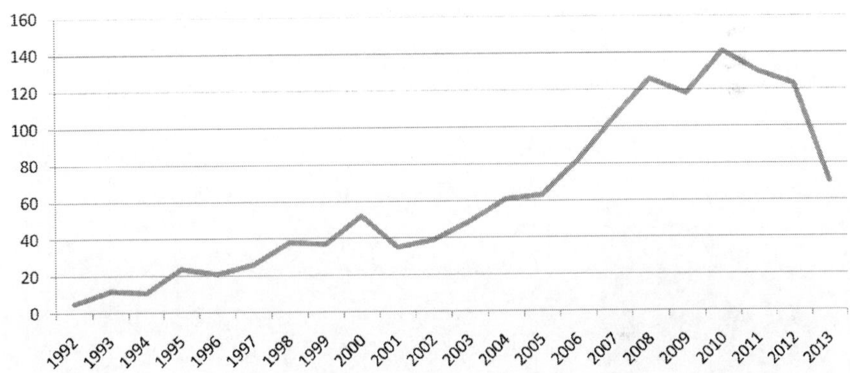

FIGURE 1.2 Growth of overall number of citations, 1992–2013

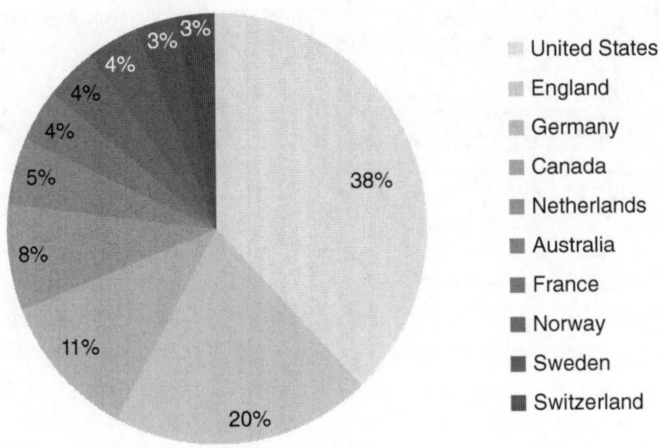

FIGURE 1.3 Percentage of total citations by country

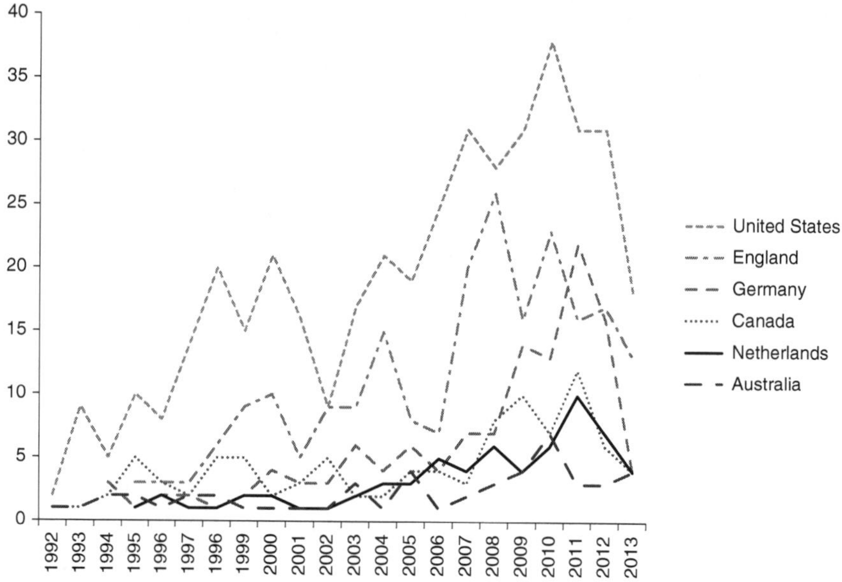

FIGURE 1.4 Diffusion of citations by country

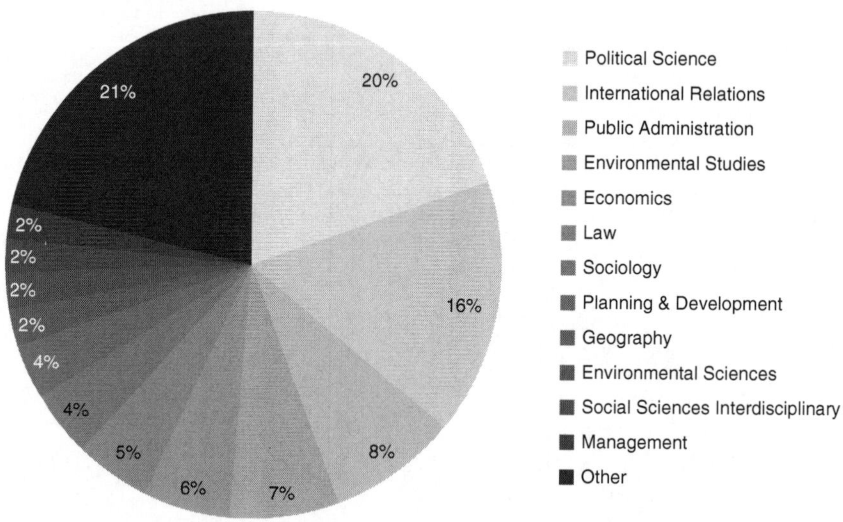

FIGURE 1.5 Citations by discipline

8 Empirical confirmation

Efforts to apply the epistemic communities' research program have spanned a variety of topics, at times blurring the distinction between studying the role of expertise and studying experts' ideas. With hindsight, many arguments that did not directly invoke epistemic communities do reflect the general argument, such as much of the literature on the triumph of free trade in the 19th century and the repeal of the Corn Laws.

Epistemic communities have been causally implicated in the form and persistence of international regimes, including 19th-century free trade (Bhagwati 1989), Bretton Woods (Ikenberry 1993), the New International Economic Order, (Murphy 1984; Rothstein 1984) a liberal post World War II trading order (Walter and Sen 2009), and international environmental regimes (Andresen, Skodin et al. 2000; Miles, Underdal et al. 2002); and the formulation of development strategies in Latin America (Adler 1987; Sikkink 1991).

Epistemic-community members have played a strong role in negotiating international environmental agreements. The hypotheses about the distinctive patterns of epistemic collective action were confirmed by numerous focused, comparative case studies of a wide array of multilateral environmental regimes. Members of the epistemic community that has dominated technical discussions in environmental regimes have subscribed to holistic ecological beliefs about the need for policy coordination subject to ecosystemic laws. Thus, they promote international environmental regimes that are grounded on policies that offer coherent plans for the management of entire ecosystems, and that are sensitive to interactions among environmental media (such as air and water), sources of pollution, and contending uses of the common property resource, rather than being limited to more traditional policies for managing discrete activities or physical-resources spaces within fairly short-term time horizons. Epistemically informed collective action has a distinctively comprehensive form and is politically resilient, yielding patterns that are not associated with political dynamics involving any other sets of political actors. Moreover, epistemically informed treaties are more likely to be effective, because the regulatory standards fit the behavior of the socio-ecosystem being managed. More generally, as discussed in Part II of this volume, the ecological view of the world has become institutionalized, leading to efforts to capture the links between issue areas.

Further studies of international institutions refined the characteristics of formal international organizations that are likely to absorb the lessons of epistemic communities and to disseminate them. Design principles for international scientific panels were also identified (Haas 2004).

The EU has been a particularly ripe area for epistemic-community studies, not surprisingly, given the relatively high socio-economic status of the countries and the multi-level governance structures that should be hospitable to epistemic community influences (Sabel and Zeitlin 2010). To mention a few, epistemic communities have been associated with negotiated policies for drug policy, monetary policy and the Economic and Monetary Union (EMU), environmental policy, agricultural policy, and legal convergence.

Transnational diffusion of national policies and regulatory standards clearly bear the imprint of epistemic communities. Such studies include Keynesianism (Galbraith 1971a; Galbraith 1971b), public-health doctrines, accounting standards, legal doctrines and interpretations, pension reform, privatization, human development, and the Washington Consensus and macroeconomic responses to financial crises (Williamson and Haggard 1994; Chwieroth 2010).

National policy choice has been analyzed in terms of epistemic-community involvement, in particular epistemic communities' independence from broad political goals of the state and from domestic interest groups. Studies have looked at pension reforms (Weyland 2005; Orenstein 2008) and credit ratings, financial management, and the sovereign debt crisis (Paudyn 2013).

Some of the seminal sea changes of world politics may be told in epistemic terms. The repeal of the Corn Laws in England and the foundations for international free trade during the 19th century relied on the involvement of an epistemic community of free trade believers organized by the Manchester School, which ultimately persuaded Robert Peele to sacrifice his political career on the platform of free trade (Bhagwati 1989; McKeown 1989). The domestic coalition in England for free trade required a component of actors whose interests were not materially defined, but rather depended on their epistemically informed understanding of the social benefits of the country as a whole (Kindleberger 1975).

The end of the Cold War can be told in part by focusing on Gorbachav's understanding of the policy choices available to him. He seems to have understood that the problem of economic decline was informed by political causes, and he received advice on the origins of it and possible responses from transnational sources (Evangelista 1997; Checkel 1998).

Epistemic communities have been subject to some conceptual stretching, including applications to include diplomacy and "appropriate behavior" (Cross 2011), religious actors and conflict resolution (Sandal 2011), and futurists (Cinquegrani 2002). In the EU Jurists have been treated as parts of epistemic communities (van Waarden and Drahos 2002).

Epistemic communities and their ideas do not always contribute to socially beneficial outcomes, although members of the epistemic community may well believe that their beliefs are socially beneficial. For instance, the Cult of the Offensive in World War I led to massive deaths. The rapid spread of Prohibition in the 1920s bore the imprint of epistemic influence (Schrad 2010). Imperialism and Colonialism, according to Schumpeter, were also the result of epistemic beliefs by a "warrior class," or an elite cadre of state policy makers (Schumpeter 1942).

More recently a host of new generative ideas have developed and spread with the assistance of epistemic communities. Human development (Fukuda-Parr 2011; Fukuda-Parr and Hulme 2011) and sustainable development are currently being considered as framing ideas for the international development agenda.

The ideational focus was absorbed into the broader constructivist research program developed in international relations and comparative politics, which looked at the role of beliefs and ideas in shaping state interests and practices, with epistemic communities serving as one of the mechanisms by which new ideas are developed and circulated (Katzenstein, Keohane et al.1998). Analyzing epistemic communities continues to provide a key group of actors who are associated with a distinctive set of processes by which causal understandings shape actor interests and regime dynamic, amid constructivists' broader focus on other ideational forces, including norms, linguistic usage, and the like.

9 Conclusion and research frontiers

More recent publications in the epistemic-communities research program have been looking more carefully at the context in which epistemic communities arise and operate. Works published in the first decade of the 21st century try to clarify the institutional factors that shape or amplify ideational consensus and dissemination, and the factors that influence the creation of relatively impartial and usable scientific advice. They try to separate the causal mechanisms of social learning from other causal mechanisms that drive collective action. Further work can help specify the distinct features of different social mechanisms in world politics, their interplay, and each one's distinctive contribution.

With the proliferation of non-state actors involved in world politics, more attention should be paid to the interplay between epistemic communities and other groups of actors.

The social construction of catalyzing crises remains under-theorized. While a crisis can easily be identified *ex post* as a catalyst of change, we do not have a strong understanding *ex ante* of how or why actors recognize that a set of events constitute a crisis.

More attention can be paid to explaining which epistemic community will prevail and when. Most work has looked at instances in which one epistemic community was superseded by another, following repudiation by events. Little attention has been paid to reconciling contending knowledge claimants. While a number of hypotheses have been raised about the factors likely to influence the attractiveness of a particular epistemic community and its ideas to decision makers at moments of crisis, these are all grounded on fixed, unitary state preferences. They do not take account of differential pressures within the state, and the case of the US decision to radically attack ozone depletion in 1985 in the face of steadfast opposition by top levels of the US administration serves as a powerful anomaly to such conventional hypotheses. One could ask why environmental scientists have tended to prevail over economists in all environmental issues other than climate change; or how Chicago School and Washington Consensus approaches have varied in their national uptake.

Additional research frontiers include a closer study of anomalous cases of failed epistemic communities—such as desertification and whaling—as well as the larger-scale dynamics by which social learning leads to broader understandings of world politics, interdependence, and identifying formation and reformation. Empirical efforts to create a more systematic population of current and historical epistemic communities will facilitate more comparative and systematic theory building about epistemic communities and their influence.

Notes

1 I am grateful to Sheila Jasanoff for this title.
2 For an elaboration of the distinctions between understanding and explanation, see Weber 1949; Weber 1968.
3 I use the term research program advisedly. While the Lakatosian formulation provides a convenient set of metrics to describe a collective research enterprise, I do not subscribe to Lakatos' sociology of knowledge and faith in cumulative knowledge through progressive research programs. With regards to my preferred stance on cumulative knowledge and progress in the social sciences, see Haas and Haas 2002; Sil and Kazenstein 2010.
4 I am eternally grateful to the late Peter Thacher, as well as Stjepan Keckes and Peter Sand.
5 For a summary of some of these influences, see Haas 1992.
6 The discussion of identifying epistemic communities and seeking to establish their causal influence in the previous section was provided in part to respond to this concern.

References

Adler, E. (1987). *The Power of Ideology*. Berkeley, University of California Press.

Adler, E. (1997). "Seizing the Middle Ground: Constructivism in World Politics." *European Journal of International Relations* **3**(3): 319–363.

Adler, E. (2013). Constructivism in International Relations. *Handbook of International Relations*. W. Carlsnaes, T. Risse and B. A. Simmons. Beverly Hills, CA, Sage: 31–64.

Adler, E. and P. M. Haas (1992). "Conclusion." *International Organization* **46**(1): 367–390.

Akerlof, G. A. and R. J. Schiller (2009). *Animal Spirits: How Human Psychology Drives the Economy and Why it Matters for Global Capitalism*. Princeton, NJ, Princeton University Press.

Andresen, S., T. Skodin, et al. (2000). *Science and Politics in International Environmental Regimes: Between Integrity and Involvement*. Manchester, Manchester University Press.

Arnheim, R. (1969). *Visual Thinking*. Berkeley, University of California Press.

Aronova, E., K. S. Baker, et al. (2010). "Big Science and Big Data in Biology." *Historical Studies in the Natural Sciences* **40**(2): 183–224.

Barnes, B. and D. Edge, Eds. (1982). *Science in Context*. Cambridge, MA, MIT Press.

Beck, U. (1986). *Risk Society: Towards a New Modernity*. London, Sage.

Belenky, M. F., C. B. McVicker, et al. (1986). *Women's Ways of Knowing*. New York, Basic Books.

Berger, P. L. and T. Luckmann (1967). *The Social Construction of Reality*. New York, Doubleday Anchor.

Bhagwati, J. (1989). *Protectionism*. Cambridge, MA, MIT Press.

Boulding, K. (1956). *The Image: Knowledge in Life and Society*. Ann Arbor, University of Michigan Press.

Braybrooke, D. and C. E. Lindblom (1963). *A Strategy of Decision: Policy Evaluation as a Social Process*. New York, Free Press of Glencoe.

Burton, J. (1972). *World Society*. Cambridge, Cambridge University Press.

Campbell, D. (1993). Evolutionary Epistemology. *Evolutionary Epistemology, Rationality, and the Sociology of Knowledge*. G. Radnitzky, W. W. Bartlet and K. Popper. Chicago, Open Court Publishing Company: 47–90.

Capra, F. (1975). *The Tao of Physics*. Berkeley, CA, Shambhala.

Checkel, J. (1998). "The Constructivist Turn in IR Theory." *World Politics* **50**(2): 324–348.

Choucri, N. and R. C. North (1975). *Nations in Conflict*. San Francisco, W.H. Freeman.

Chwieroth, J. M. (2010). *Capital Ideas: The IMF and the Rise of Financial Liberalization*. Princeton, NJ, Princeton University Press.

Cinquegrani, R. (2002). "Futurist Networks." *Futures* **34**(8): 779–783.
Cross, M. a. K. D. (2011). *Security Integration in Europe*. Ann Arbor, University of Michigan Press.
Dasgupta, P. and P. A. David (1994). "Toward a New Economics of Science." *Research Policy* **23**: 487–521.
David, P. A. (1995). "Reputation and Agency in the Historical Emergence of the Institutions of 'Open Science'." Paper presented to the National Academy of Sciences Colloquium on the Economics of Science and Technology, held at the Beckman Center, UC Irvin, 20–21 October 1995.
Evangelista, M. (1997). Domestic Structure and International Change. *New Thinking in International Relations Theory*. M. Doyle and G. J. Ikenberry. Boulder, CO, Westview Press: 202–228.
Ferguson, M. (1980). *The Aquarian Conspiracy*. New York, Houghton Mifflin.
Feyeraband, P. (1978). *Against Method*. London, Verso.
Fukuda-Parr, S. (2011). "Theory and Policy in International Development." *International Studies Review* **13**: 122–132.
Fukuda-Parr, S. and D. Hulme (2011). "International Norm Dynamics and the 'End of Poverty'." *Global Governance* **17**: 17–36.
Galbraith, J. K. (1971a). *Economics Peace and Laughter*. Boston, Houghton Mifflin Company.
Galbraith, J. K. (1971b). How Keynes Came to America. *Economics Peace and Laughter*. Boston, Houghton Mifflin Company: 43–59.
Garrett, G. and B. Weingast (1993). Ideas, Interests and Institutions. *Ideas and Foreign Policy*. J. Goldstein and R. O. Keohane. Ithaca, NY, Cornell University Press: 173–207.
Giddens, A. (2000). *Runaway World*. London, Routledge.
Gieryn, T. (1999). *Cultural Boundaries of Science*. Chicago, University of Chicago Press.
Gill, S., Ed. (1993). *Gramsci, Historical Materialism and International Relations*. Cambridge, Cambridge University Press.
Gilligan, C. (1982). *In a Different Voice*. Cambridge, MA, Harvard University Press.
Goldstein, J. (1993). *Ideas, Interests and American Trade Policy*. Ithaca, NY, Cornell University Press.
Google Scholar. From http://scholar.google.com/.
Haas, P. M. (1989). "Do regimes matter? Epistemic Communities and Mediterranean Pollution Control." *International Organization* **43**(03): 377–403.
Haas, P. M. (1990). *Saving the Mediterranean: The Politics of International Environmental Cooperation*. New York, Columbia University Press.
Haas, P. (1992). "Introduction: Epistemic Communities and International Policy Coordination." *International Organization* **46**(1): 1–37.
Haas, P. M. (2001). Epistemic Communities and Policy Knowledge. *International Encyclopedia of Social and Behavioral Sciences*. New York, Elsevier: 11578–11586.
Haas, P. M. (2004) "When Does Power Listen to Truth." *Journal of European Public Policy* **11**(4): 569–592.
Haas, P. M. (2007). Epistemic Communities. *The Oxford Handbook of International Environmental Law*. D. Bodansky, J. Brunnee and E. Hey. New York, Oxford University Press: 791–806.
Haas, P. M. (2012). Epistemic Communities. *The Oxford Companion to Comparative Politics*. J. Krieger. Oxford, Oxford University Press: 351–359.
Haas, P. M. (2014). Ideas, Experts and Governance. *The Role of 'Experts' in International Decision-Making: Advisors, Decision-Makers or Irrelevant?* M. Ambrus, K. Arts, E. Hey and H. Raulus. Cambridge, Cambridge University Press: 19–43.
Haas, P. M. and E. B. Haas (1995) "Learning to Learn." *Global Governance* **1**(3): 255–284.

Haas, P. and E. B. Haas (2002). "Pragmatic Constructivism and the Study of International Institutions." *Millennium* **31**(3): 573–602.

Haas, P. M. and C. Stevens (2011). Organized Science, Usable Knowledge and Multilateral Environmental Governance. *Governing the Air*. R. Lidskog and G. Sundqvist. Cambridge, MA, MIT Press: 125–161.

Hacking, I. (1990). *The Taming of Chance*. Cambridge, Cambridge University Press.

Hayek, F.A. (1945). "The Use of Knowledge in Society." *The American Economic Review* **35**(4): 519–530.

Held, D., A. McGrew, et al. (1999). *Global Transformations*. Stanford, CA, Stanford University Press.

Hofstadter, D. (1979). *Godel, Escher, Bach*. New York, Basic Books.

Holzner, B. and J. H. Marx (1979). *Knowledge Application: The Knowledge System in Society*. Boston, Allyn & Bacon.

Ikenberry, G. J. (1993). Creating Yesterday's New World Order. *Ideas & Foreign Policy*. J. Goldstein and R. O. Keohane. Ithaca, NY, Cornell University Press: 57–86.

Jasanoff, S. (1996). Science and Norms in Global Environmental Regimes. *Earthly Goods*. F. O. Hampson and J. Reppy. Ithaca, NY, Cornell University Press: 173–197.

Jervis, R. (1970). *The Logic of Images in International Relations*. New York, Columbia University Press.

Jervis, R. (1976). *Perception and Misperception in International Politics*. Princeton, NJ, Princeton University Press.

Jervis, R. (1997). *System Effects*. Princeton, NJ, Princeton University Press.

Katzenstein, P., Ed. (1977). *Between Power and Plenty*. Madison, University of Wisconsin Press.

Katzenstein, P. J., R. O. Keohane, et al. (1998). "International Organization and the Study of World Politics." *IO* **52**(4): 645–686.

Keohane, R. and J. Nye (1977). *Power and Interdependence*. Boston, Little, Brown.

Keohane, R. O. N., Joseph S. et al., Eds. (1971). *Transnational Relations and World Politics*. Cambridge, MA, Harvard University Press.

Kindleberger, C. (1975). "The Rise of Free Trade in Western Europe." *The Journal of Economic History* **35**(1): 20–55.

Kingdon, J. W. (1984). *Agendas, Alternatives, and Public Policies*. Boston, Little, Brown.

Knorr Cetina, K. (1991). "Epistemic Cultures: Forms of Reason in Science." *History of Political Economy* **23**(1): 105–122.

Kuhn, T. S. (1970). *The Structure of Scientific Revolutions*. Chicago, Chicago University Press.

Landes, D. S. (1998). *The Wealth and Poverty of Nations*. New York, Norton.

Latour, B. (2004). "Why has Critique Run out of Steam?." *Critical Inquiry* **30**: 225–248.

Litfin, K. (1994). *Ozone Discourses: Science and Politics in Global Environmental Cooperation*. New York, Columbia University Press.

Litfin, K. T. (1995). "Framing Science: Precautionary Discourse and the Ozone Treaties." *Millennium* **24**(2): 251–277.

McIntosh, R. P. (1985). *The Background of Ecology*. Cambridge, Cambridge University Press.

McKeown, T. (1989). "The Politics of Corn Law Repeal and Theories of Commercial Policy." *British Journal of Political Science* **19**(3): 353–380.

McNeill, J. R. (2000). *Something New Under the Sun: An Environmental History of the Twentieth-Century World*. New York, W. W. Norton.

McNeill, J. R. and W. H. McNeill (2003). *The Human Web*. New York, W. W. Norton.

Mannheim, K. (1936). *Ideology and Utopia*. London, Routledge.

Manning, P. (2003). *Navigating World History*. London, Palgrave.

March, J. G. (1994). *A Primer on Decision Making*. New York, The Free Press.

Maturana, H. R. and F. J. Varela (1987). *The Tree of Knowledge*. Boston, Shambhala.

Miles, E. L., A. Underdal, et al., Eds. (2002). *Environmental Regime Effectiveness: Confronting Theory With Evidence*. Cambridge, MA, MIT Press.
Miller, C. A. and P. N. Edwards (2001). *Changing the Atmosphere: Expert Knowledge and Environmental Governance*. Cambridge, MA, MIT Press.
Milner, H. (1992). "International Theories of Cooperation Among Nations." *World Politics* **44**(3): 466–496.
Mukerji, C. (1989). *A Fragile Power: Scientists and the State*. Princeton, NJ, Princeton University Press.
Mulkay, M. (1979). *Science and Sociology of Knowledge*. London, Allen and Unwin.
Murphy, C. N. (1984). *The Emergence of the NIEO Ideology*. Boulder, CO, Lynn Rienner Publishers.
Murphy, C. N. (1994). *International Organization and Industrial Change*. Oxford, Polity.
Nelson, R. R. (2005). *Technology Institutions and Economic Growth*. Cambidge, MA, Harvard University Press.
Odell, J. S. (1990). "Understanding International Trade Policies." *World Politics* **43**: 139–167.
Orenstein, M. (2008). *Privatizing Pensions: The Transnational Campaign for Social Security Reform*. Princeton, NJ, Princeton University Press.
Osterhammel, J. and N. P. Petersson (2005). *Globalization: A Short History*. Princeton, NJ, Princeton University Press.
Paudyn, B. (2013). "Credit Rating Agencies and the Sovereign Debt Crisis." *Review of International Political Economy* **20**(4): 788–818.
Perrow, C. (1984). *Normal Accidents: Living With High Risk Technologies*. New York, Basic Books.
Ponting, C. (1992). *A Green History of the World*. New York, St. Martin's Press.
Porter, T. M. (1986). *The Rise of Statistical Thinking 1820–1900*. Princeton, NJ, Princeton University Press.
Pouliot, V. (2007). "'Sobjectivism': Towards a Constructivist Methodology." *International Studies Quarterly* **51**: 359–384.
Risse-Kappen, T. (1994). "Ideas Do not Float Freely: Transnational Coalitions, Domestic Structures, and the End of the Cold War." *International Organization* **48**(2): 185–214.
Rose, R. (1993). *Lesson Drawing in Public Policy*. Ottawa, IDRC.
Roth, C., S. Obiedkov, et al. (2008). "Towards Concise Representation for Taxonomies of Epistemic Communities." *Lecture Notes in Computer Science* **4923**: 240–255.
Rothstein, R. (1984). "Consensual Knowledge and International Collaboration." *International Organization* **38**(4): 733–762.
Ruggie, J. G. (1975) "International Responses to Technology." *International Organization* **29**(3): 557–583.
Ruggie, J. G. (1998). "What makes the World Hang Together? Neo-Utilitarianism and the Social Constructivist Challenge." *International Organization* **52**(4): 855–885.
Sabel, C. F. and J. Zeitlin, Eds. (2010). *Experimentalist Governance in the European Union*. Oxford, Oxford University Press.
Sandal, N. (2011). "Religious Actors as Epistemic Communities in Conflict Transformation: The Cases of South Africa and Northern Ireland." *Review of International Studies* **37**(03): 929–949.
Schelling, T. C. (1978). *Micromotives and Macrobehavior*. New York, W. W. Norton.
Schrad, M. L. (2010). *The Political Power of Bad Ideas*. Oxford, Oxford University Press.
Schumpeter, J. (1942). *Capitalism, Socialism and Democracy*. New York, Harper & Row.
Sen, A. (1977). "Rational Fools." *Philosophy and Public Affairs* **6**(4): 317–344.
Shepsle, K. and M. Laver, Eds. (1994). *Cabinet Ministers and Parliamentary Government*. Cambridge, Cambridge University Press.
Sikkink, K. (1991). *Ideas and Institutions*. Ithaca, NY, Cornell University Press.

Sil, R. and P. J. Kazenstein (2010). *Beyond Paradigms: Analytic Eclecticism in the Study of World Politics*. London, Palgrave Macmillan.

Simon, H. (1978). "Rationality as a Process and as a Product of Thought." *American Economic Review* **68**: 1–16.

Solingen, E., Ed. (1994). *Scientists and the State*. Ann Arbor, University of Michigan Press.

Stein, J. G. and R. Tanter (1980). *Rational Decision-Making: Israel's Security Choices, 1967*. Colombus, Ohio State University Press.

Stone, D. A. (1989). "Causal Stories and the Formation of Policy Agendas." *Political Science Quarterly* **104**(2): 281–300.

Tilly, C. (1984). *Big Structures, Large Processes, Huge Comparisons*. New York, Russell Sage Foundation.

van Waarden, F. and M. Drahos (2002). "Courts and (Epistemic) Communities in the Convergence of Competition Models." *Journal of European Public Policy* **9**(6): 913–934.

Walsh, V. (2004). *Global Institutions and Social Knowledge: Generating Research at the Scripps Institution and the Inter-American Tropical Tuna Commission, 1900s–1990s*. Cambridge, MA, MIT Press.

Walter, A. and G. Sen (2009). *Analyzing the Global Political Economy*. Princeton, NJ, Princeton University Press.

Weaver, R. K. and B. A. Rockman, Eds. (1993). *Do Institutions Matter?* Washington DC, The Brookings Press.

Web of Knowledge. From http://apps.webofknowledge.com.silk.library.umass.edu/

Web of Science. From http://wok.mimas.ac.uk/

Weber, M. (1949). "Objectivity" in Social Science and Social Policy. *The Methodology of the Social Sciences*. E. A. Shils and H. A. Finch. New York, The Free Press: 24–37.

Weber, M. (1958a). Bureaucracy. *From Max Weber: Essays in Sociology*. H. H. Gerth and C. W. Mills. New York, Oxford University Press: 196–244.

Weber, M. (1958b). Science as a Vocation. *From Max Weber*. H. H. Gerth and C. W. Mills. New York, Oxford University Press: 129–156.

Weber, M. (1968). Objectivity. *Social Science and Social Policy Readings in the Philosophy of the Social Sciences*. M. Martin and L. C. McIntyre: 535–545

Weyland, K. (2005). "Theories of Policy Diffusion." *World Politics* **57**(January): 262–295.

Williamson, J. and S. Haggard (1994). The Political Conditions for Economic Reform. *The Political Economy of Reform*. J. Williamson. Washington DC, Institute for International Economics: 525–596.

Worster, D. (1987). *Nature's Economy*. Cambridge, Cambridge University Press.

PART I

Ontology and historical background

The central problem of our age is how to act decisively in the absence of certainty.
(Bertrand Russell)

These historically and ontologically oriented pieces present the principal historical forces which account for the problems confronting actors and the nature of actors in the contemporary world. The contemporary system of world politics is fundamentally different in a number of ways than has traditionally been depicted in most of the international relations literature.

"Transnational flow and the transformation of international relations" (Chapter 3) describes the major changes in world politics since the Industrial Revolution. Decisions makers are subject to a confusing array of challenges as a consequence of the growth in numbers and types of political actors, the diversity of functional issue areas subject to global governance, and the accelerating rates of technological change.

The analytic consequences of such changes in material forces are laid out in "The capacity of international institutions to manage Bhopal-like problems" (Chapter 3). "Bhopal-like" problems increasingly characterize international problems. These are problems which emerge from tightly coupled complex systems: such as environmental threats, climate change, sustainability challenges, nuclear proliferation, international financial coordination, macroeconomic management, and energy policy, among others. The policy challenges for decision makers are to make decisive decisions in the face of pervasive and profound uncertainty, as Bertrand Russell had presciently recognized early in the 20th century.

The full epistemic implications for studying world politics are elaborated in Chapter 3, the original introduction to the Epistemic Communities *IO* special issue. The most important implication of these ontological changes is that actors' perceptions of their policy space are socially constructed, and analysts' *ex post* assignments of actor preferences are untenable. Actors' perceptions do not correspond fully to a realist depiction of the world, in a positivist sense, because actors cannot fully comprehend the complexity of the policy problems they encounter. Thus actors' interests, preferences and policies (or strategies) depend upon the framing they apply to the problems they confront. To speak the language of game theory, pay-off matrices are made, not given. Actors have to be told that they face collective action problems. Of course these points have been recognized by a number of authors who have applied game theoretic formulations to world politics (Snyder and Diesing 1977; Mitchell 1979; Stein 1990, ch. 4; Elster 2007, ch. 12). In turn, Waltz's strict systemic evolutionary logic or systemic reproduction through the disciplining of state choices makes no sense when the systemic effects do not promptly eliminate actors making policy errors because the system is more plastic than he allows, and because actors seldom learn through reflecting on their experiences. Empirically there is a far lower mortality rate for states (Fazal 2004; Fazal 2007) and international organizations (Shanks, Jacobson et al. 1996) than Waltz's strict Darwinian logic would expect. Nor are states able to follow rational decision-making strategies.

"Constructing environmental security" (Chapter 5) provides an example of how an ambiguous new policy domain—environmental politics—has been understood in a variety of different ways as alternate theoretical frames are projected onto the issue by different actors. Different frames and actors offer different understandings of the nature of security threats from environmental scarcity, and the different attendant collective responses.

References

Elster, J. (2007). *Explaining Social Behavior.* Cambridge, Cambridge University Press.

Fazal, T. (2004). "State Death in the International System." *IO* 58(2): 311–344.

Fazal, T. M. (2007). *State Death.* Princeton, NJ, Princeton University Press.

Mitchell, R. C. (1979). "National Environmental Lobbies and the Apparent Illogic of Collective Action". *Collective Decision Making.* C. S. Russell. Baltimore, MD, Johns Hopkins University Press.

Shanks, C., H. Jacobson, et al. (1996). "Intertia and Change in the Constellation of International Organizations." *International Organization* 50(4): 593-627.

Snyder, G. and P. Diesing (1977). *Conflict Among Nations: Bargaining, Decision Making, and System Structure in International Crises.* Princeton, NJ, Princeton University Press.

Stein, A. A. (1990). *Why Nations Cooperate: Circumstance and Choice in International Relations.* Ithaca, NY, Cornell University Press.

2

TRANSNATIONAL FLOWS AND THE TRANSFORMATION OF INTERNATIONAL RELATIONS

Peter M. Haas

1.1 Introduction: The problem of new transnationalism

Yet another specter haunts the contemporary world: new international conditions obtain, as a result of a number of conjunctural and simultaneous international forces. As Alex Inkeles has written: "our future welfare, perhaps our survival, will depend on our ability initially to understand and subsequently to guide the processes of change in which we are caught up" (Inkeles and Smith 1975; OECD 1982; Drucker 1986).[1]

The flows of a number of goods (such as resources, commodities, money and people) and bads (such as pollutants) across national borders have become increasingly important for shaping relations among nations. Although the nature of the flows themselves is not necessarily new, their global nature and their extensive, possibly synergistic inter1inkages *are* novel.

This paper describes these new conditions which stem from the "transnational flows" discussed in this paper, the ensuing pressures on states, and their implications for international behavior. The basic position of this paper is that the *conjuncture* of the increased importance of such transnational *flows* across national borders, with changes in the *actors* commonly involved in international relations, and the *context* in which international relations occur has modified the way the international political and economic system is conventionally viewed and the modes through which policy coordination among nations is undertaken.

International political, economic and even ecological interdependence is by now a virtual truism (OECD 1982).[2] Historically, some aspects of interdependence have been cyclical (such as the openness of the international trading order, and the degree of national reliance on foreign trade), while others *are* secular (such as monetary relations; the environment, where pollution is transmitted longer distances than ever before). The contemporary system is characterized by a conjuncture of national sensitivity to both cyclical and secular flows. Cyclical flows are near their peaks, and secular flows are increasing. Because these forces have emerged at roughly the same time, their influence is difficult to weigh individually. Many, now interdependent trends have led formerly disparate issue-areas to become increasingly interlinked.

In general, transnational flows can be characterized along three dimensions. First, the effects of national actions are manifested across physical barriers (spatially); second, their effects are realized in the future (temporally); and third, their impact can be observed in other issue-areas where their effects are transmitted causally (functionally). Spatial flows have generally received the most attention by analysts. Now, a generally acceptable way of characterizing such a complex system is Jay Forrester's description of "multi-loop, nonlinear feedback systems."

Policy introduced to "improve" conditions may have unanticipated consequences which span all three dimensions of transnational flows. For instance, marine pollution in the Mediterranean illustrates all three dimensions. Temporally, past emissions of inorganic industrial wastes influence current and future public health in the region. Spatially, one country's emissions are transmitted by currents to its neighbors. And they may be related causally (or functionally). One individual policy space may not be effectively managed without concurrent efforts over functionally related issues. Functionally, industrial activities impede the uses of the Mediterranean by other economic sectors: by killing fish or polluting beaches, decisions made within the industrial sector are transmitted to decisions in the tourist and fishing sectors, in other countries as well. Establishing common water quality standards can give rise to political conflict between the numerous users of the Sea who desire competing quality standards. Navigation interests are pitted against fishery interests, tourist (or tourism) interests, agricultural interests and industrial interests in the establishment of common water quality standards.

In other arenas, managing international economic flows generally can raise conflicts between macroeconomic policy, monetary policy, trade policy, and setting exchange rates. Issues impinge on one another, since managing one discrete area requires establishing norms and rules for associated issues. As an illustration, it is becoming evident that exchange rates respond to short-term monetary flows between banks as well as the traditional settlement of international trade accounts. Thus, managing exchange rates requires coordinated trade policies as well as monetary policies.

From an international political perspective, it is evident that transnational flows are increasingly making governments sensitive to decisions made elsewhere. The more reliant countries are on foreign sources for the satisfaction of their domestic needs, the more sensitive they are to decisions made elsewhere. Thus, a consequence of transnational flows is that governments lose exclusive control over policy making within their territories as they lose control over discrete issues. In general, then, decision makers become more uncertain about what policies should be adopted to promote their interests within individual issue-areas, as they are unable to fully master the issues themselves, as well as complementary ones.

This paper adopts a statist approach. In other words, it takes the nation as the level of analysis and presumes that it is only national governments that have the information, experience and resources to act effectively and manage transnational issues. Although the origins of many of these global forces occur well beyond the range of traditional views of the state, governments remain responsible for formulating responses and managing problems which cross boundaries. It is also assumed that governments act to defend their autonomy, and value power resources insofar as they may contribute to such an objective. Appropriate resources vary by issue-area.

While analysts generally agree that much is "new," they dispute the origins of different strands of current conjunctures. The origins are attributed to diverse causes dating from the early Christian era to the Industrial Revolution to World War II. These differences tend to vary by the disciplinary training and politics of the analyst. Marxists and Neomarxists attribute the causes to an international capitalist system which generates externalities (Amin 1985; Redclift 1987). Liberals, drawing from a more sociological tradition blame "modern" modes of industrial production which rely on traded inanimate sources of energy (Boulding 1964; Levy 1966; Inkeles and Smith 1975; Morse 1976). A historian of science identifies the values inherent in the dominant Judeo-Christian tradition, with its focus on mastery over nature as the source of externalities (White 1966; White 1968). Structural Marxists point to the preeminence of instrumental objectives inherent in scientific and technological thinking, which distances policy makers from the ultimate effects of their policies (Ellul 1964; Habermas 1971; Horkheimer, Adorno et al. 1974). More pragmatic or eclectic writers point to styles of consumer demand in the West or local land use practices as the source of increased transnational flows (Schumacher 1973; Bahro and New Left Review 1984; Maguire and Brown 1986). This paper addresses the characteristics and consequences of this new transnationalism, leaving debates of underlying "causes" to others.

1.2 Dimensions of international change

The international political and economic system has been modified by the addition of new units and processes. Before discussing these particular changes in detail (see section 2 below), some definitions are in order. Oran Young defines a system as:

> a group of actors (i.e. units) standing in characteristic relationship to each other (structure), interacting on the basis of recognizable patterns (processes), and subject to various contextual limitations.
>
> *(Young 1968; Giddens 1984)*[3]

Among American students of international relations, contemporary Realist and Neorealist writers commonly characterize the world in terms of frequently observed patterns of force deployed to preserve the sovereignty of nation-states in the politico-military sphere (Gilpin 1975; Waltz 1979; Krasner 1983; Keohane 1984). Marxists, Neomarxists, Dependency Theorists, and radicals characterize the world in terms of unequal exchange in economic relations between nation-states (Emmanuel 1972; Amin 1977; Wallerstein 1979).

These older more traditional attributes of the international system still obtain. States retain their authoritative power in the international system. The characteristic structures of diplomacy, unequal distribution of power resources, and international organizations within which states stand in fixed relationships to one another have not been altered.

However, the addition of *new* actors, the increased *importance* to states of transnational flows, and *contextual shifts* described below contribute to a significant modification of the conventional competing views of international political and economic systems. Because many issue-areas are now highly interdependent, and their interplay is complex and unclear, relations among states are now conducted in a setting of heightened uncertainty.

1.2.1 Evidence of systemic change and its origins

The conjuncture of forces has led to an increase in the forces, and interlinkages between forces, which operate on states and decision makers. For convenience, we shall refer to this ensemble of pressures as "complexity." Complexity means:

> a set of objects or events whose description involves many variables, among which there are strong mutual interdependencies, so that the resulting system of equations cannot be solved "piece-meal", as in the case of classical celestial mechanics, where perturbations can be imposed on two-body problems.
>
> *(Rapoport 1966)*

Herbert Simon notes that:

> In such systems, the whole is more than the sum of the parts, not in an ultimate, metaphysical sense, but in the important pragmatic sense that, given the properties of the parts and the laws of interaction, it is not a trivial matter to infer the properties of the whole. In the face of complexity an in-principle reductionist may be at the same time a pragmatic holist.
>
> *(Simon 1969)*

This implies that "dependent variables often cannot be explained in terms of a single independent variable without imposed extremely far-reaching conditions concerning the state of other variables" (Young 1968). Todd La Porte operationalizes this notion according to the number of system components (actors), the relative differentiation between them, and the degree of interdependence among them (Emery and Trist 1972; La Porte 1975).

International relations are becoming more complex according to all these dimensions. More actors are involved in world politics, and policy formulation has grown increasingly technical. Higher volumes of interstate exchange are quantitative measures of such complexity. The increased importance of transnational flows to the satisfaction of national needs is the best measure of La Porte's third indicator of interdependence among actors. Changes in the rate of technical change, increasing complexity of issues, and increasing recognition of causal interlinkages between issues suggest qualitative changes.

Many of the data flows are incomplete. For some, this is simply because detailed data is not available for long periods. For others, it is because the most striking changes are recent. Of course, there are gaps in the data presented here. Of transnational flows, flows of population have received the least scholarly attention (Tietelbaum 1984; Weiner 1985).[4] No systematic inventory of interlinkages between issues exists. Most assertions are based on anecdotes, which may not be reflective of broader trends. Few easily usable techniques exist for identifying or studying such interlinkages.

1.2.2 Actors in the international system

The number of nation-states has expanded in the 20th century: from 52 countries after World War I to 67 after World War II to the present number of over 180. The simple number of bilateral relations among so many nation-states greatly exacerbates the administrative difficulties of conducting foreign policy. The number of bilateral relationships which must be monitored by foreign services grew by 8,607 in the last 70 years: from 1,326 before World War I to 2,211 in World War II to 12,720 currently (according to the equation (n)(n-l)/2). Particularly for smaller civil services, such as in many developing countries, the number of duties expected of individuals grow quite overwhelming (Obasanjo 1987).[5] Attention spans are short, and follow up is difficult.

The fora in which these actors meet have also proliferated, also indicating a growth in the range of collective concerns, such that new bodies have had to be created in which to manage them. Harold Jacobson reminds us that the number of intergovernmental organizations has grown apace: from 5 in 1865 to 49 during 1910–1914 to 84 from 1935–1939 and 621 in 1980 (Jacobson 1984).

Non-state actors have grown in numbers as well. International non-governmental organizations (NGOs) have flourished. The total number of international NGOs grew from 176 in 1909 to 973 in 1956, 1,675 in 1966 and 2,381 in 1976. Twenty-six percent of these are international scientific and professional associations, of which the largest proportion (65%) deal with health, medicine, science and technology (Crane 1972).[6] In addition to constituting new actors, these groups reflect a mounting international concern with economic and technological issues, which have a more technical nature than do the more traditional interstate concerns of security. As such, they indicate qualitative as well as quantitative systemic modifications. The growth and expansion of multi-national corporations further litters the international arena.

With the increase in saliency of technical scientific issues, the number of scientists involved in international decision making, both within governments, in delegations, and on international secretariats has grown. Such new actors are responsible for solving, or at least managing, technical issues, as well as uncovering new problems. Between 1972 and 1982 ministries or departments responsible for the environment and natural resources were established in 118 countries. In 1972, 15 developed countries and 11 developing countries had such bodies. By 1982, 34 DCs and 110 LDCs had them (Baker, Bassett et al. 1985).

In the United States the number of scientists and engineers in government increased dramatically after World War II, as has the number of regulatory agencies and government expenditures on regulation and research and development. Between 1970 and 1975 the number of major US economic regulatory agencies increased from 8 to 10, and major social regulatory agencies from 12 to 17. Economic agencies' expenditures grew by 157% for (from 166 million to 427 million), and social agencies' expenditures grew by 193% (from 1.5 billion dollars to 4.3 billion dollars) (Majone 1984).

In particular, the total number of professional scientific and technical personnel in the United States federal government grew from 123,927 in 1954 to 189,491 in 1964 to 238,041 in 1983. This mere rough doubling obscures the more pertinent change in individual expertise. From 1974 to 1983 the proportion of scientists and engineers with PhDs increased by 51%, and those with masters degrees increased by 44% (National Academy of Sciences 1964; National Academy of Sciences 1985).[7] The size of the Congressional staff has also soared. From 1955 to 1974 the House staff grew from 3,623 members to 12,444, while the Senate staff grew from 1,962 to 5,284 (Pastor 1980).

Taken together, these data suggest the growing technical nature of international relations, as well as an increasing sophistication and technical competence in the United States government, and presumably elsewhere as well. Insofar as these individuals introduce new ideas to the regulatory arena, the professionalization of government would lead to the greater prevalence of new ideas, as well as a greater separation between elected officials and their technically more competent staffs.

1.2.3 Transnational flows

Issues are becoming increasingly international. Transnational flows have been growing in absolute *size*, as well as in their *significance* to particular countries. The volume with which various goods and bads flow across borders is at a historical peak. From 1945 to the early 1980s the growth of world trade outpaced the rate of growth of national production, as is evident in Table 2.1. World imports rose from 60 billion dollars in 1950 to 1,847 billion in 1984 for an annual growth rate of 3.2%. World GDP grew by 2.8% during the same period. World merchandise trade continued to grow by about 1% more than total output from 1984 through 1987 (The Economist 1987).[8]

TABLE 2.1 Growth of world production and trade

	1950	*1960*	*1965*	*1970*	*1975*	*1980*	*1984*	*1985*
World imports (billion $US)	60.1	124.1	187.5	300.4	826.7	1,928.6	1,846.8	1,879.0
Index of world imports (1980=100)	3.1	6.4	9.3	15.6	42.8	100	95.6	97.5
Index of world GDP (1980=100)		43.7	56.4	70.7	83.2	100	109.2	
World population in billions, mid-year data	2.50	3.01	3.32	3.68	4.08	4.45	4.76	

Source: International Monetary Fund 1986; United Nations Statistical Office 1987.

The increase of world trade over world production suggests that countries are increasingly dependent on foreigners for products and markets. Table 2.2 indicates that the largest industrial countries' exposure is relatively high by historical standards. It also suggests that openness and closure of the international economic system may well be a cyclical condition (Katzenstein 1971; Stein and Rosecrance 1973; Krasner 1976; Stein 1984).[9] Current levels of trade interdependence, or sensitivity to foreign market decisions, are similar to those of the 1970s, the period before World War I, and even the inter-war period before the Great Depression.

The *nature* of these flows has changed as well. Manufactured products now account for over 60% of world trade, up from 39% in 1928. Trade in services has also flourished recently, now accounting for almost 30% of world trade (The Economist 1987).

The global *distribution* of such growth has not been symmetrical. From 1963 to 1980 the United States' share of world manufacturing output dropped from 40.37% to 29.47%. Japan's rose from 5.5% to 15.7% (from #5 to #2 ranking). The Federal Republic of Germany moved from 9.7% to 12.4% (although falling from #2 to #3 ranking) (Dicken 1986). Historically, shifts of dominance in manufacturing trade have occurred as well, as is seen in Table 2.3.

The share of developing countries in world exports of manufactures increased from 5% in 1970 to 9% in 1980 to just under 12% in 1983 (UNCTAD Bulletin 1986). Developing countries' share of world manufacturing output grew from 12.8% in 1970 to 15.3% in 1980, although this growth was largely accounted for by rapid growth in the Newly Industrialized Countries. The ten NICs' (as identified by the OECD) share of world manufacturing exports grew from 2.59% in 1963 to 7.12% in 1976, and from 5.4% of total manufacturing production to 8.87% (Dicken 1986). Their total share of world manufacturing output grew from 5.4% to 10.5% (Dicken 1986).[10]

Poorer non-oil-exporting developing countries have been increasingly marginalized from this trend towards globalization of markets. Eighty-seven percent of LDC exports of manufactures came from the industrializing

TABLE 2.2 Exports plus imports as percentage of national product

	United Kingdom		*France*		*Germany*		*Italy*		*Japan*		*Soviet Union*		*United States*	
	Years	*%*	*Years*	*%*	*Years*	*%*	*Years*	*%*	*Years*	*%*	*Years*	*%*	*Years*	*%*
Late 19th century	1877–85	49	1875–84	52	1880–89	34	1889–90	26	1878–87	13			1879–88	14
	1887–95	45	1885–94	50	1890–99	30	1891–00	25	1888–97	26			1889–98	14
Pre-World War I	1897–05	41	1895–04	49	1901–09	34	1901–10	31	1898–07	34			1899–08	11
	1909–13	52	1905–13	54	1910–13	38	1911–13	34	1908–13	33			1904–13	12
Inter-war	1924–28	38	1920–24	51	1925–29	31	1925–29	30	1918–27	41			1919–28	12
	1929–33	28	1925–34	42	1930–34	22	1930–34	20	1928–37	41			1929–38	8
	1934–38	24	1935–38	33	1935–38	12	1935–38	15	1938–42	31			1939–48	10
Post-World War II	1949–53	37	1950–54	38	1950–54	26	1950–52	26	1950–56	21	1955	4	1944–53	10
	1960	32	1960	23	1960	30	1960	26	1960	20	1960	5	1960	7
	1965	30	1965	22	1965	32	1965	26	1965	19	1965	5	1965	7
	1970	33	1970	25	1970	34	1970	30	1970	20	1970	5	1970	9
	1975	41	1975	32	1975	39	1975	41	1976	23	1975	7	1975	14
	1980	42	1980	38	1980	46	1980	44	1980	26	1980	–	1980	18
	1984	47	1984	41	1984	53	1984	46	1984	25	1984	–	1984	15

Source: Waltz 1979; International Monetary Fund 1986.

TABLE 2.3 Countries' share of world exports of manufactures (percentages based on values in US$ at current prices)

Year	*UK*	*US*	*Germany*	*Italy*	*France*	*Japan*	*Sweden*
1881–85	43.0	6.0	16.0	2.0	15.0	0.0	1.0
1899	34.5	12.1	16.6	3.8	14.9	1.6	0.9
1913	31.8	13.7	19.9	3.5	12.8	2.5	1.5
1929	23.8	21.7	15.5	3.9	11.6	4.1	1.8
1937	22.3	20.5	16.5	3.7	6.2	7.4	2.8
1950	24.6	26.6	7.0	3.6	9.6	3.4	2.8
1964	14.0	20.1	19.5	6.2	8.5	8.3	3.4
1973	9.1	15.1	22.3	6.7	9.3	13.1	3.3

Source: Hall 1986.

Republic of Korea, Hong Kong, Brazil, Mexico, Singapore, Malaysia, Yugoslavia, Philippines, India, Thailand, Indonesia and Argentina (Dicken 1986). The low income LDCs (per capita income under 300$ in 1975) have been growing more reliant on external sources for markets, while their trade partners have been growing less interested in their markets. In 1970 17.3% of all industrial country exports went to non-oil-exporting LDCs. In 1977 this figure was down to 15.8%. Conversely, the reliance of the low income LDCs on foreign trade grew from 13.8% of GNP in 1960 to 15.7% in 1976 (Streeten and World Bank 1981). According to the classical formulation of dependency, these countries are growing more dependent on the international economic system and less autonomous, as they come to rely more heavily on the international system for goods, whereas their partners find the LDC markets increasingly less important for their own sales.

World manufacturing trade as a whole has become stratified. Whereas NICs have gained a growing share of the world market, exports of the most sophisticated goods became increasingly concentrated in the United States, Japan, Federal Republic of Germany, and France. These four countries' share of the 14 industrialized countries' high-technology exports, which account for about 80% of world trade in manufactured products, rose from 59% in 1965 to 62% in 1980, as may be seen in Table 2.4.

TABLE 2.4 Share of 14 industrialized-countries' high-technology exports as percentage of total such exports

Year	*USA*	*Japan*	*F.R. Germany*	*France*
1965	28.0	5.9	17.4	7.8
1970	27.4	8.9	18.3	7.4
1975	24.8	9.9	18.1	8.7
1980	23.9	12.3	17.5	9.0

But manufacturing also gives rise to persistent toxic pollutants. The development and diffusion of toxic substances accompanied the growth and spread of production and trade in manufactured goods. Barry Commoner notes that "most pollution problems made their first appearance, or became very much worse, in the years following World War II" (Commoner 1971; White, Kates et al. 1986). Annual production of synthetic materials in the United States grew from less than 1 billion pounds prior to World War II to 172 billion pounds in 1978 (Hughes 1985). The volume of synthetic organic chemicals produced per year grew from 10 million in the 1940s to more than 10 billion pounds in 1980. Gilbert F. White, a renowned geographer, estimates that "perhaps as many as 1000 new chemicals are introduced annually." Elevated levels of atmospheric carbon dioxide have accompanied the increased use of fossil fuels for energy since the 19th century (Hughes 1985). Because the carbon dioxide in the atmosphere reflects heat back to the earth, the increasing level of carbon dioxide is feared to contribute to global warming.

The effects of many environmental problems are now felt globally, whereas in the past they were felt only locally or regionally.[11] For example, extensive logging and grazing in the 3rd century BC led to erosion in Greece, and overall deforestation in North Africa (Hughes 1975; Braudel 1977; Thirgood 1981), but such practices today may change the global climate. Seventeenth century London suffered heavily from air pollution from the burning of coal (Eisenbud 1978). The Scandinavian countries now complain that British public utilities' use of fossil fuels produces acid rain which destroys Scandinavian lakes and forests. The fallout of pollution from the 1986 Chernobyl disaster; the extensive downstream impact of the 1986 toxic chemical spill in the Rhine River; and the long-range transfer of acid rain in Europe are but a few recent examples of how pollutants may flow across long distances and borders.

The growth of transnational flows of pollutants illustrates another new dimension of transnational flows. In addition to moving spatially, these flows also have indirect effects on other economic activities, as the impact of pollutants is transmitted by natural biological systems.

The growth of the Eurocurrency market is a clear example of recent increases in transnational flows in money. The size of the market has grown from 9 billion dollars (USA) (net) in 1964 to 702 billion dollars in 1982 (net) (Cohen 1986). Deposits in the Eurocurrency market rose from 200 billion dollars (USA) in 1972 to over 3 trillion dollars (USA) in 1986 (Hormats 1987). Total international bank lending—cross-border claims plus domestic loans in foreign currencies—rose from 2.2 trillion $ in 1984 to 3.4 trillion in1986 (The Economist 1987). Effectively unregulated by domestic monetary authorities, banks have no reserve requirements against their loans. Thus, the market is potentially inflationary, but also highly sensitive to market volatility, as it lacks the safety-nets which domestic monetary markets have. With the high rates of flows between countries, a short-term disequilibrium could oscillate into a global collapse in confidence. Charles Kindleberger anxiously notes that "before World War II, an enormous run on a national currency involved as much as $100 million a day. In the 1980s, a similar shock can produce movements running to several billion dollars a day" (Kindleberger 1987). With the establishment of such a volatile source of international liquidity, national authorities lose control over domestic monetary and economic policy.

The transnational flow of information has also grown apace. The number of international transactions per average working day grew from 13,000 in 1979 to an estimated 116,000 in 1987 (Sauvant 1983).

1.2.4 Context of new transnationalism

The number of overlapping policy spaces has grown as well. Economic and welfare issues have risen to the international agenda, as developing states increasingly demand international economic equality, and the redistribution of economic resources. The United Nations convened a number of international conferences in the 1970s dedicated to a broad range of "global" problems, at times emphasizing their interactions (Biswas and Biswas 1985). Underlying the prolonged economic growth in the international system, concern has developed about the possibility of "resource exhaustion" in the face of population growth and international change so rapid as to impede mid-term market adjustments (Caldwell 1984; Biswas and Biswas 1985).

Domestically, the Keynesian Revolution has led governments to assume more responsibility over a wider variety of tasks. Most western governments are now responsible for the provision of full employment, growth, and stable prices, in addition to the conventional tasks of defense and printing money. There is also greater public expectation of governmental control over society.

The growing stock of scientific knowledge has led the public to presume that governments may be better able to efficiently and effectively implement a wider range of policies.

More importantly, change itself has now become a familiar aspect of the contemporary system. An increased rate in inventions and innovations, combined with a reduced interval between invention and commercialization contribute to a cascading effect of new problems, as well as foreshortening the time to deal with existing ones. Table 2.5 indicates broadly the accelerating pace of scientific and technological advances.

TABLE 2.5 Scientific discoveries and technological advances, 700–1900

Dates	*Scientific discoveries in period*	*Important inventions in period*	*Total number of discoveries and inventions in period*
701–1200	12	21	33
1201–1450	82	51	133
1451–1600	278	153	431
1601–1750	852	306	1,158
1751–1900	5,611	3,859	9,470

Source: Sprout and Sprout 1971.

2 years	Solar Battery (1953–1955)
2 years	High Speed Grinding
5 Years	Transistors (1948–1953)
6 years	Atomic Bomb (1939–1945)
10 years	Nuclear Reactor (1932–1942)
12 years	Television (1922–1934)
15 years	Radar (1925–1940)
18 years	X-Ray Tube (1884–1915)
35 years	Radio (1867–1902)
33 years	Vacuum Tube (1884–1915)
56 years	Telephone (1820–1876)
65 years	Electric Motor (1821–1886)
85 years	Steam Engine
112 years	Photography (1727–1839)

1..............1..............1..............1
1700 1800 1900 2000

FIGURE 2.1 Discovery to development gap

Source: McHale 1969; Sprout and Sprout 1971; Mensch 1979.

In short, the rate of discovery has grown from .066/year in the Black Ages to .887/year in the Middle Ages to 2.87/year in the Renaissance to 7.72/year during the antecedents to the Industrial Revolution to 63.13/year in the Modern Era.

The interval between inventions (or innovations) and their commercialization has shrunk as well. Figure 2.1 indicates this diminishing gap between invention and commercialization for a selected group of technological innovations.

Technological advances illuminate new problems as well as creating new problems. The development of new, more finely calibrated, scientific equipment reveals new problems which could not be measured or identified with older technology. It is now possible to sensitively monitor global trends from satellites, and evaluate water quality to much finer tolerances than in the past. Moreover, in the 1970s global modeling offered a framework in which observations could be systematically evaluated, complete with alternative scenarios of what could happen without human intervention, and in conjuncture with other global trends.

While problems are becoming more international and require a broader and interdisciplinary approach for their resolution, scientific advances have become increasingly more specialized. With the exponential growth of publicly available scientific knowledge, scientists are hard pressed to keep abreast of developments within their own disciplines (De Solla Price 1975), much less at their interstices.

World population has also grown at a high rate. Accelerating rates of growth exacerbate the pressures of governments, as they must satisfy the burgeoning demand of such a rapidly growing population (Choucri 1984). Table 2.6 indicates the relatively high rates of aggregate population growth occurring recently.

TABLE 2.6 Historical world population growth rates

Period	*Annual percentage growth*
1750–1800	0.4
1800–1850	0.5
1850–1900	0.5
1900–1920	0.6
1920–1930	1.0
1930–1940	1.1
1940–1950	1.0
1950–1960	1.9
1960–1970	2.0
1970–1980	1.8
1980–1981	1.7

Source: Hughes 1985.

Although rates of growth may have been accelerating, the distribution of global population has remained fairly constant, as is seen in Table 2.7. Differential rates of growth between more and less developed regions still persist.

TABLE 2.7 Distribution of world population, 1750–2000

Region		*Population (in millions)*				
		1750	*1850*	*1950*	*1983*	*2000*
Asia (except USSR)		498	801	1,281	2,730	3,564
	China	200	430	560	1,023	1,244
	India and Pakistan	190	233	434	826	1,108
	Japan	30	31	83	119	130
	Indonesia	12	23	77	156	199
	Rest of Asia (except USSR)	67	87	227	606	883
Africa		106	111	222	513	851
	North Africa	10	15	53	120	188
	Rest of Africa	96	96	169	393	663
Europe (except USSR)		125	208	392	484	511
USSR		42	76	180	272	309
America		18	64	328	649	866
	USA and Canada	2	26	166	259	302
	Latin America	16	38	162	360	603
Oceania		2	2	13	24	29
World total		791	1,262	2,515	4,677	6,130

(Continued)

TABLE 2.7 (Continued)

Region		*Population (in millions)*				
		1750	*1850*	*1950*	*1983*	*2000*
As percentage of world total						
	Asia	63	63	55	58	58
	Africa	13	9	9	11	14
	Europe	16	17	16	10	8
	USSR	5	6	7	6	5
	USA/Canada	–	2	7	6	5
	Latin America	2	3	6	8	10
	Oceana	–	–	–	–	–

Source: Hughes 1985.

Although analyses are indeterminate about the effects of such growth (i.e. at what levels will the world or states be unable to satisfy aggregate demand?) it is a "fact" that population growth increases stress on the international system by creating an escalating baseline of demand for resources most governments are called upon by their citizens to provide.

1.3 Implications of systemic change for international relations

The implications of this conjuncture of trends are ambiguous. Analysts are unused to speculating about the effects of the interplay of issues, rather than the implications of the issues or forces themselves. Moreover, such efforts have been heavily shaped by their author's presuppositions about the reactive capacity of nation-states, markets and international institutions. Many of such implications for behavior are necessarily highly speculative, as they hinge upon assumptions that leaders will recognize the greater uncertainty and act in some way to reduce it.

Early reports to the Club of Rome and subsequent global modeling exercises predicted global collapse, as did the 1980 Report to the United States Congress by the President (Council on Environmental Quality 1982; Meadows, Richardson et al. 1982; Onuf 1983; Woodwell 1985).[12] On the other hand, technological optimists such as Julian Simon and Herman Kahn refer to such projections as mere "globaloney," pointing to the adaptive capacity of modern societies (Simon and Kahn 1984). In the middle lie the projections by international diplomats and civil servants.

A less apocalyptic prediction is offered by Maurice Strong:

> … the prospect of slow but probably accelerating slide into chaos due to social limits on our ability to cope with the complexity inherent in a high-technology society. Political, psychological and institutional limitations could

> condemn the world to a vicious cycle of interlocking crises, with the institutional structure of society breaking down or becoming paralyzed by the sheer weight and complexity of problems it cannot handle. And this could happen well before resource limitations put physical restraints on man's activities.
>
> *(Strong 1973)*

Strong fears that the convergence of forces may create a systemic load which will exceed the capacity of the system to manage. As early as 1975 a Trilateral Commission Report warned that increasing demands on the state might lead to a possible "overload," and the ensuing collapse of democratic institutions (Crozier, Huntington et al. 1975). However, as mentioned earlier, it is not clear that current trajectories necessarily lead to the transcendence of some "natural limits." Rather, they need to be examined in light of the new forces brought to bear by these trends, and the effects of the uncertainty which results from their treatment in tandem.

At the heart of the conjuncture of forces is an ensuing condition of international uncertainty, described below. Two sets of behavior are affected by uncertainty. Domestically, leaders face new pressures driving them into the international arena to respond to such demands. Internationally, there are more pressures to cooperate as states must cooperate to manage shared resources, or to handle the externalities resulting from flows across borders. In addition, prevailing uncertainty may lead leaders to confer greater decision-making authority on scientists or other experts who are able to reduce such uncertainty.

1.3.1 The problems of uncertainty

Two broad trends have been described. One is the increasing difficulty of mastering any particular issue-area. Its increasingly technical nature, and the number of actors whose actions must be understood and coordinated in order to effectively manage them, combine to make the management of many international issues problematic. Second is the interdependence between issue-areas, which further impedes a country's ability to unilaterally formulate policies. This trend towards greater complexity gives rise to an associated condition of uncertainty. Faced with such new "objective" forces, decision makers may experience greater cognitive states of uncertainty. As the causal links between issues become unclear, and temporal and spatial flows in turn impede national autonomy, leaders lose certainty about planning. Charles Lindblom observes that "we shall have to come to some kind of conclusion long before we can achieve any kind of comprehensive or broad mastery of a plan" (Lindblom 1973).

The addition of the natural environment to these international issues further confounds policy making. Harvey Brooks writes:

> The ecosystems which comprise the biosphere are incredibly complex interacting systems, and we have little present capacity to predict the effect of disturbances of these systems engendered by human activity. The fear is that

> we may do irreversible damage inadvertently through the fact that our capacity to alter nature to our purpose progresses much more rapidly than our understanding of nature's complexities and hence of the ultimate effects of our alterations. There is no agreement, for example, on how robust the system as a whole is with respect to the size of the perturbations we are capable of making in the natural biogeochemical cycles. Are we dealing with a highly resilient system, with many internal compensating mechanisms, or are we dealing with a system that is in very delicate balance, where a relatively small change might trigger a shift to an entirely new and less benign stable state?
>
> *(Brooks 1982)*

Uncertainty persists with respect to all three dimensions of transnational flows and most international issues. Policy makers everywhere are unsure about the time frame in which a problem may be manifested. The dispersion between sectors or issue-areas may be murky. And geographically, the spread of a problem is often unknown. Lastly, the actual distribution of costs and benefits are often unclear from policies or events which actors strive to avoid.

Decision makers are uncertain about the precise effects of actions. More saliently, it also means that they often lack enough information about the likelihood of an event to estimate risks in order to calculate decision models of alternative outcomes, probability valuing or ranking the likelihood of different outcomes (Steinbruner 1974).[13] Writ large, most planning becomes an ungainly optimization exercise requiring linear programming models based on weak causal models with incomplete data.

Decision makers as well as analysts are often uncertain about the likely effects of policies, and the significance of evidence which is publicized as being extremely high-risk. For example, presented in 1972 with DDT in penguins, observers were unsure whether this constituted a minor anomaly, or evidence of much deeper global disruptions (Thacher 1985). The unanticipated negative side effects of many large-scale Third World development projects have come as rude shocks (Farvar and Milton 1972). We simply do not know if the appropriate systemic metaphor is one of punctuated equilibrium, exponential change, or catastrophe. There is a corresponding increase in uncertainty about the effects of public and private policies in economic issues as well as scientific ones (Strange 1986). There is also widespread ignorance about the effects of policies resulting from the inability to acquire a comprehensive understanding about the functional inter1inkages of issues. The wide diffusion of information, and rapid change of relevant information about markets and other actors' activities further impede the acquisition of such an understanding (Strange 1986).[14]

With this probing sense of the need to get the "whole picture," the deepest concern is that our approaches to governing the inter1inkages between issues will be grounded on an incomplete understanding of the world. But, we are not sure if various fields of action are linked in terms of one underlying comprehensive logic, and therefore must all be solved jointly (Lovelock 1979; Myers 1985; Myers, Nath et al. 1986), or if there is merely increased interlinkages between a host of still partially independent forces in the world, and issue-areas manifesting them. If the former is true, global multi-disciplinary institutions are necessary for problem management. If the latter is the case, many issues may become prematurely conflated, because of an underlying presumption that all issues are interrelated. If they are actually independent, then a strategy of "functional eclecticism" (Ruggie and Haas 1975) would be sufficient, given the existence of appropriate institutions. The final section addresses this last question about the degree of holism in environmental issues, and the availability of international institutions to coordinate policy making for them.

1.3.2 Domestic implications

Domestically, governments are faced with an interpenetrated set of forces which demand government action. Many of these problems stem from population growth; yet government responses are constrained by the array of international forces which buffet them.

Population growth aggravates the pressures on governments to adapt to increasing demands. Subsequent problems ensue from the various ways in which governments respond to these demands. Responses are mitigated by technological innovations, which may create new resources or enable societies to convert existing resources more efficiently.

Population growth per se is not determinative for resource deterioration or depletion. It does not directly correlate to violence, or even resource depletion (Choucri 1974; Durham 1979; Kleinman 1980; Boserup 1981; Boserup 1981; Repetto and Holmes 1983; Choucri 1984).[15] However, it is a shaping factor for governmental policy, and when a broad variety of governments worldwide face population growth rates in excess of 2% per year, pressure is placed on the underlying economic system to provide education, healthcare, food and employment (World Bank 1984). In countries where half the population is under the age of 15, governments are faced with providing employment for the burgeoning cohort which will enter the economy within the next ten years. With lower rates of growth, when governments have more time to respond, it may merely offer a market signal for the development of more efficient production to respond to increased demand.

Governments have a wide area of discretion in how they choose to respond; from forced relocation as in Indonesia and Kampuchea to energy conservation strategies, as in the US in the late 1970s. It is only when domestic responses to increased demand are blocked that population pressures become determinative, and countries are forced to rely more heavily on the international system for the provision of scarce commodities. The conjunctural elements from other historically transmitted pressures provide the constraints on responses to population pressures, and drive countries to more interactions with one another. To respond to these pressures, LDCs may escalate their demands at international meetings for global economic redistribution. Internationally, migration problems may become even more politicized, as burgeoning populations will seek to emigrate to areas offering more economic opportunities.

Domestically, governments may face challenges from disgruntled elements of the population (Azar and Moon 1984). Governments' legitimacy may be challenged by uprooted domestic groups whose aspirations are not satisfied, or who suffer from localized pollution. Food riots resulting from food shortages or food subsidy reforms have seriously threatened the legitimacy of the standing governments in Poland (1980), Sudan (1981, 1985), Tunisia (1983–1984), Morocco (1984), and Egypt. Their range of acceptable domestic policies are constrained by their involvement with international organizations, such as the International Monetary Fund. Internationally, weak Third World governments may be made more vulnerable to internationally transmitted forces as they become more reliant on international markets and aid for the provision of food and markets for domestically produced products. Disruptions in these complex networks of interrelated issues may have severe consequences for national legitimacy and international security of governments. As weak LDC governments are confronted with resource scarcity and allocation decisions, they may have to become increasingly authoritarian to constrain domestic dissent. Conversely, if the resource bases in LDCs are allowed to deteriorate further, these countries may be unable (or refuse) to generate sufficient export earnings to repay their international debts, and exacerbate the plight of the international financial system.

1.3.3 International institutional implications

As transnational flows become more prevalent and states become more interdependent, governments lose control and autonomy over domestic policies. They are more vulnerable to decisions made abroad, as well as to unanticipated shocks or crises whose origins are outside their borders. Leaders are in a pickle. Pushed by traditional nationalist and bureaucratic forces to insulate themselves from the world, they are also forced by external events to coordinate their policies with neighbors (The New York Times 1987).[16]

Logically, governments have three options to manage their sensitivity to external forces. Leaders have tried all three strategies. Governments may seek to extricate themselves from the international system, through pursuing more self-reliant strategies, including protectionism, currency zones, capital controls, and immigration quotas.

In cases where extrication is not feasible, such as with many environmental issues, as well as many short-term international economic relations, they may seek to coordinate their policies, in order to achieve jointly beneficial outcomes. Governments may need to coordinate policies in order to directly manage the spatial aspect of transnational flows. Research and policy coordination is also needed to manage the functional interlinkages between issues, so that governments will not inadvertently interfere with each other's pursuit of policy objectives. Lastly, governments may seek to defer action through temporarily transferring policy or research tasks to new groups of experts who may be better able to master the uncertainty of the issue-areas.

1.3.4 Cooperation and conflict

Interdependence gives rise to non-violent conflict as well as cooperation. Seeking to retain autonomy, governments attempt to make each other bear the adjustment costs of managing transboundary flows and reducing uncertainty.

Conflict may arise at a number of different levels of interaction. It may arise between states which disagree over values, objectives, or means. Disagreements may occur between bureaucratic policy makers, as well as experts consulted for advice on problems and more traditional foreign service officers. It may even arise for individual experts encountering cognitive dissonance between their professional views and their foreign policy instructions.

Current patterns of practice would lead observers to predict that cooperation will seldom occur, states will frequently ride freely, and everyone will discount the future highly (Ruggie 1972; Oye 1986).

However, with the added conditions of greater complexity and uncertainty, even less cooperative behavior would be forthcoming because of the lower likelihood of reciprocity either from the system or from partners (Kadushin 1981; Krasner 1983).

Actual cooperation is inhibited by three factors. No country is presently willing or capable of taking a strong leadership position. Few international institutions or governments are designed to handle the multiple dimensions of transnational flows. Lastly, it often is not clear what policy is most efficacious at achieving a desired outcome.

Many authors have related international order and cooperation to a stable balance of international power. Presently, such a distribution is in flux, as the United States' economic hegemony is widely regarded as declining, certainly relative to its unusual monopoly over international reserves, trade and production immediately following World War II (Strange 1983; Russett 1985; Strange 1985; Strange 1987). In the issue of international trade, this decay is evident in Table 2.8.

Historically, stable trading orders have closely correlated with one country's domination over world markets, notably the United Kingdom in the late 19th century and the United States following World War II.[17]

Governments or international agencies are seldom organized to cope with cross-cutting issues. They are designed according to geographic or functional logics, and thus are insensitive to recognizing or managing natural resource and environmental quality issues which transcend or cut across such institutional boundaries. For instance, in the United States the Departments of Agriculture, Interior, Energy and the Environmental Protection Agency all deal with portions of international environmental problems. The Departments of State and the

TABLE 2.8 Exports plus imports as percentage of world trade for selected countries

Years	*United Kingdom*[a]	*France*	*Germany*	*Italy*	*Japan*	*Soviet Union*	*United States*[b]
1870–80	24.0	10.8	9.7	3.0	n.a.	4.5	8.0
1880–89	22.4	10.2	10.3	2.9	n.a.	3.9	9.8
1913	15.5	7.3	12.1	2.9	n.a.	12.8	12.9
1928	13.7	6.1	9.3	2.8	n.a.	8.3	17.3
1937	14.1	4.8	8.3	2.3	5.1	7.4	16.0
1955	9.7	5.3	6.6	2.4	2.3	3.3	14.6
1960	8.7	4.9	8.1	3.1	3.3	4.2	13.4
1965	8.0	5.7	9.4	4.0	4.2	4.8	14.4
1970	6.9	6.3	11.0	4.9	6.2	3.9	15.0
1975	5.8	6.4	10.0	4.5	6.6	4.0	13.0
1980	5.9	6.6	10.0	4.7	7.1		12.6
1984	5.5	5.6	9.0	4.4	8.5		15.5

[a] United Kingdom and Ireland through 1937.
[b] North America for years 1913–1937.

Source: Waltz 1979; International Monetary Fund 1986.

National Security Council have trouble incorporating the subject matter, the knowledge base and the domestic political considerations into foreign policy to manage such issues. International organizations are designed similarly, with relatively narrow functional mandates. Nor is scientific knowledge organized by a holistic logic.

1.3.5 Empowerment of scientists and experts

Facing a bewildering array of uncertain steps between immediate policy alternatives and desired objectives, governments are increasingly faced with the need to resort to expert advice. Scientists and experts are likely to have authority conferred on them to the extent that they can mitigate uncertainty (Wilensky 1967; Sapolsky 1971; Nelkin 1975; Benveniste 1977; Mulkay 1979; Nelkin 1979).[18] In addition to being able to reduce uncertainty, resorting to expert advice is valuable domestically as a political device to avoid or postpone short-term conflict by deferring policy responsibility to the experts, even though they often lack sufficient information to provide fully bounded, coherent policy. Internationally, we may expect more cooperation and policy convergence according to the degree to which such new actors all provide concurrent and convergent policy advice, and can gain access to the decision-making process.

With a greater involvement of experts in decision making, persuasion rather than force may become a more prevalent pattern of conflict resolution. However, with asymmetric distributions of access to information, and technical competence in interpreting the information, analysts must be ever sensitive to the latent coercive effects of an uninformed acceptance of a new language of discourse underlying shared meanings. If technical competence is not shared by most cooperating parties, a technical agreement may mask coercion of weaker parties unfamiliar with the technical subtleties of the issue.

1.4 Summary and conclusions

As a result of a number of interdependent conjunctural forces, the world has become a more complex place, in turn making the effects of decisions more uncertain. These conditions pose fundamental problems for the traditional manner of conducting relations between states. The systemic principles of sovereignty and anarchy are challenged by these recent systemic conditions, as traditional national policies addressed at preserving sovereignty become increasingly difficult to sustain in the face of increasing vulnerability to international political and economic forces. The need for increased coordination of national policy making and the use of international institutions to develop collective strategies fundamentally questions the adequacy of such traditional, unilateral or "self-help" strategies. The threefold ways in which the effects of policies are transmitted internationally—spatially, functionally, and temporally—compound the uncertainty surrounding the effects of most decisions and policies.

Speculation about accommodation to these new international conditions is difficult. Countries may respond in a variety of ways, from increasing conflict to coordinating policies to delegating authority to new actors. This paper has attempted to indicate some of the traditional patterns of behavior which will be difficult to sustain in the light of such new conditions.

The apocalyptic predictions commonly heard in the 1970s do not appear to be immediately forthcoming. How states will actually respond to these recent conditions remains to be seen, but the conditions evoke a variety of problems for decision makers. Domestically, governments face greater demands from their populations, which can only be met by resorting to sources available abroad. Internationally, policy makers are caught in a web of interconnected flows and forces. The future promises to be interesting, if uncertain.

Notes

1 Similar conclusions, although drawn from an analysis of different aspects of world politics, are drawn by (Drucker 1986; Brooks 1982; OECD 1982).

2 Similar phrases coming from a variety of traditions *are* "globalization," "complexity," "complex interdependence," the "global problematique" and "structural change." The systemic changes and trends discussed in this paper are similar to those expressed by these notions. The primary thrust here is their simultaneity.

3 (Biddens 1984) follows a similar notion of system which encompasses patterns of behavior.

4 Various exceptions are (Weiner 1985) and (Teitelbaum 1984). These pieces tend to be more schematic, and lack the historical scope presented in discussions of trade, money, and pollution.

5 A former Nigerian Head of State bemoans the fact that "in most African countries it is difficult, if not impossible, to gain access to relevant and timely information on key national, regional, and global issues." He was referring to the technical aspects of issues as well as the number of potential actors. These aspects are discussed below.
6 Groups devoted to labor, commerce and industry made up only 15% of the total.
7 All numbers include computer specialists.
8 However, previously since the 1960s trade growth had exceeded production growth by 2.5%.
9 See for more refined interpretations which draw a similar conclusion.
10 These are, with their 1980 percentage shares Brazil (3.01), Spain (2.24), Mexico (1.95), Yugoslavia (.89), S. Korea (.66), Taiwan (.46), Portugal (.4), Greece (.31), Hong Kong (.27), Singapore (.16). The rate of growth for S Korea, Singapore and Taiwan was dramatic for the 1960s and 1970s, with annual rates of growth in excess of 10%.
11 In some cases, such as with inorganic compounds, this may be due to the fact that the pollutants themselves have longer residency times in the environment, and hence may affect people further away. In other cases it may be due to the greater density of habitation.
12 For a thorough overview of the Reports to the Club of Rome see (Onuf 1983), and for the global modeling exercises see (Meadows 1982).
13 For a discussion of limited response to such an environment.
14 Strange notes that international institutions are unable to collect and disseminate such information.
15 For a good study of how population pressures exacerbated existing socio-economic conditions in Central America, and contributed to the Soccer War between Honduras and El Salvador.
16 James A. Baker III, Secretary of the Treasury of the United States, puts it trenchantly. "One of the most difficult jobs in the world is coordinating economic policy. Of necessity, a country's domestic agenda has to come first unless you're going to cede sovereignty, and we're not going to do that."
17 It is striking, however, that although the United States' share of world trade is relatively unchanged since 1950, its willingness to support and comply with the norms and rules of international trade regimes has changed dramatically.
18 Scientists' effect on American domestic politics has been at best equivocal. Internationally, scientists may have a greater role insofar as there are greater incentives for them to maintain solidarity. To assess their effects one must better specify their collective characteristics.

References

Amin, S. (1977). *Imperialism and Unequal Development.* New York, Monthly Review Press.

Amin, S. (1985). "Modes of Production, History and Unequal Development." *Science and Society* 49(2): 194–207.

Azar, E. and C.-i. Moon (1984). "Third World National Security: Toward a New Conceptual Framework." *International Interactions* 11(2): 103–135.

Bahro, R. and New Left Review (1984). *From Red to Green: Interviews with New Left Review.* London, Verso.

Baker, M., L. Bassett, et al. (1985). *The World Environmental Handbook.* New York, World Environment Center.

Benveniste, G. (1977). *The Politics of Expertise.* San Francisco, Boyd and Fraser.

Biswas, A. and M. Biswas (1985). "The Global Environment: Past, Present and Future." *Resources Policy* 11(1): 25–42.

Biswas, M. and A. Biswas (1985). "Replenishing IFAD." *Food Policy* 10(1): 2–2.

Boserup, E. (1981). *Population and Technological Change: A Study of Long-Term Trends.* Chicago, University of Chicago Press.

Boserup, E. (1981). *Population and Technology.* Oxford, Blackwell.

Boulding, K. E. (1964). *The Meaning of the Twentieth Century: The Great Transition.* New York, Harper & Row.

Braudel, F. (1977). *Afterthoughts on Material Civilization and Capitalism.* Baltimore, Johns Hopkins University.

Brooks, H. (1982). "Can Technology Assure Unending Material Progress?" *Progress and Its Discontents.* G. Almond, M. Chodorow and R. H. Pearce. Berkeley, University of California Press.

Caldwell, L. K. (1984). *International Environmental Policy.* Durham, Duke University Press.

Choucri, N. (1974). *Population Dynamics and International Violence; Propositions, Insights, and Evidence.* Lexington, Lexington Books.

Choucri, N. (1984). *Multidisciplinary Perspectives on Population and Conflict.* Syracuse, Syracuse University Press.

Cohen, J. (1986). "An Epistemic Conception of Democracy." *Ethics* 97(1): 26–38.

Commoner, B. (1971). *The Closing Circle: Nature, Man, and Technology.* New York, Knopf.

Council on Environmental Quality (1982). *The Thirteenth Annual Report of the Council on Environmental Quality.* Washington D.C.

Crane, D. (1972). *Invisible Colleges: Diffusion of Knowledge in Scientific Communities.* Chicago, University of Chicago Press.

Crozier, M., S. P. Huntington, et al. (1975). *The Crisis of Democracy: A Report on the Governability of Democracies to the Trilateral Commission.* New York, New York University Press.

De Solla Price, D. J. (1975). *Science Since Babylon.* New Haven, Yale University Press.

Dicken, P. (1986). *Global Shift: Industrial Change in a Turbulent World.* London, Harper & Row.

Drucker, P. F. (1986). *The Frontiers of Management: Where Tomorrow's Decisions are Being Shaped Today.* New York, Truman Talley Books.

Durham, W. H. (1979). *Scarcity and Survival in Central America: Ecological Origins of the Soccer War.* Stanford, Stanford University Press.

Eisenbud, M. (1978). *Environment, Technology, and Health: Human Ecology in Historical Perspective.* New York, New York University Press.

Ellul, J. (1964). *The Technological Society.* New York, Knopf.

Emery, F. E. and E. L. Trist (1972). *Towards a Social Ecology: Contextual Appreciation of the Future in the Present.* London, Plenum Press.

Emmanuel, A. (1972). *Unequal Exchange: A Study of the Imperialism of Trade.* New York, Monthly Review Press.

Farvar, M. T. and J. P. Milton (1972). *The Careless Technology.* Garden City, Natural History Press.

Giddens, A. (1984). *The Constitution of Society.* Berkeley, University of California.

Gilpin, R. (1975). *U.S. Power and the Multinational Corporation: The Political Economy of Foreign Direct Investment.* New York, Basic Books.

Habermas, J. (1971). *Knowledge and Human Interests.* Boston, Beacon Press.

Hall, P. A. (1986). *Governing the Economy.* Cambridge, Polity Press.

Horkheimer, M., T. W. Adorno, et al. (1974). *Aspects of Sociology.* London, Heinemann.

Hormats, R. D. (1987). *Reforming the International Monetary System: From Roosevelt to Reagan.* New York, Foreign Policy Association.

Hughes, B. (1985). *World Future: A Critical Analysis of Alternatives.* Baltimore, Johns Hopkins University Press.

Hughes, J. D. (1975). *Ecology in Ancient Civilizations.* Albuquerque, University of New Mexico Press.

Inkeles, A. and D. H. Smith (1975). *Becoming Modern: Individual Change in Six Developing Countries.* London, Heinemann Educational.

International Monetary Fund (1986). *International Financial Statistics Yearbook*, UN.

Jacobson, H. K. (1984). *Networks of Interdependence.* New York, Knopf.

Kadushin, C. (1981). "Notes on Expectations of Rewards in N-Person Networks." *Continuities and Structural Inquiry.* P. Blau and R. K. Merton. Beverly Hills, Sage.

Katzenstein, P. J. (1971). "Hare and Tortoise: Race Toward integration." *International Organization* 25(2): 290–295.

Keohane, R. O. (1984). *After Hegemony: Cooperation and Discord in the World Political Economy.* Princeton, NJ, Princeton University Press.

Kindleberger, C. P. (1987). *Marshall Plan Days.* Boston, Allen & Unwin.

Kleinman, D. S. (1980). *Human Adaption and Population Growth: A Non-Malthusian Perspective.* Montclair, Allanheld, Osmun.

Krasner, S. D. (1976). "State Power and Structure of International Trade." *World Politics* 28(3): 317–347.

Krasner, S. D., Ed. (1983). "Structural Causes and Regime Consequences: Regimes as Intervening Variables." *International Regimes.* Ithaca, NY, Cornell University Press.

La Porte, T. R. (1975). *Organized Social Complexity: Challenge to Politics and Policy.* Princeton, Princeton University Press.

Levy, M. J. (1966). *Modernization and the Structure of Societies: A Setting for International Affairs.* Princeton, Princeton University Press.

Lindblom, C. E. (1973). *Incrementalism and Environmentalism.*Environmental Protection Agency. EPA-600/5-73-010. Washington D.C.

Lovelock, J. (1979). *Gaia: A New Look at Life on Earth.* Oxford, Oxford University Press.

Maguire, A. and J. W. Brown (1986). *Bordering on Trouble: Resources and Politics in Latin America.* Bethesda, Adler & Adler.

Majone, G. (1984). "Science and Trans-Science in Standard Setting." *Science, Technology, & Human Values* 9(1): 15–22.

McHale, J. (1969). *The Future of the Future.* New York, Braziller.

Meadows, D. H., J. M. Richardson, et al. (1982). *Groping in the Dark: The First Decade of Global Modelling.* Chichester, Wiley.

Mensch, G. (1979). *Stalemate in Technology: Innovations Overcome the Depression.* Cambridge, Ballinger Pub. Co.

Morse, E. (1976). *Modernization and the Transformation of International Relations.* New York, Basic Books.

Mulkay, M. J. (1979). *Science and the Sociology of Knowledge.* London, Allen & Unwin.

Myers, N. (1985). *The Gaia Atlas of Planet Management.* London, Pan Books.

Myers, N., U. R. Nath, et al. (1986). *Gaia: An Atlas of Planet Management.* Garden City, NY, Anchor Press/Doubleday.

National Academy of Sciences (1964). *Annual Report—National Academy of Sciences.* Ulan Press.

National Academy of Sciences (1985). *Annual Report—National Academy of Sciences.* Ulan Press.

Nelkin, D. (1975). "The Political Impact of Technical Expertise." *Social Studies of Science* 5: 35–54.

Nelkin, D. (1979). *Controversy Politics of Technical Decisions.* Beverely Hills, Sage.

Obasanjo, O. (1987). *Africa in Perspective: Myths and Realities.* New York, Council of Foreign Relations.

OECD (1982). *Total Net Official Flows from DAC Member Countries and Multilateral Agencies by Type of Flow*, OECD Publishing.

Onuf, N. G. (1983). "Reports to the Club of Rome." *World Politics*: 121–146.

Oye, K., Ed. (1986). *Cooperation Under Anarchy*. Princeton, Princeton University Press.

Pastor, R. A. (1980). *Congress and the Politics of U.S. Foreign Economic Policy, 1929–1976*. Berkeley, University of California Press.

Rapoport, A. (1966). *Two-Person Game Theory: The Essential Ideas*. Ann Arbor, University of Michigan Press.

Redclift, M. (1987). *Sustainable Development: Exploring the Contradictions*. London, Methuen.

Repetto, R. and T. Holmes (1983). "The Role of Population in Resource Depletion in Developing Countries." *Population and Development Review* 9(4): 609–632.

Ruggie, J. G. (1972). "Collective Goods and Future International Collaboration." *American Political Science Review* LXVI(3): 888–892.

Ruggie, J. G. and E. B. Haas (1975). "Environmental and Resource Interdependencies." *Commission on the Organization for the Conduct of Foreign Policy, Appendices*. Washington, D.C., US Government Printing Office. 1: 218–230.

Russett, B. (1985). *World Politics: The Menu for Choice*. New York, W.H. Freeman.

Sapolsky, H. M. (1971). *Creating the Invulnerable Deterrent: Programmatic and Bureaucratic Success in the Polaris System Development*. Cambridge, Massachusetts Institute of Technology.

Sauvant, K. P. (1983). "Transborder Data Flows and the Developing Countries." *International Organization* 47: 359–371.

Schumacher, E. F. (1973). *Small is Beautiful: A Study of Economics as if People Mattered*. London, Blond and Briggs.

Simon, H. (1969). "The Architecture of Complexity". *The Sciences of the Artificial*. Cambridge, MIT Press.

Simon, J. L. and H. Kahn (1984). *The Resourceful Earth: A Response to Global 2000*. New York, Blackwell.

Sprout, H. and M. Sprout (1971). *Toward a Politics of Planet Earth*. New York, Van Nostrand Reinhold Company.

Stein, A. (1984). "The Hegemon's Dilemma: Great Britain, the United States, and the International Economic Order." *International Organization* 38(2): 355–386.

Stein, A. and R. Rosecrance (1973). "Interdependence: Myth or Reality." *World Politics* 26(1): 1–27.

Steinbruner, J. D. (1974). *The Cybernetic Theory of Decision*. Princeton, Princeton University Press.

Strange, S. (1983). "*Cave Hic Dragones*: A Critique of Regime Analysis." *International Regimes*. S. D. Krasner. Ithaca, Cornell University Press.

Strange, S. (1985). "Protectionism and World Politics." *International Organization* 39(2): 233–259.

Strange, S. (1986). *Casino Capitalism*. New York, Blackwell.

Strange, S. (1987). "The Persistent Myth of Lost Hegemony." *International Organization* 41(4): 551–574.

Streeten, P. and World Bank (1981). *First Things First: Meeting Basic Human Needs in the Developing Countries*. New York, Oxford University Press.

Strong, M. (1973). "One Year After Stockholm." *Foreign Affairs* 51(4): 690–707.

Thacher, P. S. (1985). "Learning to Cope with Multi-Dimensional Problems." *Social Education* 49(3).

The Economist (1987). *The Economist*. September 26.

The Economist (1987). *The Economist*. June 20.

The Economist (1987). *The Economist*. October 24.

The New York Times (1987). *The New York Times.* May 31.

Thirgood, J. V. (1981). *Man and the Mediterranean Forest: A History of Resource Depletion.* London, Academic Press.

Tietelbaum, M. (1984). *The British Fertility Decline: Demographic Transition in the Crucible of the Industrial Revolution.* Princeton, Princeton University Press.

UNCTAD Bulletin (1986). *UNCTAD Bulletin.* May.

United Nations Statistical Office (1987). *Statistical Yearbook, 1982–1984,* UN.

Wallerstein, I. (1979). *The Capitalist World-Economy.* Cambridge, Cambridge University Press.

Waltz, K. (1979). *Theory of International Politics.* Reading, MA, Addison Wesley.

Weiner, B. (1985). *Human Motivation.* New York, Springer-Verlag.

White, G., R. Kates, et al. (1986). *Geography, Resources, and Environment.* Chicago, University of Chicago Press.

White, L. (1966). *The Transformation of the Roman World: Gibbon's Problem After Two Centuries.* Berkeley, University of California Press.

White, L. (1968). Machina ex deo*: Essays in the Dynamism of Western Culture.* Cambridge, MIT Press.

Wilensky, H. L. (1967). *Organizational Intelligence: Knowledge and Policy in Government and Industry.* New York, Basic Books.

Woodwell, G. (1985). "On the Limits of Nature." *The Global Possible: Resources, Development, and the New Century.* R. Repetto and World Resources Institute. New Haven, Yale University Press.

World Bank (1984). *World Bank Development Report: Population Change and Development,* World Bank, Oxford University Press.

Young, O. R. (1968). "A Systemic Approach to International Politics." Princeton, Center of International Studies, Woodrow Wilson School of Public and International Affairs: 61.

3

THE CAPACITY OF INTERNATIONAL INSTITUTIONS TO MANAGE BHOPAL-LIKE PROBLEMS[1]

Peter M. Haas

The Bhopal disaster is an example of a broader class of problems that involve managing transboundary risks associated with the production of potentially hazardous substances. International institutions have been coming to grips with such problems only since the 1972 United Nations Conference on the Human Environment (UNCHE). This chapter investigates the extent to which the United Nations system has responded to such new risks and the extent to which international institutions have learned, or are capable of learning, to capture the linkages that produce disasters like Bhopal. This chapter begins with a description of the major transboundary dimensions of industrial risk management, describes institutional responses to them, and evaluates these responses in light of their relationship to the relevant transboundary dimensions and the special needs of developing countries. Thus, the chapter has a normative concern with institutional design: I specify the important features of Bhopal-like problems whose treatment I regard as essential for effective public management. However, this exercise also serves an analytic purpose in identifying the major parameters against which a policy change can be measured, and in assessing whether appropriate lessons can be drawn by the responsible institutions and the parameters within which such institutional learning may occur.

In using the term "institutional learning," I follow Ernst Haas's definition of a "change in the definition of the problem to be solved by a given organization" on the basis of new information (Haas 1990). Information is a key factor in institutional change to environmental problems because of the high uncertainty and risk involved with managing highly technical and unfamiliar issues. In this chapter I focus on institutional learning, although references may be made in passing to the relation between institutional learning and the broader process of social learning by which states improve their capacity to deal with industrial risks and environmental problems.

Accidental pollution, such as the MIC release at Bhopal, is but one of many forms of low probability–high risk disruptions that unfold within interactive and tightly coupled systems in which consequences of action occur in a nonlinear manner (Perrow 1984; Fabiana and Theys 1987; Smets 1988). Industrial risk, in short, is but one of many forms of transboundary environmental risk that public authorities seek to manage. Industrial risks include accidents, water and air pollution resulting from operational discharges, hazardous and toxic wastes disposal, and the risks associated with the transport and handling of such substances.

Risky activities occur in most countries with industrialized sectors of the economy. Such activities often have transboundary effects, and are causally related to other activities in multiple subtle ways. The Organization for Economic Cooperation and Development (OECD) has recorded 178 major industrial accidents from 1970 to 1989 (OECD 1990)[2]. Of these, 65 (38%) occurred in developing countries. India alone accounts for 9 percent of the world's major industrial accidents occurring worldwide during this period and 25 percent of those occurring in developing countries.

The problem for international management is that the often irreversible effects of human decisions are felt in places separated by time and distance from the place where those decisions were initially made. Had Bhopal not been located in the heart of the Indian subcontinent, it could easily have posed direct transboundary threats to environmental quality and public health as Chernobyl did. Present choices threaten to circumscribe a future range of action, often in ways that are not fully understood. When there is potential for transboundary effects, no one country is able to guard against insults to its environment through its own actions, unaided by other countries' concurrent actions. Solutions may be imported from abroad—or imposed, as when U.S. lawyers descended on Bhopal. Yet the state of understanding about the nature and extent of these interlinkages remains sketchy at best, and the full extent of possible environmental disruptions and resilience are seldom known in advance.

Bhopal, as we saw in earlier chapters, also implicated problems specific to managing industrial risks in developing countries. Third World governments often lack the administrative experience or capacity to assess environmental risks, to design effective institutions, or even to monitor and enforce decisions. With little local participation in formulating decisions, public policies seldom reflect local knowledge or preferences. Even more than in the North, governments are obsessed with promoting short-term economic objectives, often at the expense of public health or longer-term environmental quality.

The Bhopal disaster revealed that the private sector, including even experienced multinationals, is incapable of effectively managing such problems alone. Investment decisions relating to the construction and siting of industrial facilities such as the Bhopal plant are commonly taken without regard to distinctive local contexts or conditions, such as poorly trained workforces, shortage of capital, and proximity to highly concentrated population centers. Established operational practices used in industrialized societies may not be easily transferred to developing countries. With only rudimentary regulatory frameworks in many developing countries, transnational corporations, such as Union Carbide at Bhopal, have no incentive to apply the strict rules they follow in developed countries. Moreover, as the United Nations Center on Transnational Corporations notes, "External constraints, like required foreign minority ownership or inadequate local infrastructure, sometimes impede the use of up-to-date technologies in developing countries" (1991).

International institutions may be of help in coping with the growing transboundary dimensions of risk associated with industrial activities. I speak here of institutions in the broad sense of "established rules, norms and conventions" mediated by formal international organizations that "prescribe behavioral roles, constrain activity, and shape expectations" (Keohane 1989; Haas, Keohane et al. 1993). Through their activities, international institutions may influence or modify patterns of state behavior within a given area of activity, such as the management of transboundary risks associated with industrial projects in developing countries. They can help provide information about the transnational effects of industrial projects, and thus contribute to their better design (Laird 1994). They can provide facilities through which member states may coordinate measures to manage the transboundary externalities from their activities and their responses to actual disasters. They may also help to develop and provide the kind of integrative policy models whose application is necessary to manage the tightly coupled nature of these problems. Through training they may also help to build national administrative and technical capabilities to deal with such problems in the future.

There are reasons to doubt the ability of international organizations to help countries manage such a challenge adequately. Few organizations have the interdisciplinary competence to formulate and disseminate policy relevant to the cross-cutting nature of these problems, and the difficulties of coordination between international organizations are well known. Future generations, who are likely to suffer most heavily from environmental degradation, are represented poorly or not at all in present decision making fora (Streeten 1986; Weiss 1989). Is the decentralized United Nations system, in particular, able to respond to the interdisciplinary and nonlinear (and hence nondiscrete) nature of such problems? (Seynes 1972; Ruggie 1979)

Analytic dimensions of transboundary environmental problems

International environmental problems, in particular hazard exports and transboundary pollution, are widely regarded as a composite of tightly coupled relationships between a variety of social activities. The environmental effects of policy choices are transmitted between countries along three analytic dimensions, reflecting the interdependencies that characterize industrial policy in a global economy.

The first dimension is intersectoral. Choices affecting one sector or area of activity are felt in other sectors as well. Effects are transmitted from one issue to another by the functional linkages between issues. Functional connections may be determined by whether activities or issues are tightly or loosely coupled to other activities or issues, and to which ones (Simon 1969; Lindblom 1973; Clark and Munn 1987).

Transboundary effects are also felt over time, as when the consequences of immediate choices are felt by future generations. The temporal dimension may be short term (such as under six months), medium term (such as six months to 5 years), and long term (over five years). With most ordinary uses of economic discounting techniques, planners generally disregard or heavily discount long-term effects, thereby downplaying the latent consequences of present industrial activity.

Finally, issues are related spatially, both when hazardous products such as pesticides are shipped across boundaries and when pollution travels over long distances, as in the case of radiation from Chernobyl. This dimension may be scaled according to the number of countries producing and consuming a given problem: from global (countries all around the world contribute to and are affected by the problem) to regional (the transboundary effects are limited to a regional scope) to widely distributed (a number of countries are affected by a problem but the problem itself is generated purely within national boundaries).

There are also three modes by which such effects are actually transmitted or mediated. First-order effects are the physical effects of action or inaction. Second-order effects are the social effects that occur as a consequence of physical degradation of the environment. Third-order effects are the consequence of actions taken to protect the environment or defer deterioration, including the opportunity costs of varying policies and mobility of investment and resources as environmental quality changes (Siebert 1987). Not all of these can be covered in this brief chapter.

Functional linkages

At Bhopal the siting of a potentially hazardous facility was tightly coupled with a number of other activities. Decisions regarding the production and management of MIC at Bhopal were closely affected by such conditions as population growth, urbanization, agricultural production, transportation and energy use, and economic policy. The industrial siting decision in turn had effects on public health, environmental quality, and agricultural production. While regularized use of MIC-based pesticides was intended to increase agricultural production, excessive use and ecosystemic build-up of such materials were destined to lead to longer-term declines.

In Bhopal, a potentially hazardous substance was produced in a densely populated area with weak technological and administrative controls. Accordingly, the accident had widespread and unanticipated consequences. Yet the proximate factors associated with the magnitude of the disaster cannot be separated from the broader social factors that laid the groundwork for such an event. Locating industrial sources nearest to the major inputs, such as incidental labor supply, reflects deeper political choices that developing countries can seldom evade: in particular, a potentially tragic choice of economic efficiency over serious public health or environmental considerations.

Events like Bhopal must be seen as part of a broader web of tightly coupled decisions that cut across many areas of public policy. In a country like India, high rates of urban growth and urbanization and limited availability of transportation and infrastructural support relate in turn to patterns of resource and energy consumption. The Union Carbide factory would not have been producing insecticides if it were not for India's broader need for agricultural chemicals to expand food production to satisfy a growing population and to obtain foreign exchange through exports. Throughout the world, but particularly in developing countries, planning and managing urban systems and improving the environment of human settlements pose a daunting challenge, particularly when urban populations are often growing at 6–8 percent per year from natural increase and rural migration.

Thus one can argue that the siting decision was preceded by (possibly unavoidable) social choices regarding centralized and large-scale industrial production, which entailed the location of factories near the major industrial inputs—labor, energy, and intermediate products—with an attendant rise in the potential for disaster. This industrial development strategy is endemic throughout the world, and surely continues in India, where concentrated growth zones are planned for the future (Cherian 1991; Cherian 1991). For the present, at least, the inadequacy of infrastructure rules out less centralized industrial development that would keep centers from being located in such vulnerable areas.

Spatial linkages

Many environmental problems emerge as a result of flows of polluting facilities and pollutants transmitted spatially between countries. That is, policy effects are felt across international borders, either physically, through more or less complex biogeochemical cycles, or by markets that transmit the price effects of local disruptions. Had Union Carbide built its factory closer to a border we could well have been discussing a physical transboundary case similar to Chernobyl or the pollution of the Rhine from the accident at Sandoz (Eijndhoven 1994). With greater reliance on international trade for enhancing national welfare, market channels, too, are becoming increasingly important for the transnational dispersal of environmental risks.

Except for its legal ramifications, the most salient spatial dimension of the Bhopal disaster was ultimately national. The accident involved purely private goods which appeared within the boundaries of one territorial state and whose effects remained within those boundaries. Nonetheless, Bhopal called attention to the transnational dimension inherent in technology transfer. Decisions developed within one region's economic or cultural context were transplanted elsewhere, such as the engineering design features of the Bhopal plant, including infrastructure, management practices, and labor skills, which were originally created for market conditions in the United States. Thus, technological decisions that appear at first to be purely national can in reality produce widely distributed effects in countries throughout the world.

Temporal linkages

Temporally, the effects of an environmental mishap may be felt in the future both by victims of that episode and in areas that suffer longer-term impacts through multiple feedback loops. The effects of toxic substances are often not fully understood when they are originally developed (as, for example, in the case of methyl isocyanate), and when populations are exposed, the full effects may not be felt for as much as twenty to fifty years. Environmental detriments, too, may be irreversible, at least in the short to medium run. Adverse effects in functionally related areas may take even longer to make themselves felt. An accident of significant proportions may cause significant changes in food and agricultural production, energy use, and population level, but such effects are unlikely to occur for at least five to ten years. Formulating better policies to address such temporal linkages is difficult because of the poor information generally available about these issues.

Institutional design

While no existing international organizations have competency to deal with all three dimensions by which Bhopal-like problems acquire transnational status, we may investigate how such institutions have responded to environmental problems within their traditionally delimited domains. Which institutions have developed new activities to take account of all three dimensions of policy linkages? The most effective institution for responding to new environmental challenges would be one that is sufficiently resilient to avoid making irreversible decisions and expert enough to respond to unfolding understanding about environmental conditions, while possibly even contributing to a better understanding of environmental threats and dynamics. Its institutional design is best seen as that of a switchboard operation, rather than a centralized or strictly hierarchically designed body. Such a design is most likely to improve the ability of governments and societies to deal with industrial and environmental risks because the institution is able to encourage social learning by ensuring that policy-relevant knowledge claims and new ideas about environmental management are provided to the organization and its members and are widely disseminated.

At a minimum, institutional learning requires an institutional design that provides for the provision of nonpartisan scientific information about the physical environment, the regularized feedback of information regarding activities by governments and firms, the support of developing countries' capacity to conduct the environmental monitoring and research and apply it indigenously to their policy process, and the widespread dissemination of such information in a readily usable manner, so that private groups can keep track of each other's activities and hold governments accountable for enforcing their environmental commitments. By soliciting input from a variety of different actors with different experiences and concerns, including the scientific community, multinational corporations (MNCs), governments, and grassroots nongovernmental organizations (NGOs), such an institution is able to help the international system respond more effectively to environmental issues.

Tightly coupled issues that give rise to events such as Bhopal can best be treated by institutions with a broad range of competencies for studying and coordinating policies and for initiating and coordinating research and monitoring activities by other agencies. The wider the functional interlinkages underlying environmental problems, the broader the range of functional skills that are required for policy making and a coordinative and, at times, catalytic agency is needed to avert disjointed policies. While analysts may regard international standard-setting and rulemaking as necessary to assure adequate management of transboundary environmental problems, such functions are more commonly performed through coordination among nations than through supranational administration.

Most importantly, institutions with the capacity to develop useful policy guidance must offer integrated management procedures that simultaneously address the multiple interlinkages that characterize transboundary problems. The United Nations Environment Programme (UNEP) notes that "what must be borne in mind, however, is that to be effective the activities that make up the environment programme demand a systems approach to their conceptualization and a multidisciplinary input to their planning and execution" (UNEP 1988a, b).

Memberships must be appropriate to include participants across the likely spatial spread of possible effects. Thus, problems arising out of physical flow of pollutants should be relegated to regional or small bodies with limited membership for most efficient responses. Regional bodies are likely to have more experience and localized knowledge about the context of such flows. Widely distributed national problems, by contrast, may be better treated by multilateral development banks and aid agencies, with global or regional membership, which may provide information about management styles and technology as well as financial assistance.

Unlike environmental problems originating with physical transnational flows, widely distributed problems do not need coordinative institutions for their resolution. Rather, their effective management relies more heavily on providing timely information about the scope of the problem, expert advice on alternative policies, training of national officials, and financial support for projects. Many initiatives have already been undertaken within such institutions as the World Bank and multilateral development agencies, which possess the appropriate membership and institutional resources. Preventing or mitigating such problems in the future will require the involvement of institutions with regional to global membership, as well as technical capabilities to identify problems, gather data, monitor and evaluate environmental quality, estimate risks, assess projects' impacts, disseminate information, provide consultants and expert advice, conduct training, and coordinate national and international programs (Skolnikoff 1972; Kay and Jacobson 1983)

Institutional responses

International institutions were alerted to issues of environmental management in the late 1960s by a host of widely publicized disasters. Two years of preparations for the 1972 United Nations Conference on the Human Environment made national governments sensitive to a variety of new environmental policy concerns. The conference adopted a declaration with 26 principles and 109 recommendations aimed at getting UN agencies to integrate major environmental issues within their programmatic activities (Skolnikoff and Kay 1972).

Since 1972, numerous existing institutions created new programs to cope with newly revealed policy needs. Some dimensions of these problems corresponded to existing institutional channels for dealing with international issues. Some others, however, because they cut across the narrower functional authority of existing international agencies, required the establishment of new bodies or the creative coordination of ongoing programs by intergovernmental and nongovernmental organizations. As different institutions modified their activities to cope with the environmental aspects of their domains of action, their efforts were orchestrated by UNEP, which was created at UNCHE with a "catalytic" mission to spur and coordinate environmental action throughout the United Nations system. Since its inception in 1973, UNEP has expanded its focus to incorporate the activities of nongovernmental organizations as well.

Even with the creation of UNEP and the Environment Coordination Board in 1972, efforts to coordinate and catalyze environmental activities within the United Nations were fraught with interorganizational jealousies and conflicts. The consolidation in 1978 of coordinating responsibilities within a single senior-level working committee of the heads of the major UN agencies with environmental responsibilities, known as the Designated Officials for Environmental Matters (DOEM), was but a modest institutional reform.

In attempting to instill comprehensive environmental policies throughout the UN system, UNEP tried to link decision making across a host of functionally interlinked issues, including population growth, agricultural production, industrial siting and general industrial practices, and energy use (Caldwell 1990). UNEP officials hoped to capture the interconnected nature of all environmental policy problems, including industrial wastes and the environmental aspects of development planning, through a master plan by which the Program would coordinate its own projects with those of other UN agencies. This approach, UNEP officials hoped, would lead to more coherent planning, as well as imbuing the other UN agencies with a more thoroughgoing environmental orientation (UNEP 1988a, b). In 1988 UNEP adopted the System-Wide Medium-Term Environment Programme for 1990–1995. Overall, however, UNEP's experience has been disappointing, as it lacks the financial resources to induce other agencies or governments to adopt environmental programs and the bureaucratic clout to compel them to do so (Hold-Gate 1984; McCormick 1989; Thacher 1991).

In 1988 the World Bank developed dramatically new environmental assessment procedures for many of its projects, including industrial siting for chemical and fertilizer factories. A manual was published in 1988 specifying techniques to assess industrial hazards in new projects (Technica 1988). Environmental assessments are now required for new Bank projects and components that have "diverse and significant components that may have specific environmental impacts" (World Bank 1989). Cross-sectoral guidelines were issued in 1991 to alert project designers to heed possible cross-sectoral links associated with all industrial projects, as well as possible physical transboundary effects. The new guidelines also called for strengthening the administrative capabilities of local environmental authorities (World Bank 1991a; World Bank 1991b).

The Bank approved programmatic changes as well. Over the last two years the Bank has begun to consider ecological factors in its traditional development-oriented activities. New funds were allocated to support projects that would promote environmental protection or compensate for environmental degradation. About 25 percent of all World Bank projects were assessed for their environmental impact during 1990. While structural adjustment loans are excluded from environmental assessment, four of the 14 structural adjustment loans administered in 1989 had explicit environmental objectives; agriculture and the environment in the Gambia, natural resources management in Ghana, and forestry in Guinea-Buissau and Laos. Five sector adjustment loans specifically addressed environmental objectives. Loans for free-standing environmental projects increased in number from two during the 1989 fiscal year to eleven in fiscal 1990, and 107 of 222 total approved sectoral loans contained environmental components. The Bank announced in 1990 that from 1990 to 1992 it planned to conduct environmental assessments for 376 projects in the pipeline, although only 58 (15%) will receive intensive assessment. Of intended projects, 47 percent will receive environmental assessments: 40 percent of African projects; 59 percent of Asian projects; 53 percent of European, Middle East, and North African projects; and 41 percent of Latin American and Caribbean projects. Most of the work relates to investments in power, energy, and agriculture, followed by transportation and industry, reflecting the Bank's overall portfolio (Goodland 1990; World Bank 1990a; World Bank 1990b). The $3.5 billion Global Environment Fund was established by the World Bank, UNEP, and the United Nations Development Program (UNDP) in 1990 to finance projects in developing countries with global or regional environmental benefits.

These new operational policies also created an expanded role for NGOs, who are now consulted and involved in Bank operations, particularly in providing advice on consultants and in making available local information on development sites. While these initiatives can be seen as an effort to co-opt NGO critics by bringing them inside and investing them with a stake in continued Bank lending, they have also had the effect of establishing NGOs as watchdogs. U.S. and Indian NGOs, for example, now work together in exchanging and publicizing information about World Bank activities. They also provide local information to the Bank that officials do not otherwise receive.

Other dimensions of industrial and environmental risk have been confronted by institutions directly involved with industrial risk problems and with functionally tightly coupled issues. Population problems have fallen under the purview of the World Bank and the United Nations Fund for Population Assistance (UNFPA). By 1987 UNFPA had provided more than $1.5 billion to nearly 150 countries for support of family planning, population information, education, data collection and analysis, establishment of research institutes, and formulation and implementation of population policies (UN Chronicle 1987). The United States, however, cut off its contributions to UNFPA in 1986 to protest its support for abortion. Public health efforts more generally are supported by the World Health Organization (WHO) and national aid agencies.

Several institutional efforts exist to deal with agricultural production and the second- and third-order effects of agrochemicals such as were produced at Bhopal. The thirteen institutes in the Consultative Group on International Agricultural Research conduct research on various new high-yield agricultural crops. More resistant strains are expected to reduce the need for chemical pesticides. The Food and Agricultural Organization (FAO), the World Food Council (WFC), and the International Fund for Agricultural Development (IFAD) support other efforts aimed at expanding agricultural production.

Information on the management, transport, and use of farm chemicals is collected and disseminated by many institutions. Information about agrochemicals is circulated through FAO's and WHO's Codex Alimentarius Commission, which has also adopted some two hundred international standards for food quality standards and maximum residue levels for pesticides and agrochemicals. Some forty codes of practice, guidelines, and other texts covering food products have also been developed. UNEP works with WHO and FAO to create demonstration projects on correct use of pesticides, now in place in more than thirty countries.

FAO's Code of Conduct on the Distribution and Use of Pesticides recently imposed a voluntary "prior informed consent" provision on international trade in pesticides. FAO administers technical assistance projects and training programs on the correct use and safe handling of pesticides in developing countries. Future plans include efforts to prevent groundwater contamination by pesticides and to promote correct use and safe handling of pesticides. While ambitious in conception and well intentioned in practice, such projects all too frequently lack the resources and host country commitment to affect workplace practices significantly (Paarlberg 1993).

Some steps have been taken since 1984 to improve disaster response procedures. UNEP sponsors a program for training local officials to respond to industrial accidents (UNEP 1988a, b). Since 1986 UNEP has conducted training workshops for officials from developing nations in managing hazardous wastes and controlling disasters. Draft laws and regulations are also provided to developing countries for possible national adoption. The World Bank, too, is currently developing training programs in disaster response.

In 1987 UNEP developed the London guidelines for the exchange of information on chemicals in international trade, and, with the FAO, began developing and distributing lists of chemicals banned and severely restricted by more than ten countries. With the ILO and WHO, UNEP is involved in the International Programme on Chemical Safety (IPCS) intended to conduct and disseminate assessments of the risk to human health from exposure to chemicals, as well as to support future research. Started in 1980, the program by 1992 involved twenty-seven countries and sixty-seven institutions (Mercier and Draper 1984; Garbino 1990) In 1984 the General Assembly approved the continuing update of a Consolidated List of five hundred potentially dangerous products banned, restricted, or unapproved in sixty countries, covering pharmaceuticals, agricultural chemicals, industrial chemicals, and consumer products.

UNEP is helping to design and promote hazardous waste management strategies, particularly for developing countries. UNEP provides information on the management of toxic substances, and national standards through its International Registry of Potentially Toxic Chemicals (IRPTC), which keeps a file on approximately 450 chemicals with national and international recommendations and legal requirements related to their control (see Laird, this volume, Chapter 10, for an appraisal of IRPTC). UNEP's Paris-based Industry and Environment Office circulates information on hazardous waste management techniques. UNEP developed in 1987 the Cairo Guidelines and Principles for the Environmentally Sound Management of Hazardous Wastes, and is preparing with WHO a technical manual for the safe disposal of hazardous wastes with special emphasis on the problems and needs of developing countries.

To date, developing policies for routine industrial practices largely remains an activity of the industrialized countries. Research, information transfer, and policy advice come from the Organization for Economic Cooperation and Development (OECD), the United Nations Economic Commission for Europe (ECE), and the European Community (EC). The EC responded to industrial disasters by adopting the 1986 Seveso Directive, which requires much more extensive reporting from companies and public notification of industrial siting decisions (Eijndhoven 1994). EC directives have been issued on toxic wastes, waste handling and disposal, sewage sludge, used oils, PCBs and PCTs, titanium-dioxide wastes, and packaging (Johnson and Corcelle 1989). Few infractions of these directives have been reported (International Environment Reporter 1990).

The International Labor Organization (ILO) has proposed a convention concerning safety in the use of chemicals in the workplace, stressing chemical hazard evaluation and information provision to workers, along with preventive measures and programs (Preparatory Committee for the UNCED 1990). Institutional measures also exist for the transport of hazardous wastes. The OECD adopted the first Decision on the Export of Hazardous Waste from its member countries in 1986, and the EC has a directive on the transboundary transfers of hazardous wastes. No comparable specialized programs exist as yet for managing toxic and industrial wastes in the Newly Industrializing Countries, although with high rates of industrialization and economic growth, these countries will surely soon be encountering such problems at home and in their relations with other Third World countries.

Most recently, some efforts have been taken by international organizations to capture the cross-cutting dimensions of environmental problems. Programs have been developed that are sensitive to the intersectoral nature of environmental threats and that contain elements for monitoring and linking sectors previously considered to be discrete. UNEP and WHO have developed global monitoring networks that are capable of tracing the multiple sources of environmental degradation. These illustrate for decision makers the extensive functional interlinkages among environmental issues by identifying chemical inputs into the environment and relating their introduction to observed health effects in various regions (Kjellstrom 1988; UNEP and WHO 1988a; UNEP and WHO 1988a; UNEP, WHO et al. 1988; UNEP 1989). UNEP's Global Environment Monitoring System (GEMS), run in conjunction with other specialized agencies, coordinates monitoring stations in 142 countries including some 30,000 scientists and technicians. The modeling efforts and systems analysis underway at the International Institute for Applied Systems Analysis (IIASA) in Vienna lends further scientific credibility to the linkages identified between such issues.

Institutional efforts have also shifted somewhat from the physical transmission of hazards to the social factors that promote the spread of environmental hazards. Whereas the physical effects were, in most cases, limited to the regional or widely distributed scale, the globalization of international economic relations has made the second- and third-order consequences of environmental decisions also of a global nature. International financial institutions, such as the World Bank and the International Monetary Fund, have attempted to harmonize the way in which second-order market effects are transmitted by promoting the Polluter Pays Principle and through efforts to reduce state subsidies for activities that contribute to environmental degradation. The OECD and EC also support the Polluter Pays Principle as a way to promote environmental protection consistently with their formal responsibility for harmonizing economic policies and eliminating market barriers. Economists are trying to revise the United Nations System of National Accounts in order to account for the environmental stocks from which countries generate wealth and which were traditionally not factored into systems of national accounting (IMF 1990).

UNEP trains officials in developing countries to apply economic planning techniques that pay closer attention to the multiple connections between economic development and environmental protection. UNEP has been influential in helping draft and administer multilateral treaties regulating transboundary dimensions of environmental risk. The 1989 Basel Convention on Trade in Hazardous Substances and the 1973 Convention on International Trade in Endangered Species constrain the use of markets that promote environmentally damaging activity. The Basel Convention, which closely regulates the export of hazardous wastes and demands extensive reporting from exporting to importing states, may be seen in part as a response to the worldwide outcry that greeted Bhopal.

International institutions and environmental learning

Despite bureaucratic inertia and a generalized reluctance to change fixed patterns of action, the programmatic changes described above indicate a moderate shift in international institutions' approach to managing industrial risks. Three themes are striking in this twenty-year period of institutional responses to shocks such as Bhopal and to the growing mass of new information about the complex nature of environmental risks.

First, interagency coordination to manage environmental programs systematically within the UN system failed, due to a lack of bureaucratic leverage by UNEP and by the uniform reluctance among funding countries to support such a massive overhaul of the UN system.

Second, while most institutional efforts do not deal explicitly with the underlying social causes of such problems, many institutions modified their traditional activities in order to integrate environmental considerations that directly affected their operational missions. Virtually all organizations developed projects to deal with industrial risks, and also supplemented their activities to deal with some of the direct environmental linkages to their missions, such as population for UNFPA, public health for WHO, and pesticide use for FAO.

Third, a variety of more important, functional linkages were recognized by UNEP and the World Bank. Each organization developed new programs, and the World Bank modified old programs to take account of the spatial, temporal, and functionally related environmental effects of their activities. Recent IMF and World Bank efforts that have begun to integrate long-term environmental considerations into shorter-term structural adjustments are also consistent with this pattern of more comprehensive programmatic accommodations (Ruggie 1986; Ruggie 1989).

Recent writing in political science about institutional learning helps to explain this variation in institutional responses to environmental challenges. Institutional learning is a political process through which collective behavior is modified in light of new collective understanding (Argyris and Schon 1978; Nye Jr. 1987; Haas 1990; Haas 1990; Haas 1990; Breslauer and Tetlock 1991) Such learning may take one of two forms: single-loop or double-loop. Single-loop learning refers to modifications in organizational practices that take into account new problems. Double-loop learning refers to the modification of an institutional mission, as members and officials reflect on new understanding about the world to modify their mission in light of what they now understand to be desirable or possible. New institutional objectives and means of control are adopted out of a recognition of the complex pattern of causal relations in the issue that the policy addresses. Ernst Haas regards only this second form as "true" learning; the first is adaptation to changed circumstances, which does not entail the scope that institutional responses require to manage industrial and environmental issues effectively within the context of the new ecological "problematique."

Most of the organizational responses documented above reflect merely single-loop learning. Institutions applied new information about environmental linkages to modify their existing patterns of activity incrementally. While numerous institutions developed new programs for environmental monitoring, circulating guidelines and standards, and coordinating state policies, such responses by the UNFPA, ILO, and FAO were merely adaptive: new means or instruments were appended to preexisting ends. Nor were the results comprehensive enough to be deemed adequate for designated policy purposes.

Most of the single-loop resources neglect the different context of industrial siting in the Third World. The new projects continue to reflect an overarching concern and experience with the environmental problems of the industrialized countries. The training projects sponsored by international institutions tend to retain the technocratic perspective of public planning in the North. Few provisions exist to improve administrative capabilities in developing countries, to include indigenous participation in project planning, or to inject traditional knowledge and local experience into new projects.

The World Bank and UNEP demonstrated double-loop learning by virtue of their systematic efforts to redefine their activities. Questions about the institutions' prior programmatic ends led to substantially new, comprehensive programs for dealing with industrial siting and environmental planning. These institutions' programs were also more sensitive to the different conditions in developing countries. The institutional changes of greatest potential value to developing countries are those that stress the provision of environmental information and model policies, as well as training in new technology and techniques. Examples include measures for training LDC officials in planning for emergency responses and managing hazardous wastes. Recent efforts to include NGOs in international planning may also provide a channel for local groups that lack such standing within their own states to express their concerns about industrial planning and to provide an understanding that is often absent from development projects.

Not all international institutions are equally capable of learning. Ernst Haas notes that two necessary conditions for institutional, or double-loop, learning are "a relatively stable coalition of like-minded states … and sufficient consensual knowledge … to provide the rationale for the novel nesting of problems and solutions" (Haas 1990). Since the Second Industrial Revolution in the late twentieth century, policy makers have accorded respect to specialists in science and technology because of the growing technical nature of international relations and because national power and wealth have become intimately tied to a country's scientific and technological resources. Where true institutional learning has occurred, it has often been spurred by the inclusion of transnational networks of experts in the operations of international institutions. The policy lessons learned by the institutions reflect the technical understanding and beliefs of these experts. These groups are known as epistemic communities. They are composed of professionals who share common causal beliefs, common values, common truth tests, and a common policy enterprise (Haas 1992a). Because of their shared command of potentially instrumental technical knowledge, national decision makers often defer to their advice, and they are hired by international institutions as officials and as consultants. Thus, they are responsible for articulating the precise knowledge relevant to the functional responsibilities of international institutions.

During the 1980s, members of a global ecological epistemic community emerged as key actors in international environmental protection. They were trained in the new sciences of systems ecology and informed with holistic beliefs about the nature of social and physical systems. Their preferred policies and programs reflected their causal beliefs regarding the widespread existence of tightly coupled systems, including the choices associated with industrial siting, which required comprehensive environmental planning and participation from local citizens' groups in areas where development projects were planned. Many of them were hired by international organizations and helped to draft the new environmental programs intended to cope with industrial accidents and other environmental threats. Epistemic communities imparted new information to national decision makers and to the staffs of international institutions. In turn they helped redirect the programmatic activities of their institutions through a combination of measures. They persuaded their colleagues of the need for new approaches. They helped identify projects and ideas that served as focal points around which other groups could mobilize. Finally, they gained positions of influence in national administrations and international institutions, from which they projected their desired programs. Significant policy change has been limited to those institutions—the World Bank and UNEP for the problem of industrial accidents—in which ecological epistemic community members have greatest bureaucratic influence.

While information about the magnitude of industrial problems and environmental threats has been widespread for at least a decade, only UNEP and the World Bank took significant steps to modify their operations in light of this new perspective. Double-loop learning occurred only in the institutions in which epistemic communities existed and were able to consolidate bureaucratic power within the institution.

In UNEP, for instance, a staff committed to holistic environmental planning helped design a number of programs that integrated environmental considerations into the activities of other institutions, as well as expanding UNEP's role in training developing country officials in more comprehensive planning techniques. The 1988 World Bank reforms were drafted by a like-minded group (Haas 1992b). Once inscribed in the Bank's policy, these procedural reforms exercised far more leverage than had been anticipated by the Bank's major critics in the United States Treasury Department, Congress, and the Washington NGO community, each of which had initially used environmental arguments expediently to promote its own aims. Such moves even exceeded the wishes of major donors, the Bank's primary clientele, who opposed any massive rethinking that would disrupt the flow of development finance. Still more profound and long-lasting impacts of the programmatic changes at the World Bank may be experienced through the new perspectives on economic development and environmental management that the Bank imparts in its training seminars and publications for developing country officials.

Inhibitions to institutional learning surely exist as well. Governmental opposition can inhibit the effectiveness of new institutional programs. The conceptual policy changes and programmatic reforms conceived in international institutions were not easily converted to changes in practice in developing countries that lacked resources for institutions, equipment, or training programs. As noted earlier, U.S. opposition to abortion after 1985 undercut the UNFPA budget, just as U.S. opposition to projects that would provide concessionary technology transfer to developing countries led to the limiting of U.S. financial support to UNEP after 1977.

Other general inhibitions to institutional learning may be easily identified. Institutions are unlikely to adopt programs that run counter to the interests of the stronger countries on their governing boards who contribute the majority of their budgets and who vote on programmatic matters. The FAO was long regarded as a captive of agricultural producers, and not surprisingly has limited its environmental activities to increasing agricultural yield rather than to limiting the use of farm chemicals. IMO has long been the bailiwick of tanker owners, and its focus on navigation and safety over stringent discharge standards reflects this distribution of power. The most extensive management strategies have addressed problems encountered in the North, for which it is easier to mobilize financial support from multilateral institutions which are funded primarily by the Northern countries. There are few institutional efforts of comparable scope for problems specific to developing countries, and the programs that exist remain underfunded.

Few institutions, moreover, have the organizational capacity for sustained institutional learning. They are typically staffed by people with applied skills, rather than with the interdisciplinary and holistic expertise that leads staff to recognize intersectoral linkages and act on them. Moreover, organizational habits and inertia lead bodies to avoid making dramatic or transformational programmatic shifts. Most responses remained consistent with the organizations' traditional responsibilities.

In the absence of proactive epistemic communities, institutional responses have generally been of the adaptive variety. The functionally most complex issues, for which the least amount of learning has occurred, are the ones that continue to pose the greatest challenges to international organizations responsible for environmental programs.

Conclusion

International institutions responded in a variety of ways to Bhopal and similar environmental challenges. Management of discrete aspects of the transboundary risks associated with industrial accidents was assumed by institutions within the United Nations system whose traditional responsibilities most closely resembled the analytic dimensions of the new problems. Little effective interagency coordination occurred, as most UN agencies developed new projects consistent with their traditional responsibilities. But institutions in which ecological epistemic communities gained control developed more holistic and comprehensive programs. In short, international institutions, and countries participating in them, recognized the importance of a new set of multidimensional and interconnected problems. A new, though still inadequate, pattern of practices appears to be emerging as these actors adjust their behavior to capture some of the most obvious dimensions of environmental interconnectedness.

TABLE 3.1 Major accidents involving hazardous substances, 1970–1989

Year	*Number*	*Year*	*Number*
1970	3	1980	11
1971	3	1981	8
1972	4	1982	6
1973	3	1983	6
1974	10	1984	12
1975	5	1985	15
1976	7		
1977	7	1987	12
1978	9	1988	10
1979	14	1989	9

Source: OECD 1990.

The spatial and temporal dimensions of many environmental issues were fairly well addressed by existing international bodies. New programs were directed at compensating for administrative weaknesses in developing countries that inhibit them from effectively preventing or managing such problems. However, some of the complex functional interlinkages disclosed by the Bhopal disaster elude satisfactory treatment, because they exceed the limited functional resources of current organizations and because epistemic communities capable of articulating the new scientific information to the institutions were unable to obtain access.

Major gaps therefore remain in the institutional treatment of international environmental problems, and many of the programs developed by these institutions are still only incipient. A new wave of environmental consciousness, such as the one that propelled UNCHE, may be needed in order to push the UN system and other international bodies to expand their search for relevant knowledge into a deeper phase of institutional learning. This responsibility will now lie with the Commission for Sustainable Development, established in the United Nations to coordinate institutional and international activities for sustainable development after the United Nations Conference on Environment and Development. By facilitating access of the scientific community and grassroots organizations to international discussions, and by encouraging the UN agencies to more fully integrate environmental dimensions into their activities the Commission may further promote institutional learning about the management of industrial risks.

Acknowledgements

I wish to acknowledge the valuable research assistance of Jenifer Conrad in this enterprise. My thanks to Ernst B. Haas, Sheila Jasanoff, Gene Rochlin, Peter Sand, and Oran Young for their comments on previous drafts.

Notes

1 Copyright © 1994 the University of Pennsylvania Press. This chapter was first published in Sheila Jasanoff (ed.) *Learning from Disaster: Policy Consequences of Bhopal*, Philadelphia: University of Pennsylvania Press, 1994, pp. 225–247. Reprinted with permission of the University of Pennsylvania Press.

2 Data cover incidents with at least 24 deaths, 125 injuries, 10,000 people evacuated or deprived of water, or $10 million (US) in damages. The data do not include oil spills at sea from ships or mining accidents.

References

(1991). "Companies Think Globally on the Environment." *Transnationals* 6.

Argyris, C. and D. Schon (1978). *Organizational Learning*. Reading, Addison-Wesley.

Breslauer, G. and P. Tetlock (1991). *Learning in U.S. and Soviet Foreign Policy*. Boulder, CO, Westview.

Caldwell, L. K. (1990). *International Environmental Policy*. Durham, NC, Duke University Press.

Cherian, A. (1991) "Chembur: A Dead-End for Safety." *Times of India*. August 1, 1991.

Cherian, A. (1991) "What Happened to the Model Plan?" *Times of India*. July 31, 1991.

Clark, W. C. and R. E. Munn (1987). *Sustainable Development of the Biosphere*. Cambridge, Cambridge University Press.

Eijndhoven, J. V. (1994). "Disaster Prevention in Europe". *Learning from Disaster: Risk Management After Bhopal*. S. Jasanoff. Philadelphia, University of Pennsylvania Press.

Fabiana, J.-L. and J. Theys (1987). *La Societe Vulnerable: evaluer Et Maitriser Les Risques*, Paris, Presses de l'Ecole normale superieure.

Garbino, J. d. P. (1990) "Do We Need all These Chemicals?" *World Health*: 13–15.

Goodland, R. (1990) "New Environmental Assessment Unit Promotes Operational Directive to Share 'Best Practice' Experience." *Environmental Bulletin*.

Haas, E. B. (1990). *When Knowledge is Power: Three Models of Change in International Organizations*. Berkeley, University of California Press.

Haas, P. M. (1990). *Saving the Mediterranean: The Politics of International Environmental Cooperation*. New York, Columbia University Press.

Haas, P. M. (1992a). *From Theory to Practice: Ecological Ideas and Development Policy*. Center for International Affairs; Harvard University.

Haas, P. M. (1992b). "Knowledge, Power, and International Policy Coordination." *International Organization* 46(1).

Haas, P. M., R. O. Keohane, et al., Eds. (1993). *Institutions for the Earth: Sources of Effective International Environmental Protection*. Global Environmental Accords Series. Cambridge, MA, MIT Press.

Hold-Gate, M. (1984). "UNEP: Some Personal Thoughts." *Mazingira*: 17–20.

IMF (1990). "The Economy and the Environment: Revising the National Accounts." *International Monetary Fund Survey*: 161–169.

International Environment Reporter. (1990) "Alleged Environmental Infractions Among European Community Members." *International Environment Reporter*.

Johnson, S. P. and G. Corcelle (1989). *The Environmental Policy of the European Communities*. London, Graham and Trotman.

Kay, D. and H. Jacobson, Eds. (1983). *Environmental Protection: The International Dimension*. Totowa, NJ, Allanheld Osmun.

Keohane, R. O. (1989). *International Institutions and State Power: Essays in International Relations Theory*. Boulder, CO, Westview Press.

Kjellstrom, T. (1988). "Hazards of the Environment: Measuring the Harm." *World Health*: 2–5.

Laird, F. (1994). "Information and Disaster Prevention". *Learning from Disaster: Risk Management After Bhopal*. S. Jasanoff. Philadelphia, University of Pennsylvania Press.

Lindblom, C. E. (1973). *Incrementalism and Environmentalism*. Environmental Protection Agency. EPA-600/5-73-010. Washington D.C.

McCormick, J. (1989). *Reclaiming Paradise: The Global Environmental Movement*. Bloomington, Indiana University Press.

Mercier, M. and M. Draper. (1984) "Chemical Safety: The International Outlook" *World Health*.

Nye Jr., J. S. (1987). "Nuclear Learning." *International Organization* 41(3): 378–382.

OECD (1990). *State of the Environment 1990*. Paris, OECD Publishing.

Paarlberg, R. (1993). "Managing Pesticide Use in Developing Countries". *Institutions for the Earth: Sources of Effective International Environmental Protection*. P. M. Haas, R. O. Keohane and M. A. Levy. Cambidge, MA, MIT Press.

Perrow, C. (1984). *Normal Accidents: Living with High Risk Technologies*. New York, Basic Books.

Preparatory Committee for the UNCED. (1990). *Annex I to the Report of the Secretary-General of the Conference*./CONF. 151/PC/5.

Ruggie, J. G. (1979). "On the Problems of 'the Global Problematique': What Roles for International Organizations?" *Alternatives* 5: 517–550.

Ruggie, J. G. (1986). "Social Time and International Policy: Conceptualizing Global Population and Resource Issues". *Persistent Patterns and Emergent Structures in a Waning Century*. M. P. Karns. New York, Praeger.

Ruggie, J. G. (1989). "International Structure and International Transformation: Space, Time, and Method". *Global Changes and Theoretical Challenges*. J. N. Rosenau and E.-O. Czempiel. Lexington, MA, Lexington Books.

Seynes, P. d. (1972). "Prospects for a Future Whole World." *International Organization* 26(1): 1-17.

Siebert, H. (1987). *Economics of the Environment*. Berlin, Springer-Verlag.

Simon, H. (1969). "The Architecture of Complexity". *The Sciences of the Artificial*. Cambridge, MA, MIT Press.

Skolnikoff, E. (1972). *The International Imperatives of Technology*. Berkeley: Institute of International Studies, University of California.

Skolnikoff, E. and D. Kay, Eds. (1972). *World EcoCrisis*. Madison, University of Wisconsin Press.

Smets, H. (1988). "The Cost of Accidental Pollution." *UNEP Industry and Environment*: 28–33.

Streeten, P. (1986). "What do We Owe the Future?" *Resources Policy* 12(1).

Technica (1988). *Techniques for Assessing Industrial Hazards: A Manual*. Washington D.C., World Bank.

Thacher, P. S. (1991). *Global Security and Risk Management*. Geneva, World Federation of United Nations Associations.

UN Chronicle. (1987) "Population Commission Recommends Continued Monitoring of World Population Trends and Policies." *UN Chronicle*.

UNEP. (1988a). *APELL Awareness and Preparedness for Emergencies at Local Level* No. E.88.111.D.3. New York, UNEP.

UNEP (1988b). *The United Nations Systems-Wide Medium-Term Environment Programme 1990–1995*. Nairobi, UNEP.

UNEP (1989). *Environmental Data Report*. London, Basil Blackwell.

UNEP and WHO (1988a). *Assessment of Freshwater Quality*. London, Monitoring and Assessment Research Centre.

UNEP and WHO (1988b). *Assessment of Urban Air Quality*. London, Monitoring and Assessment Research Centre.

UNEP, WHO, et al. (1988). *Assessment of Chemical Contaminants in Fund*. London, Monitoring and Assessment Research Centre.

Weiss, E. B. (1989). *In Fairness to Future Generations*. Ardsley-on-Hudson, NY, Transnational Publishers.

World Bank. (1989). *World Bank Operational Manual*. A. A. Operational Directive 4.00. Washington D.C., World Bank.

World Bank (1990a). *Annual Report 1990*. Washington D.C., World Bank.

World Bank (1990b). *The World Bank and the Environment: First Annual Report. Fiscal Year 1990*. Washington D.C., World Bank.

World Bank. (1991a). *Environment Assessment Sourcebook, vol. 3: Guidelines for Environmental Assessment of Energy and Industry Projects*. Environment Department. World Bank Technical Paper no. 154. Washington D.C., World Bank.

World Bank. (1991b). *Environmental Assessment Sourcebook, vol. 1: Policies, Procedures, and Cross-Sectoral Issues*. Environment Department. World Bank Technical Paper No. 139. Washington D.C., World Bank.

4

INTRODUCTION: EPISTEMIC COMMUNITIES AND INTERNATIONAL POLICY COORDINATION[1]

Peter M. Haas[2]

The growing technical uncertainties and complexities of problems of global concern have made international policy coordination not only increasingly necessary but also increasingly difficult. If decision makers are unfamiliar with the technical aspects of a specific problem, how do they define state interests and develop viable solutions? What factors shape their behavior? Under conditions of uncertainty, what are the origins of international institutions? And how can we best study the processes through which international policy coordination and order emerge?

While a variety of analytic approaches have been used to address the problems of international cooperation, the approaches have yielded only fragmentary insights. At its core, the study of policy coordination among states involves arguments about determinism versus free will and about the ways in which the international system is maintained and transformed. Among the overlapping topics of debate are whether national behavior is determined or broadly conditioned by system-level factors, unit-level factors, or some complex interplay between the two; whether state policymakers can identify national interests and behave independently of pressures from the social groups they nominally represent; and whether states respond consistently to opportunities to create, defend, or expand their own wealth and power, to enhance collective material benefits, or to promote nonmaterial values (Katzenstein 1977; Gourevitch 1978; Archer 1982; Krasner 1984; Lentner 1984; Benjamin and Duvall 1985; Evans, Rueschemeyer et al. 1985; Katzenstein 1986; Nordlinger 1987; Wendt 1987; Putnam 1988; Dessler 1989). A related question of debate is the extent to which state actors fully recognize and appreciate the anarchic nature of the system and, consequently, whether rational choice, deductive-type approaches or interpretive approaches are most appropriate for the study of international cooperation (Keohane 1988).

In focusing on the structure of international or domestic power in their explanations of policy coordination, many authors ignore the possibility that actors can learn new patterns of reasoning and may consequently begin to pursue new state interests. While others mention this possibility, few investigate the conditions that foster a change in state interests and the mechanisms through which the new interests can be realized (Gilpin 1981; Krasner 1983; Keohane 1984; Krasner 1985).

We acknowledge that systemic conditions and domestic pressures impose constraints on state behavior, but we argue that there is still a wide degree of latitude for state action. How states identify their interests and recognize the latitude of actions deemed appropriate in specific issue-areas of policymaking are functions of the manner in which the problems are understood by the policymakers or are represented by those to whom they turn for advice under conditions of uncertainty. Recognizing that human agency lies at the interstices between systemic conditions, knowledge, and national actions, we offer an approach that examines the role that networks of knowledge-based experts—epistemic communities—play in articulating the cause-and-effect relationships of complex problems, helping states identify their interests, framing the issues for collective debate, proposing specific policies, and identifying salient points for negotiation. We argue that control over knowledge and information is an important dimension of power and that the diffusion of new ideas and information can lead to new patterns of behavior and prove to be an important determinant of international policy coordination.

An epistemic community is a network of professionals with recognized expertise and competence in a particular domain and an authoritative claim to policy-relevant knowledge within that domain or issue-area (Polanyi 1962; Kuhn 1970; Fleck 1979; Holzner and Marx 1979).[3] Although an epistemic community may consist of professionals from a variety of disciplines and backgrounds, they have (1) a shared set of normative and principled beliefs, which provide a value-based rationale for the social action of community members; (2) shared causal beliefs, which are derived from their analysis of practices leading or contributing to a central set of problems in their domain and which then serve as the basis for elucidating the multiple linkages between possible policy actions and desired outcomes; (3) shared notions of validity—that is, intersubjective, internally defined criteria for weighing and validating knowledge in the domain of their expertise; and (4) a common policy enterprise—that is, a set of common practices associated with a set of problems to which their professional competence is directed, presumably out of the conviction that human welfare will be enhanced as a consequence.[4]

The causal logic of epistemic policy coordination is simple. The major dynamics are uncertainty, interpretation, and institutionalization. In international policy coordination, the forms of uncertainty that tend to stimulate demands for information are those which arise from the strong dependence of states on each other's policy choices for success in obtaining goals and those which involve multiple and only partly estimable consequences of action. Examples include the uncertainties about strategies to avert nuclear destruction and the uncertainties about how to respond to the hypothesized threats to an invisible layer of ozone located seven to fifteen miles above the earth's surface. These forms of uncertainty give rise to demands for particular sorts of information. The information needed does not consist of guesses about others' intentions, about the probability of discrete events occurring, or about a state's own ability to pursue unilaterally attainable goals that are amenable to treatment by various political rules of thumb. Rather, it consists of depictions of social or physical processes, their interrelation with other processes, and the likely consequences of actions that require application of considerable scientific or technical expertise. The information is thus neither guesses nor "raw" data; it is the product of human interpretations of social and physical phenomena.

Epistemic communities are one possible provider of this sort of information and advice. As demands for such information arise, networks or communities of specialists capable of producing and providing the information emerge and proliferate. The members of a prevailing community become strong actors at the national and transnational level as decision makers solicit their information and delegate responsibility to them. A community's advice, though, is informed by its own broader worldview. To the extent to which an epistemic community consolidates bureaucratic power within national administrations and international secretariats, it stands to institutionalize its influence and insinuate its views into broader international politics.

Members of transnational epistemic communities can influence state interests either by directly identifying them for decision makers or by illuminating the salient dimensions of an issue from which the decision makers may then deduce their interests. The decision makers in one state may, in turn, influence the interests and behavior of other states, thereby increasing the likelihood of convergent state behavior and international policy coordination, informed by the causal beliefs and policy preferences of the epistemic community. Similarly, epistemic communities may contribute to the creation and maintenance of social institutions that guide international behavior. As a consequence of the continued influence of these institutions, established patterns of cooperation in a given issue-area may persist even though systemic power concentrations may no longer be sufficient to compel countries to coordinate their behavior.

By focusing on the various ways in which new ideas and information are diffused and taken into account by decision makers, the epistemic communities approach suggests a nonsystemic origin for state interests and identifies a dynamic for persistent cooperation independent of the distribution of international power. It assumes that state actors are uncertainty reducers as well as power and wealth pursuers. It also seeks to explain the substantive nature of coordinated policy arrangements, a subject on which many structural analysts are notably silent. Yet to some extent, the approach supplements structural theories of international behavior: in response to new knowledge articulated by epistemic communities, a state may elect to pursue entirely new objectives, in which case outcomes may be shaped by the distribution of information as well as by the distribution of power capabilities. Table 4.1 presents a schematized outline of the epistemic communities approach and compares it with other approaches to the study of policy change that have been advanced by international relations scholars.

Pursuing the epistemic communities approach, contributors to this volume analyze the impact of five epistemic-like communities on decision making in a variety of issues concerning the international political economy, international security, and the environment. In analyzing the processes leading to policy coordination in a specific issue-area, each author describes the membership and shared

TABLE 4.1 Distinguishing epistemic communities from other groups

Approach	*Level of analysis and area of study*	*Factors that influence policy change*	*Mechanisms and effects of change*	*Primary actors*
Epistemic communities approach	Transnational; state administrators and international institutions	Knowledge; causal and principled beliefs	Diffusion of information and learning; shifts in the patterns of decision making	Epistemic communities; individual states
Neorealist approaches (Waltz 1979; Gilpin 1981; Krasner 1985)	International; states in political and economic systems	Distribution of capabilities; distribution of costs and benefits from actions	Technological changes and war; shifts in the available power resources of states and the nature of the game	States
Dependency theory-based approaches (Cardoso and Faletto 1979; Evans 1979; Wallerstein 1979; Evans 1989)	International; global system	Comparative advantage of states in the global division of labor; control over economic resources	Changes in production; shifts in the location of states in the global division of labor	States in the core, periphery and semiperiphery; multinational corporations
Poststructuralist approaches (Derian and Shapiro 1989; Ashley and Walker 1990)	International; discourse and language	Usage and meanings of words	Discourse; the opening of new political spaces and opportunities	Unclear

beliefs of an expert community, traces the community's actions, and discusses its impact. By comparing the beliefs and the behaviors of policymakers in one country over time and by comparing them in countries in which expert communities were active versus those in which they were not, the authors try to specify the extent to which decision-making processes were influenced by the community as opposed to the political factors and actors emphasized in other approaches to international relations.

The articles by William Drake and Kalypso Nicolaidis, Emanuel Adler, M.J. Peterson and Peter Haas investigate the ways in which epistemic communities initially framed the issues for collective debate, thereby influencing subsequent negotiations and bringing about their preferred outcomes to the exclusion of others in the case involving trade in services, nuclear arms control, management of whaling, and protection of stratospheric ozone. In the whaling and ozone cases, the authors also outline the role that epistemic communities played in identifying specific policies for national and collective adoption. In the study regarding the principles and practice of food aid, Raymond Hopkins traces the changes in the beliefs and understandings of the epistemic community that had a hand in the food aid regime and links these changes to regime reforms. Ethan Kapstein's analysis of banking regulators and G. John Ikenberry's analysis of economists involved in Anglo-American postwar economic settlement both shed light on the epistemic communities approach by discussing the factors that differentiate these groups from the epistemic communities discussed in the other case studies included here. And James Sebenius adds an additional viewpoint in his commentary on the commonalities and differences between the epistemic communities approach and negotiation analysis.

While all of the case studies in this volume consider the array of political and systemic constraints within which expert communities operate, Ikenberry focuses in particular on how political factors can impede the application of the consensual views of specialists. In his analysis of postwar economic management, he thus offers a limiting case, indicating that epistemic agreement was possibly in those areas removed from the political whirl. One of the conclusions that can be drawn from Ikenberry's study, as well as from earlier studies of epistemic-like communities presented elsewhere (King 1973; Wooster 1973; Russell 1974; Johnson 1975; Odell 1982; Ascher 1983; Evera 1984; Weir and Skocpol 1985; Adler 1986; Hall 1986; Crane and Finkle 1988; Hodgson 1988; Haas 1990),[5] is that while the form and specific policy choices are influenced by transnational knowledge-based networks, the extent to which state behavior reflects the preference of these networks remains strongly conditioned by the distribution of power internationally. Thus, the range of impact that we might expect of epistemic communities and epistemic-like communities remains conditioned and bounded by international and national structural realities. The extent of the conditioning—the amount of flexibility in the international system available for reflection and understanding in the face of power and structure—is the focus of this volume.

The international setting for epistemic communities

The modern administrative state: expansion, professionalization, and deference to the "knowledge elite"

Many of the major dimensions of contemporary international relations can be traced to the late nineteenth century, when crafts and guilds were declining and scientific and engineering expertise were increasingly applied to commercial research, development, and governance (Ford 1953; Dupree 1963; Chandler 1966; Chandler 1977; Mowery and Rosenberg 1989; Yates 1990).[6] Scientific rationality began to prevail over alternative paradigms of knowledge as a model for decision-making science as well, although it did not reach its peak until about fifty years later, when logical positivism and the ideas of the Vienna Circle were embraced and the entry of white-coated professionals into the public policy process became more widespread. As Harvey Brooks observed in 1965, "Much of the history of social progress in the Twentieth Century can be described in terms of the transfer of wider and wider areas of public policy from politics to expertise" (Brooks 1965). With the proliferation of government ministries and agencies to coordinate and handle many new tasks, regulation has become an increasingly important bureaucratic function (Wilson 1980; Aberbach, Putnam et al. 1981; Suleiman 1984; Moe 1989), and the expertise required has extended to a wider range of disciplines than ever before.

The domain of public governance has also grown correspondingly technical. Despite the fact that numerous ministries established for conducting War World II were decommissioned in subsequent years, the total number of ministries tripled during the period from the late 1940s to the mid-1970s. Around 1950, there were 70 independent countries with 850 ministries, or roughly 12 ministries per country. By 1975, there were 140 independent countries with 2,500 ministries, or nearly 18 ministries per country, indicating a strong shift toward more active social regulation (Blondell 1982a; Blondell 1982b).[7] The rapid growth of governmental agencies was particularly evident in the United States, where two economic regulatory agencies and five major social regulatory agencies were created during the five-year period from 1970 to 1975, while the federal budget allocations for economic and social regulation grew by 157 percent and 193 percent, respectively (Kistiakowsky 1976; Skowronek 1982; Gilpin and Wright 1984; Majone 1984; Galambos 1987; Maier 1987; Smith 1990).

Governments of industrialized countries also developed a greater interest in planning and began to establish futures-oriented research bodies (Dror 1986). With decolonization and the frequent emulation of the Western development models, the attitudes of these governments spread to those of the Third World as well (Nehru 1989; Perutz 1989).[8] This was reflected, for example, in the fact that the governments of 118 countries established agencies responsible for environmental and natural resources between 1972 and 1982.

The process of professionalization accompanied the expansion of bureaucracies in many countries. In the United States, for example, the number of scientific and technical personnel employed by the federal government grew from 123,923 in 1954 to 189,941 in 1976 to 238,041 in 1983. This mere doubling of the number over nearly three decades obscures other pertinent changes in individual expertise in the U.S. government employees. From 1973 to 1983 alone, the proportion of scientists and engineers with doctoral degrees grew by 51 percent and the proportion with masters degrees grew by 44 percent. During the same period, the government was increasing its staff of scientists, engineers, and computer specialists by 4 percent per year, while the increase for other personnel was only 2 percent a year. By 1983, scientists, engineers, and computer specialists comprised 15 percent of the government white-collar work-force, in contrast to 13 percent in 1973 and in contrast to 6 percent of the nongovernmental work-force in 1983 (National Science Foundation 1985).[9]

These trends contributed to the emergence of what Dorothy Nelkin has called "the policy role of the knowledge elite" (Primack and Hippel 1974; Nelkin 1975; Nelkin 1979a, b). The proliferation of new agencies and the practice of staffing them with professionals also contributed to the erosion of centralized control over public bureaucracies, which has occurred despite widespread efforts since World War II to curb the discretion of bureaucratic administrators. As Joel Aberbach, Robert Putnam, and Bert Rockman found in their survey of public servants in the major Western industrialized societies, the overwhelming majority of civil servants regard themselves as technicians, policymakers, and brokers, unlike elected officials, who primarily regard themselves as advocates and partisans (Aberbach, Putnam et al. 1981). In the case of professionals, the degree to which they are sympathetic with the missions of the agencies in which they work is influenced by a variety of factors, including the extent of their specialized training, the field in which they were trained, and their personal views (Dietz and Rycroft 1987; Gormley 1987; Hays 1987). In other words, "where they stand" is associated with factors other than "where they sit."

In international bureaucracies, such as the United Nations (UN), technical responsibilities have proliferated since the inauguration of the International Geophysical Year in 1957, yet the training of personnel within the UN system has not kept pace. Only 13 percent of the staff members have doctorates, and less than 50 percent hold more than a first university degree (Fromuth and Raymond 1987; Williams 1987). In 1986, when the UN employed 54,000 people worldwide, about 18,000 were serving "professional" functions, 4,000 to 5,000 of which were "substantive" in nature (Mango 1986).[10] Nevertheless, the budgeting of funds in the UN indicates a shift away from the more traditional political and security considerations of the General Assembly and toward the more technical concerns of specialized agencies (Feld 1971; Crane 1981).[11]

Thus, the expansion and professionalization of bureaucracies and the growing technical nature of problems have fostered an increase in the deference paid to technical expertise and, in particular, to that of scientists. "In modern societies," Barry Barnes and David Edge have argued, "Science is near to being the source of cognitive authority: anyone who would be widely believed and trusted as an interpreter of nature needs license from the scientific community" (Weinberg 1972; Barnes and Edge 1982).

As several studies have pointed out, policymakers and leaders typically expect to remain in control even when delegating authority (Moe 1984; Bendor, Taylor et al. 1987). Questions arise, then, about the effects that the interaction of experts and politicians have on policy choices. Many expected that scientists, because of their common faith in the scientific method, would make policymaking more rational. Yet even in cases involving what is regarded as a technical issue, policymaking decisions generally involve the weighing of a number of complex and nontechnical issues centering around who is to get what in society and at what cost. Despite the veneer of objectivity and value neutrality achieved by pointing to the input of scientists, policy choices remain highly political in their allocative consequences (Socolow 1976; Rich 1979; Ezhrahi 1980; DeLeon 1988). Especially in cases in which scientific evidence is ambiguous and the experts themselves are split into contending factions, issues have tended to be resolved less on their technical merits than on their political ones. That scientists working within the bureaucracy have a common faith in the scientific method does not guarantee their solidarity, nor does it make them immune to pressures from the institutions in which they work or from political temptation.

Studies of science policy and of scientist' effects on American Policy and regulation have been at best equivocal, finding only slight and transitory influence by scientists (Kornhauser 1962; Mulkay 1979; Nelkin 1979a, b; Weingart 1982). Similarly, early studies of policy coordination in technical areas have demonstrated that state decision makers were no more willing to sacrifice autonomy in these areas than in issues of security; that as their governments grew cognizant of the political costs of technical coordination, they grow more unwilling to coordinate their actions; and that many foreign ministries proved resistant to any encroachment by technical functional ministries on their sphere of responsibility (Ruggie 1972; Nau 1974; Williams 1974). Thus, in spite of the increasing involvement of technocrats in government institutions and contrary to the hopes of functionalists such as David Mitrany, outcomes in technical issues proved little different from those of more conventional high politics.

Unlike the functionalists, who turned their attention to the development of common activities and the transfer of technocratic loyalty to a superordinate authority, the concern of the contributors to this volume is with styles of policymaking and changes in the patterns of policymakers' reasoning. As argued below, the increasing uncertainties associated with many modern responsibilities of international governance have led policymakers to turn to new and different channels of advice, often with the result that international policy coordination is enhanced.

Decision-making processes: complexity, uncertainty, and the turn to epistemic communities for advice

Among the factors that have contributed to the uncertainties faced by decision makers are the increasingly complex and technical nature of the ever-widening range of issues considered on the international agenda, including monetary, macroeconomic, technological, environmental, health, and population issues; the growth in the complexity of the international political system in terms of the number of actors and the extent of interactions; and the expansion of the global economy and the modern administrative state (Polyani 1944; Levy 1966; Cooper 1968; Inkeles 1975; La Porte 1975; Morse 1976; OECD 1982; Ruggie 1983; Clark and Munn 1986; Toulmin 1990; Keohane and Nye 2001). Forced to deal with a broader range of issues than they were traditionally accustomed to, decision makers have turned to specialists to ameliorate the uncertainties and help them understand the current issues and anticipate future trends (Wilensky 1967; Winner 1975; Benveniste 1977; Spiegel-Rosing and De Solla Price 1977; Hirshleifer and Riley 1979; Slouka 1982; Ascher 1983; Brennan and Buchanan 1985; Dror 1986).

Complexity tests the limits of human understanding. Although knowledge may be better than it was in the past about the dynamics of any of the individual issues, the nature of the interactions between them is particularly difficult to grasp and deal with effectively in the policymaking process. For example, to the extent that economic interdependence and a globalized economy require policy coordination among countries to pursue domestic goals, the domestic agendas and international agendas have become increasingly linked, yet decision makers have often failed to comprehend the complex linkages. The result, as some analysts have complained, is that "to fear a greater extent than in the past, the individual who must make the difficult economic choices in Washington are in the dark" (Aho and Levinson 1988).

Similarly, in the case of international environmental issues, decision makers are seldom certain of the complex interplay of components of the ecosystem and are therefore unable to anticipate the long-term consequences of measures designed to address one of the many environmental issues under current consideration. Without the help of experts, they risk making choices that not only ignore the interlinkages with other issues but also highly discount the uncertain future, with the result that a policy choice made now might jeopardize future choices and threaten future generations.

Conditions of uncertainty, as characterized by Alexander George, are those under which actors must make choices without "adequate information about the situation at hand" or in the face of "the inadequacy of available general knowledge needed for assessing the expected outcomes of different courses of action" (George 1980). A growing number of issues and problems faced by decision makers fit this description. That this is true indeed undermines the utility of many conventional approaches to international relations, which presume that a state's self-interests are clear and that the ways in which its interests may be most efficaciously pursued are equally clear (Alchian 1950). As several authors have warned, however, misperceptions of the nature of the international setting, as well as misperceptions of others' intentions and actions, are most likely to occur under conditions of uncertainty (Jervis 1976; Snyder and Diesing 1977; Stein 1990; Vertzberger 1990).

Decision makers do not always recognize that their understanding of complex issues and linkages is limited, and it often takes a crisis or shock to overcome institutional inertia and habit and spur them to seek help from an epistemic community. In some cases, information generated by an epistemic community may in fact create a shock, as often occurs with scientific advances or reports that make their way into the news, simultaneously capturing the attention of the public and policymakers and pressuring them into action. In estimating the effect that shocks or crises have on decision makers, the contributors to this volume are influenced by two schools of thought. Those informed by organization theory presume that decision makers will seek information and defer to actors who are able to provide credible technical advice. Those applying the political literature presume that leaders will only defer to technical advice that will enable them to pursue preexisting ends and to expand political coalitions. This does not, however, rule out the possibility that leaders would defer to specialists under circumstances in which they are uncertain about what course of action is in their own political interests, nor does it exclude the possibility that their delegation of authority will persist past the initial crisis or shock.

The concept of uncertainty is thus important in our analysis for two reasons. First, in the face of uncertainty, and more so in the wake of a shock or crisis, many of the conditions facilitating a focus on power are absent. It is difficult for leaders to identify their potential political allies and to be sure of what strategies are most likely to help them retain power. And, second, poorly understood conditions may create enough turbulence that established operating procedures may break down, making institutions unworkable. Neither power nor institutional cues to behavior will be available, and new patterns of action may ensue.

Under conditions of uncertainty, then, decision makers have a variety of incentives and reasons for consulting epistemic communities (Williamson 1975),[12] some of them more politically motivated than others. First, following a shock or crisis, epistemic communities can elucidate the cause-and-effect relationships and provide advice about the likely results of various courses of action. In some cases, they can help decision makers gain a sense of who the winners and losers would be as the result of a particular action or event, as was the case in considerations about banning chlorofluorocarbon use or facing a possible environmental disaster. Decision makers seldom apply the types of decision-making heuristics that scientists apply under conditions of uncertainty (Kahneman, Slovic et al. 1982). Indeed, as Jon Elster argues, decision makers generally "are unable to assign numerical probabilities to the various answers of what will happen. They can at most list the possible answers, not estimate their probabilities" (Steinbruner 1974; Simon 1978; Elster 1983). While they may desire probability statistics and similar data for purposes of determining the gravity of a situation, they may also use the information for other purposes, such as justifying a "wait and watch" policy and deferring responsibility until the future, when other actors may be held responsible.

Second, epistemic communities can shed light on the nature of the complex interlinkages between issues and on the chain of events that might proceed either from failure to take action or from instituting a particular policy. Information is at a premium in the face of possible systemic volatility, when efforts to solve or curb a problem in one domain or issue-area may have unanticipated negative feedback effects on others.

Third, epistemic communities can help define the self-interests of a state or factions within it. The process of elucidating the cause-and-effect relationships of problems can in fact lead to the redefinition of preconceived interests or to the identification of new interests.

Fourth, epistemic communities can help formulate policies. Their role in this regard will depend on the reasons for which their advice is sought. In some cases, decision makers will seek advice to gain information which will justify or legitimate a policy that they wish to pursue for political ends. An epistemic community's efforts might thus be limited to working out the details of the policy, helping decision makers anticipate the conflicts of interest that would emerge with respect to particular points, and then building coalitions in support of the policy. If the policy is instituted and problems ensue, the decision makers have the option of pointing to the information given to them by experts and spreading the blame (Perl 1971; King and Melanson 1972). Again, however, it is important to stress that epistemic communities called in for political reasons may succeed in imposing their views and moving toward goals other than those initially envisioned by the decision makers.

In less politically motivated cases, epistemic communities have a greater hand in the various stages of the policymaking process, including the introduction of policy alternatives, the selection of policies, and the building of national and international coalitions in support of the policies. "The definition of the alternatives," as E. E. Schattschneider noted, "is the supreme instrument of power" (Schattschneider 1975). By pointing out which alternatives are not viable on the basis of their causal understanding of the problems to be addressed, the community members can limit the range of alternatives under consideration. While the actual choice of policies remains the domain of the decision makers, it can also be influenced by community members. As Herbert Simon points out, almost all organizations engage in some form of "satisficing" or procedural rationality in their consideration of policy alternatives (Simon 1983; Simon 1985). If rationality is bounded, epistemic communities may be responsible for circumscribing the boundaries and delimiting the options.

Distinguishing epistemic communities from other groups

As outlined earlier, members of epistemic communities not only hold in common a set of principled and causal beliefs but also have shared notions of validity and a shared policy enterprise. Their authoritative claim to policy-relevant knowledge in a particular domain is based on their recognized expertise within that domain. These features distinguish epistemic communities from other groups often involved in policy coordination.

Epistemic communities need not be made up of natural scientists; they can consist of social scientists or individuals from any discipline or profession who have a sufficiently strong claim to a body of knowledge that is valued by society. Nor need an epistemic community's causal beliefs and notions of validity be based on the methodology employed in the natural sciences; they can originate from shared knowledge about the nature of social or other processes, based on analytic methods or techniques deemed appropriate to the disciplines or professions they pursue. In this volume of articles, for example, while the community involved in efforts to protect the ozone layer claimed authority based on knowledge about atmospheric science, communities involved in other efforts had expertise related to disciplines and professions such as economics and engineering.

While national epistemic communities may emerge and direct their activities largely toward a single country, as in the case of the U.S. community and the Soviet community described by Adler, they may in some cases become transnational over time as a result of the diffusion of community ideas through conferences, journals, research collaboration, and a variety of informal communications and contacts. But epistemic communities need not be transnational, nor need their members meet regularly in a formal manner. Collaboration in the absence of material interests binding together actors in different countries with common policy agendas would strongly suggest the existence of an epistemic community with transnational membership. A transnational community's ideas may take root in an international organization or in various state bodies, after which they are diffused to other states via the decision makers who have been influenced by the ideas. As a result, the community can have a systemic impact. Because of its larger diffusion network, a transnational community's influence is likely to be much more sustained and intense than that of a national community.

The epistemic community members' professional training, prestige, and reputation for expertise in an area highly valued by society or elite decision makers accord them access to the political system and legitimize or authorize their activities. Similarly, their claims to knowledge, supported by tests of validity, accord them influence over policy debates and serve as their primary social power resource (Schluchter 1987).[13] At the same time, the professional pedigrees and validity tests set the community members apart from other social actors or groups (Haas 1990)[14] and not only serve as a barrier to their entry into the community but also limit the influence that these other actors or groups might have in the policy debate. In response to new information generated in their domain of expertise, epistemic community members may still engage in internal and often intense debates leading to a refinement of their ideas and the generation of a new consensus about the knowledge base.

As Figure 4.1 indicates, it is the combination of having a shared set of causal and principled (analytic and normative) beliefs, a consensual knowledge base, and a common policy enterprise (common interests) that distinguishes epistemic communities from various other groups. They differ from interest groups in that the epistemic community members have shared causal beliefs and cause-and-effect understandings. If confronted with anomalies that undermined their causal beliefs, they would withdraw from the policy debate, unlike interest groups. Peterson's case regarding the management of whaling, for example, stresses the difference between

		Causal beliefs	
Principled beliefs		Shared	Unshared
	Shared	Epistemic communities	Interest groups and social movements
	Unshared	Disciplines and professions	Legislators, bureaucratic agencies and bureaucratic coalitions

		Knowledge base	
Interests		Consensual	Disputed or absent
	Shared	Epistemic communities	Interests groups, social movements and bureaucratic coalitions
	Unshared	Disciplines and professions	Legislators and bureaucratic agencies

FIGURE 4.1 Distinguishing epistemic communities from other groups

the epistemic community of cetologists, the economic interest group of whaling industry managers, and the issue-oriented lobbying coalition of environmentalists.

Epistemic communities must also be distinguished from the broader scientific community as well as from professions and disciplines (Wilensky 1964;Vollmer and Mills 1966; Schluchter 1987).[15] Although members of a given profession or discipline may share a set of causal approaches or orientations and have a consensual knowledge base, they lack the shared normative commitments of members of an epistemic community. An epistemic community's ethical standards arise from its principled approach to the issue at hand, rather than from a professional code. Unlike members of a profession or discipline, who seldom limit themselves to work that is closely congruent with their principled values (Derber, Schwartz et al. 1990), members of an epistemic community tend to pursue activities that closely reflect the community's principled beliefs and tend to affiliate and identify themselves with groups that likewise reflect or seek to promote these beliefs. In practice, however, short-term alliances based on common research and concerns often exist between members of epistemic communities and professions (Haas 1989).[16]

The point to be stressed here is that while economists as a whole constitute a profession, members of a particular subgroup of economists—for example, Keynesians or followers of one of the schools of development economics—may constitute an epistemic community of their own and systematically contribute to a concrete set of projects informed by their preferred views, beliefs, and ideas.

The beliefs and goals of epistemic communities differ from those of bureaucratic bodies, but the approaches to analyzing epistemic communities and bureaucratic politics share a focus on administrative empowerment of specialized knowledge groups. Bureaucratic bodies operate largely to preserve their missions and budgets

(Allison 1971; Art 1973), whereas epistemic communities apply their causal knowledge to a policy enterprise subject to their normative objectives. Consequently, although members of epistemic communities may use the bureaucratic leverage they are able to acquire through obtaining key personnel slots within bureaucracies, their behavior is different from that of the individuals typically analyzed in terms of their bureaucratic constraints. Such a normative component means that epistemic community members are not merely policy entrepreneurs.

Because the behavior within and by an epistemic community is guided by various kinds of normative and causal beliefs as well as circumstance, it will differ from the behavior typically analyzed and predicted by rational choice theorists and principal-agent theorists. The combination of shared causal beliefs and shared principled beliefs held by epistemic community members would inform the advice they offer and would offset or outweigh the pressures for them to offer alternative advice which is more consistent with the preexisting political interests or preferences of high-level policymakers or which might further their individual careers (Olson 1965; Hechter 1987). Sociologist Joseph Ben-David, writing about scientific communities with tightly shared beliefs, notes that they provide "an example of an extreme case of effective social control by a minimum of informal sanctions" and demonstrate "one of the interesting instances where a group of people is held together by a common purpose and shared norms without the need of reinforcement by familial, ecological, or political ties" (Ben-David 1984).

The solidarity of epistemic community members derives not only from their shared interests, which are based on cosmopolitan beliefs of promoting collective betterment, but also from their shared aversions, which are based on their reluctance to deal with policy agendas outside their common policy enterprise or invoke policies based on explanations that they do not accept. The members' institutional ties, informal networks, and collective political practices also contribute to the persistence and solidarity of the community in several ways. They provide members with a valuable institutional structure in which to compare information and to find moral support for their sometimes socially and politically marginalized beliefs. They also strengthen the commitments of individuals and inhibit them from subsequently recanting the beliefs shared with and reinforced by their fellow community members.

Cognate literature

Numerous bodies of literature shed light on the three major dynamics—uncertainty, interpretation, and institutionalization—that are explored in the epistemic communities approach to the study of international policy coordination presented here. Insights gained from work in various disciplines appear to support our arguments that epistemic communities are not epiphenomenal; that policy is not merely determined by a consistent set of deeper economic, political, or social structures that in some way generate a preconditioned set of outcomes; and that while some political and social conditions surely penetrate all technical advice and

TABLE 4.2 Variables discussed in the literature on policy coordination

	Defining characteristics of variable				
Variable	*Principled beliefs*	*Causal beliefs*	*Validity test*	*Policy enterprise*	*Explanatory implications of variable*
Epistemic communities	X or	X	X	X	Policy coordination reflects epistemic community-induced changes in state interests and patterns of decision making
Ideas	X or	X			Policy coordination reflects the substance of ideas
Belief systems, operational codes, and cognitive maps	X	X			Belief systems orient behavior and shape perceptions
Consensual knowledge		X	X		Policy coordination reflects consensual knowledge
Policy networks		X		X	Policy outcomes reflect the collusion of interested parties
Transnational and transgovernmental channels and politics				X	Information is diffused and political alliances are forged via functional transnational and transgovernmental networks
Institutions and organizations				X	Policy outcomes reflect historically inherited preferences and styles

the outlooks of specialists, all specialists are not subject to the same set of conditioning forces. While international relations scholars have introduced many variables and concepts to help us understand policy outcomes and coordination (see Table 4.2), we argue that epistemic communities, as objects of study, are distinct from these concepts in that they may convey new patterns of reasoning to decision makers and encourage them to pursue new paths of policymaking, which may in turn lead to unpredicted or unpredictable outcomes.

Reality is socially constructed

Decision makers are most likely to turn to epistemic communities under conditions of uncertainty. While their goal is ostensibly to obtain "knowledge" that will ameliorate the uncertainty and give them some handle on the "reality" or "truth" of the situation at hand, the specialists called upon for advice bring with them their interpretations of the knowledge, which are in turn based on their causally informed vision of reality and their notions of validity.

As numerous social and cultural theorists have argued, reality is socially constructed. Epistemologically, the world and our representation of it are not isomorphic; our concept of reality is mediated by prior assumptions, expectations, and experience (Berger and Luckmann 1967; Toulmin 1970). Even knowledge "cannot mean the 'grasping' of reality itself," Burkhart Holzner and John Marx argue. "In fact," they add, "philosophical progress has produced the conclusive insight that there can be no such thing as the direct and 'true' apprehension of 'reality' itself. More strictly speaking, we are compelled to define 'knowledge' as the communicable mapping of some aspect of experienced reality by an observer in symbolic terms" (Holzner and Marx 1979).

In a similar vein, philosophers and sociologists of science have pointed to the epistemological difficulty in verifying our collective visions of the world. Radical constructivists, for example, contend that since every language we use to describe the world is socially constructed, there is no "objective" basis for identifying material reality and all claims for objectivity are therefore suspect (Hollis and Luke 1982; Mulkay and Knorr-Cetina 1983; Woolgar 1988). In other words, subject and object are mutually constitutive; no description can exist independently of the social circumstances under which description is made. Science, they argue, is no different from any other form of knowledge creation, and there is no basis for privileging "scientific" knowledge.

Alternatively, those with a more essentialist or materialist view argue that the world is a real and separate object of inquiry that exists independently of the analyst and that, although the categories in which it is identified are socially constructed, consensus about the nature of the world is possible in the long run. This limited constructivist view informs the analyses presented by most of the authors contributing to this volume. It also has implications for evaluating the validity of a given body of knowledge, pointing to the need for a consensus theory of a finite and temporally bounded notion of truth, rather than a correspondence theory (Rorty 1979; Munz 1985; Rouse 1987). Although knowledge is only accepted belief, not correct belief, correct beliefs may evolve over time, as progressively more accurate characterizations of the world are consensually formulated (Toulmin 1972; Laudin 1977; Laudin 1984; Campbell and Paller 1989; Campbell 1993). By reference to internally formulated truth tests, contending groups may collectively validate their conclusions and their beliefs may converge intersubjectively in the medium run.

The epistemic communities approach focuses on this process through which consensus is reached within a given domain of expertise and through which the consensual knowledge is diffused to and carried forward by other actors. Its primary concern is the political influence that an epistemic community can have on collective policymaking, rather the correctness of the advice given. While epistemic communities provide consensual knowledge, they do not necessarily generate truth. The epistemological impossibility of confirming access to reality means that the group responsible for articulating the dimensions of reality has great social and political influence. It can identify and represent what is of public concern, particularly in cases in which the physical manifestations of a problem are themselves unclear, such as the case involving threats to the stratospheric ozone layer.

Pursuing ontological lines of inquiry, some scholars in the fields of critical theory and sociology of science have taken a more instrumental approach to analyzing how the nature of reality is elucidated by groups. Rather than addressing questions about contending individual access to reality, they have investigated the effects of special social institutions on specialists' research. Specialists' choices of research programs, they argue, are influenced by a variety of factors, ranging from predispositions of funding agencies and the possibilities for career advancement to the deeper social forces that identify fruitful research questions and facilitate their investigation while impeding the investigation of others (Mannheim 1936; Habermas 1971; Tesh 1988). Moreover, scientists' observations are themselves the function of prior interests and are influenced by factors such as language usage (Whorf 1956; Segall, Campbell et al. 1966; Pitkin 1972; Cole and Scribner 1974; Laitin 1977; Lukes 1982).[17] This is emphasized by Joseph Rouse, who argues that "scientific observation of the world is theoretically selected and interpreted and functions only within a network of presupposed theories. Observation is very far from giving us an independent check on the accuracy of our theoretical representations" (Rouse 1987).[18] Technical advice that fixes attention on specific problems thus reflects more fundamental social and economical functional needs.

Many authors have been quick to point out that the increasing influence of specialized groups such as epistemic communities may have serious negative implications for such deep-seated political values as democracy and participation. The transfer of decision-making authority to a group of elite specialists, they argue, can further limit access to power by the public (Dickson 1981; Laird 1985; Sclove 1990). The Frankfurt school of scholars and many critics of modern society and technology, including Ivan Illich, Jacques Ellul, and Lewis Mumford, saw the turn of elite specialists as the first wave of victory of instrumental reason over fundamental interests. Others have warned that privileging the advice of specialists in a particular domain, such as engineering, may result in generation of "bad" decisions, either because it leads to a neglect of potentially valuable interdisciplinary insights or ignores the social ends to which decisions regarding specific issues are directed (Macrae 1976; Noble 1977; Gouldner 1979; Nelkin 1979a, b).[19]

These approaches lead us to ask what unspoken societal assumptions experts transmit. Is there a dominant social culture that influences the ideas developed and disseminated by scholars? Potentially, even Karl Popper's argument that theories can be evaluated according to their internal coherence rather than their correspondence to empirical reality must rest upon some social set of conditions which dictates the value of logical-deductive thought and the impossibility of "a" and "not a" being true simultaneously (Lakatos 1970; Popper 1972; Feyeraband 1978).[20] That new conditions bring about shifts in our own beliefs would appear to be supported by the fact that what Popper refers to as the "third world" of past lore, the body of works stored in libraries, is continually reinterpreted and evokes different responses in subsequent generations of scholars.

Additional questions remain. First, to what extent are specialists' theoretical edifices socially conditioned? And, second, does this conditioning reflect a systemic bias? That is, can it be found in all technical advice, or is it merely another social factor that must be considered in specific circumstances? If there is systematic bias, then the analysis of beliefs and perception omits some of the other fundamental forces in international politics and focuses on epiphenomena. Specialists will not have a salutary effect on policy conditions, since they only mask deeper-seated economic forces. Yet this argument goes too far.

Few would deny that technical advice reflects some prior social conditioning. I would not go so far as to argue, however, that all technical advice shares the same conditioning, that such conditioning is irrevocable over the medium to long term, or that all disciplines are equally burdened. Although the specialists' claim to privileged knowledge is certainly suspect, it is not irrevocably flawed. If the consensus theory of truth is valid, then the fundamental distortions to collective understanding may over time be reduced or at least be detectable by informed study. It is by no means clear that the same sets of constraints and censors operate in every instance of specialization and interpretation. For instance, although Jurgen Habermas argued that social needs dictate that centralized and instrumental bodies of knowledge are deployed for policy purposes (Habermas 1971), the growing application of ecologically informed views is more integrative and open in its orientation. Moreover, Michel Foucault ultimately failed to demonstrate a consistent source of social influences that operated on the development of disciplining beliefs and practices (Foucault 1973; Foucault 1978; Foucault 1979). To take an example presented in this volume of articles, corporate atmospheric scientists stuck to their scientific beliefs, counter to the immediate economic needs of their company.[21] That there have been pathological instances of the widespread joining of political norms and scientific means is reflected in Nazi medical science and Lysenko's evolutionary studies in the Soviet Union. But social pressures in these cases served as a form of control and led to their halt.

Ideas inform policies

Prevailing ideas may be an important determinant of policy choice and persistence. For instance, under the sway of economic liberalism, open trade policies emerged and remained prevalent in the nineteenth and twentieth centuries despite strong pressures toward protectionism. And in the case of the repeal of the Corn Laws, the popularity of economic liberalism helped overcome or preempt political resistance of agricultural interest groups (Kindleberger 1975; Odell 1982; Goldstein 1988; Goldstein 1989). Pointing to the persistent power of economic orthodoxy, John Maynard Keynes observed that "practical men, who believe themselves to be quite exempt from any intellectual influences, are usually the slave of some defunct economist. Madmen in authority who hear voices in the air are distilling their frenzy from some academic scribbler of years back" (Keynes 1936).

John Ruggie offers similar arguments with respect to the power of broader visions of reality, or *epistemes*, that provide the assumptions from which policies follow and shape the pattern of politics over the long run. He also introduces the term "epistemic communities" in keeping with this package of dominant worldviews:

> Institutionalization involves not only the institutional grid of the state and the international political order, through which behavior is acted out, but also the *epistemes* through which political relationships are visualized. I have borrowed this term from Michel Foucault, to refer to a dominant way of looking at social reality, a set of shared symbols and references, mutual expectations and a mutual predictability of intention. Epistemic communities may be said to consist of interrelated roles which grow up around an *episteme*; they delimit, for their members, *the* proper construction of social reality.
>
> *(Ruggie and Haas 1975)*

Our usage of the term "epistemic community" is at a lower level of abstraction than Ruggie's usage. We use the term to refer to a concrete collection of individuals who share the same worldview (or *episteme)* and in particular share the four aspects of it that were outlined earlier. While members of an epistemic community by definition share an *episteme* with each other, they do not necessarily share it with other groups or individuals. In practice, the number of members in the communities we describe is relatively small. It is the political infiltration of an epistemic community into governing institutions which lays the groundwork for a broader acceptance of the community's beliefs and ideas about the proper construction of social reality. The result in turn may be the creation of *the* proper construction of reality with respect to a particular issue-area as well as mutual expectations and a mutual predictability of intention. The intent of the articles in this volume is to analyze this process in numerous concrete cases and discern the extent to which the substantive content of policies was shaped by community views and the extent to which other actors and political forces played a role.

While the notion that ideas inform policies is provocative, it leaves a number of questions unanswered. Are ideas themselves socially conditioned, or do social conditions merely affect which ideas gain acceptance? How are ideas disseminated? Why do some prevail over others? What is the life cycle of ideas? How do they evolve? Even if scholars resort to a natural selection model for the evolution of ideas, as Donald Campbell does (Campbell 1993), then they must identify the mechanisms or agents of selection. A fruitful application of ideas to policy choice, at least over time, requires a greater specification of how ideas emerge and change (or evolve and are selected for). Without compelling answers to the questions that remain in this regard, it is difficult to support the argument that ideas are independent variables and not just intervening variables.

Epistemic communities are channels through which new ideas circulate from societies to governments as well as from country to country. However, an epistemic community cannot be reduced to the ideas it embodies or purveys, since these ideas are transmitted in tandem with a set of causal and principled beliefs and reflect a particular political vision. The ideas would be sterile without carriers, who function more or less as cognitive baggage handlers as well as gatekeepers governing the entry of new ideas into institutions. The influence that an epistemic community has and the ideas that it transmits may well be mutually reinforcing.

Actor's understanding of the world and the formulation of alternative actions are shaped by belief systems, operational codes, and cognitive maps

While we can draw on epistemological arguments about perceptions of and access to reality, we can also draw on the insights of cognitive psychologists, who stress the conditioning role that prior beliefs and established operating procedures play in determining how individuals will respond to new situations or events and choose a course of action when confronted with uncertainty (Nisbett and Ross 1980; Larson 1988; Lave 1988; Stein 1988; Holsti 1989). Faced with a new situation, we identify and interpret problems within existing frameworks and according to past protocols and then try to manage the problems according to operating procedures that we have applied in analogous cases. Aspects of situation that cannot be dealt with in established ways are only incompletely perceived and processed, with the result that salient dimensions of a problem or issue at hand are often ignored (Larson 1988; Stein 1988).

Investigating crises during which decisions affecting survival had to be made quickly by heads of state and high-level advisers, scholars have found that information processing was at best incremental and that decision makers tended to apply simplified images of reality which were highly resistant to modification (Goffman 1974; Nisbett and Ross 1980; Lave 1988; Holsti 1989). Examining non-crisis cases as well, other analysts have noted that decision makers are not always aware of the possible impact of the signals they send, since they tend to presume that the receivers of these signals have a worldview which mirrors their own (Jervis 1976; Jervis 1988). Similarly, decision makers' understanding of others' behavior is shaped by their own beliefs, motives, and intentions, and this sometimes leads them to misinterpret the signals they receive from others. Even the international context in which problems are to be resolved is not equally transparent to all actors (Snyder and Diesing 1977). Factors such as these can contribute to the breakdown of cooperation.

As Ole Holsti has argued, belief systems impose "cognitive restrains on rationality" (Holsti 1977). More broadly, the combination of prior belief systems, operational codes, and cognitive maps shapes decision makers' responses not only by influencing the ways in which they interpret the world but also by erecting barriers to the types of information they consider valuable (Boulding 1956; Boulding 1962; Axelrod 1976; George 1976; Carbonell 1981; Mefford 1987).

Yet policy responses to uncertainty cannot be reduced to cognitive psychology. Belief systems may be conferred by epistemic communities. As argued in an earlier section, whether decision makers turn to epistemic communities for advice depends on the level of their uncertainty about an issue-area. Failed policies, crises, and unanticipated events that call into question their understanding of an issue-area are likely to precipitate searches for new information, as are the increasing complexity and technical nature of problems. If decision makers have no strong preconceived views and beliefs about an issue-area in which regulation is to be undertaken for the first time, an epistemic community can have an even greater impact in shaping their interpretations and actions in this case and in establishing the patterns of behavior that they will follow in subsequent cases regarding the issue-area.

Consensual knowledge may contribute to policy coordination and more comprehensive policies

Before states can agree on whether and how to deal collectively with a specific problem, they must reach some consensus about the nature and scope of the problem and also about the manner in which the problem relates to other concerns in the same and additional issue-areas. Ernst B. Haas has argued that "interrelatedness may also become interdependence in the sense that new scientific knowledge will create a consensual basis for the recognition of new cause–effect links which had not been recognized before" (Haas 1975; Haas, Williams et al. 1977; Haas 1980). As the scientific consensus becomes the collective consensus of decision makers and as the nature of the problem is collectively redefined in broader and more interlinked terms, the need for more comprehensive patterns of policy coordination may also be recognized and pursued. Whether collective behavior becomes more comprehensive rather than merely ad hoc and incremental will in turn depend on the extent to which the scientists' and the decision makers' views coincide and the extent to which the negotiations reflect the pursuit of politically motivated linkages and the struggle for control among states. In general, governments and organizations may be said to learn through the evolution of consensual knowledge.

While the role of consensual knowledge in policy coordination has been the focus of numerous studies, the process by which the views of specialists are accepted and acted upon by decision makers are poorly specified. In particular, studies have not addressed the question of how recalcitrant states can be persuaded to accept new causal understandings that point to policies which are contrary to their conceptions of self-interest. Moreover, given the many examples of different states reacting in different ways to the same consensual evidence provided by specialists (Gillespie, Eva et al. 1979; Jackson 1986; Hoberg 1990), it is unclear how effective consensual knowledge is, as an independent variable, at explaining or predicting state behavior. The organizational structures through which consensual knowledge is diffused may be equally important. Epistemic communities can insinuate their views and influence national governments and international organizations by occupying niches in advisory and regulatory bodies. This suggests that the application of consensual knowledge to policymaking depends on the ability of the groups transmitting this knowledge to gain and exercise bureaucratic power. Moreover, while governments and organizations may learn through the evolution of knowledge, the learning does not necessarily lead to policy coordination. In the case of international commodity arrangements, for example, the consensus emerging from new knowledge was that cooperation would not result in joint gains; hence, the efforts at policy coordination collapsed (Rothstein 1984). This suggests that whether epistemic influence leads to policy coordination is a function of whether the causal beliefs of epistemic communities demonstrate the need for it.

Policy choices are often made by discrete networks of actors

Numerous scholars have argued that domestic regulation in cases involving complex and highly technical issues is often the result of collusion among interested parties. Decision making, rather than being centralized, occurs within an amorphous set of subgovernments. Whether the parties involved are characterized as interest groups, iron triangles, advocacy coalitions, issue networks, or policy networks (Heclo 1978; Walker 1981; Gais, Peterson et al. 1984; Kingdon 1984; Hansen 1987; Sabatier 1987), the point is the same: small networks of policy specialists congregate to discuss specific issues, set agendas, and formulate policy alternatives outside the formal bureaucratic channels, and they also serve as brokers for admitting new ideas into decision-making circles of bureaucrats and elected officials (Sundquist 1978). While much of the literature in this regard focuses on policymaking in the United States, similar technocratic subgovernments elsewhere have been discussed (Grindle 1980; Suleiman 1984). Unfortunately, however, most of the literature has remained descriptive rather than analytic. It does not identify or explore the common causal beliefs that participants may carry with them, nor does it indicate the degree to which such groups actually influence policy outcomes.

As with the transgovernmental literature discussed below, this literature on domestic networks supports several of the arguments pursued in the epistemic communities approach, among them the argument that a nonsystemic level of analysis is useful for considering decisions made in response to systemic stimuli and the arguments that networks of specialists can become strong actors at the national level. Key locations from which members of epistemic communities could gain significant leverage over policy choices include think tanks, regulatory agencies, and the type of governmental policy research bodies that are more common outside of the United States. Allied through transnational and transgovernmental networks, the specialists could have an impact on international policy coordination.

Coalitions are built transgovernmentally and transnationally

International relations scholars have also identified and pointed to the significance of transgovernmental and transnational channels through which political alliances are forged and information regarding technical issues is transmitted between government officials, international secretariats, nongovernmental bodies, and nongovernmental actors, including communities of professional scientists (Keohane and Nye 1971; Keohane and Nye 1974; Hopkins 1978; Evan 1981; Willets 1982; Jonsson 1986; Cernea 1988; Jonsson and Bolin 1988). Many of these scholars have argued that management tasks at the international level have to some extent been usurped by groups of functionally equivalent nonstate actors who act relatively independently of the policies of top leaders of their governments. The members of these groups, when operating in tandem through tacit alliances, can concurrently promote their ideas and specific policy objectives within their own countries and governments.

This approach describes the coordinating role of members of international secretariats and of governmental and nongovernmental bodies and the channels through which they interact, but it is unclear about what outcomes are likely to occur other than the formation of short-term policy coalitions among individuals who occupy similar positions or levels of responsibility and interact on a regular basis. Such channels could be used just as well by higher-level foreign policy officials to extend their own view of state interests. With respect to transgovernmental alliances, for example, the approach does not investigate the origin of the interests of the members involved. Do their interests stem from their common bureaucratic roles within their own governments, or are they based on preexisting beliefs and interests which they brought to their jobs and which are likely to be pursued even after they leave their current posts? In the absence of attention paid to any common causal beliefs and understandings among the members, we may well tend to conclude that such alliances will be short-lived (Crane 1984). Indeed, while those pursuing studies of transgovernmental alliances clearly identified channels of diplomacy and interstate interaction which had been neglected previously, they were unable to demonstrate that activities emanating from these channels had any independent influence on outcomes. Moreover, they found that as issues gained in saliency, they came to assume the characteristics of high politics, with the result that the transgovernmental linkages waned (Jacobson 1974).

As Diana Crane found in her study of transnational scientific groups, however, the shared causal beliefs of the individuals in these groups proved to be more important determinants of outcomes than did the channels through which they operated: "These studies show that it is not necessarily an ISPA [international scientific and professional association] which exerts political influence but the expert committee which may or may not be affiliated with an ISPA. The invisible college which cuts across all the organizations involved, both IGOs [international governmental organizations] and INGOs [international nongovernmental organizations], plays an important role in integrating the fragmented IGO programs" (Crane 1981). Crane's findings lend support to our argument that epistemic communities operating through transnationally applied policy networks can prove influential in policy coordination.

Organizations are not always captured

The recent literature on the "new institutionalism" brings together many of the arguments concerning the process of decision making (March and Olsen 1984; March and Olsen 1989).[22] Those pursuing the institutionalist approach emphasize the relative autonomy of political institutions, which may mediate the pressures on decision makers from international structures and domestic forces. Ultimately, they argue, institutional choices are influenced to a greater extent by historically inherited preferences and styles than by external structural factors. This means that the initial identification of interests and decision-making procedures will have a major influence on subsequent policy choices, alternatives deemed possible, and actual state behavior—as was evident, for example, in the choice of the Norwegian oil ministry to apply the same safety regulations to offshore oil rigs in the 1970s as it had applied to earlier oil tankers (March and Olsen 1989).

In a similar vein, studies based on organization theory indicate that institutions do not continually monitor their surroundings and reevaluate prior choices. Indeed, in the absence of crises, there will be little reconsideration of past choices, subject to what Peter Blau terms the elasticity of demand for advice (Blau 1967; March 1981). Yet when prior understandings lead to unexpected outcomes and uncertainty, organizations will take on the function of information seekers and solicitors of advice. As James Wilson notes in this regard, large institutions are more likely to generate new ideas but are less likely to adopt them than are smaller institutions (Wilson 1966).

In conjunction, the bodies of literature on organization theory and the new institutionalism shed light on the circumstances under which groups with new ideas are likely to emerge, offer an explanation for their enduring influence, and point to the conclusion that, once in place, a group will persist until subsequent crises challenge its ability to provide advice.

Methodology and guidance for further research

The epistemic communities approach pursued in this volume is distinct from the approaches reviewed above, although it rigorously integrates and builds on the insights gained from many of them. It distinguishes epistemic communities from other groups that seek to exert influence on decision makers, and it specifies in greater detail both the factors that lead knowledge-based groups to cohere and the mechanisms by which they gain and retain influence in the policymaking process.

The research techniques for demonstrating the impact of epistemic communities on the policymaking process are straightforward but painstaking. With respect to a specific community, they involve identifying community membership, determining the community members' principled and causal beliefs, tracing their activities and demonstrating their influence on decision makers at various points in time, identifying alternative credible outcomes that were foreclosed as a result of their influence, and exploring alternative explanations for the actions of decision makers (George 1979; George and McKeown 1985). The use of counterfactuals may be helpful in this regard.

One of the points emphasizes in the earlier section about distinguishing an epistemic community from another type of knowledge-based group (a discipline, a profession, a coalition of bureaucrats, and so forth) was that while the members of any knowledge-based group may share criteria of validity and a policy enterprise, members of an epistemic community in addition share principled (normative) and causal beliefs. Individuals in the community may be found among the respected experts whose names recur on delegation lists to intergovernmental meetings or among those responsible for drafting background reports on briefing diplomats. The process of tracing causal beliefs of a community calls for a detailed study of materials such as the early publications of community members, testimonies before legislative bodies, speeches, biographical accounts, and interviews. The process of tracing causal beliefs is obviously easier if members' backgrounds are in disciplines that make copious use of equations and models. Operationally, epistemic community members could be distinguished from nonmembers on the basis of whether the key variables and transformation equations incorporated in their models agree. Their beliefs and spread of networks could then be depicted by causal mapping and network analysis (Axelrod 1976). The extent to which epistemic beliefs mask social conditioning can be assessed through judicious use of secondary literature regarding the intellectual history of the disciplines from which the epistemic community derives its understanding of the world.

A robust study of an epistemic community's influence calls for comparative studies of countries and organizations in which the community has been active and those in which it has not. Moreover, it calls for an analysis of policies and practices pursued by governments and organizations not only in the period during which a community is active but also the periods before and after in order to determine both the emergence and the persistence of influence.

As the studies in this body demonstrate, epistemic communities have exerted their influence on decision makers in a wide variety of issue-areas. Generally called upon for advice under conditions of uncertainty, they have often proved to be significant actors in shaping patterns of international policy coordination.

Notes

1 Copyright © 1992 The MIT Press. This chapter was first published in *International Organization*, Vol. 46 No. 1 Winter, pp. 1–37, 1992. Reprinted with permission of the MIT Press.

2 I am grateful to Pete Andrews, Peter Cowhey, Barbara Crane, George Hoberg, Raymond Hopkins, Ethan Kapstein, Peter Katzenstein, Stephen Krasner, Craig Murphy, John Odell, Gail Osherenko, M. J. Peterson, Gene Rochlin, and Richard Sclove for comments on earlier versions of this chapter. Matt Lepori and Casey Stevens provided research assistance for this reader article.

3 The term "epistemic communities" has been defined or used in a variety of ways, most frequently to refer to scientific communities. In this volume, we stress that epistemic communities need not be made up of natural scientists or of professionals

applying the same methodology that natural scientists do. Moreover, when referring to epistemic communities consisting primarily of natural scientists, we adopt a stricter definition than do, for example, Holzner and Marx, who use the term "epistemic community" in reference to a shared faith in the scientific method as a way of generating truth. This ignores that such faith can still bond together people with diverse interpretations of ambiguous data. By our definition, what bonds members of an epistemic community is their shared belief or faith in the verity and the applicability of particular forms of knowledge or specific truths. Our notion of "epistemic community" somewhat resembles Fleck's notion of a "thought collective"—a sociological group with a common style of thinking. It also somewhat resembles Kuhn's broader sociological definition of a paradigm, which is "an entire constellation of beliefs, values, techniques, and so on shared by members of a given community" and which governs "not a subject matter but a group of practitioners."

4 Other characteristics of epistemic communities that were mentioned or discussed during the preparation of this volume include the following: members of an epistemic community share intersubjective understandings; have a shared way of knowing; have shared patterns of reasoning; have a policy project drawing on shared values, shared causal beliefs, and the use of shared discursive practices; and have a shared commitment to the application and production of knowledge. These phrases were not incorporated in the formal definition listed here; they are simply provided to evoke additional notions that are associated with epistemic communities.

5 A number of earlier studies focusing on the interplay between expertise, technical issues, consensual knowledge, and state power have considered the role of epistemic-like communities in the decision-making process. At the level of international organizations, such studies have been undertaken with regard to a wide variety of issue-areas and have demonstrated that webs of nonstate actors provided information and were involved in the shaping of agendas and the defining of state interest. While all of these studies cannot be listed here, a few examples show the range of areas analyzed. For examples of studies in comparative politics that discuss the role of epistemic-like communities in the development and enforcement of common policies see (King 1973; Weir and Skocpol 1985). With respect to policy coordination, it is worth stressing that even if actors believe that their common understandings will contribute to enhancing the collective good, serious unanticipated consequences are possible; see (Odell 1982; Evera 1984; Adler 1986; Hodgson 1988).

6 While the transfer of authority to the sphere of the secular and rational can be traced back to the eighteenth century and the granting of Noblesse de la Robe in France, the integration of scientists and engineers into a new rationalized corporate structure really began with the second industrial revolution in the 1880s. For background information see (Ford 1953). Regarding the acceleration of technically grounded forms of governance and decision making see (Dupree 1963; Chandler 1966; Chandler 1977; Mowrey and Rosenberg 1989; Yates 1990).

7 For data on the professional backgrounds of ministers and individuals occupying other ministerial posts see (Blondell 1982a, b). Blondell notes that 9.5 percent of the ministers serving between 1945 and 1981 could be considered "specialists," with most of this group consisting of civil engineers, electrical engineers, and agronomists.

8 Jawarhul Nehru, arguing that less developed countries must also turn toward science, offered the following rationale: "It is science alone that can solve the problems of hunger and poverty, of insanitation and illiteracy, of superstition and deadening custom and tradition, of vast resources running to waste, of a rich countries inhabited by starving people...Who indeed could afford to ignore science today? At every turn we have to seek its aid...The future belongs to science and those who make friends with science."

9 During the period from 1973 to 1978, the increase in scientists, engineers, and computer specialists occurred largely outside of the Defense Department.

10 Based on his survey of 75 percent of the UN's professional staff, Mango concluded that about 4,000 served key functions "in all areas of human endeavor from peace and disarmament to health, nutrition, industry, communications and the environment."Thus, for the full 100 percent of the staff, the figure may have reached 5,000.

11 The percentage of the UN budget allocated for specialized agencies steadily rose from 45.1 percent in 1950 to 60.5 percent in 1985. With the adoption of the Kaasebaum amendment, the percentage has remained at the 1985 level. Two specialized areas involving science and technology—that of food and agriculture and that of health—have come to control over 25 percent of the resources of the UN system. See UN documents nos. A/1312, A/3023, A/6122, A/7608, A/42/683, and A/10360, UN, New York, 1951, 1956, 1967, 1971, 1976, 1986, respectively. The highest postwar rates were in the areas of science and technology, followed by economics and finance.

12 Oliver Williamson argues that under conditions of uncertainty, organizations are likely to develop internal methods to generate more and better information instead of turning to external sources.

13 "It seems that in the case of functional authority," writes Schluchter, "it is the 'trust' institutionalized in the internal relations 'experts' that communicates to outsiders faith in the value of specialized knowledge."

14 According to the definition of epistemic communities employed in this volume, community members have intersubjective, internally defined validity tests. This contrasts with Ernst Haas's usage of the concept of epistemic communities, in which he explicitly mentions that such communities "profess beliefs in extracommunity reality tests." Although there are other differences between his and our usage, they are fairly minor. I believe that this particular difference in emphasis on intracommunity vs. extracommunity truth tests springs primarily from differing overarching historical visions. Ernst seeks to demonstrate the evolution of rationality over time, possibly through a gradual intercession of epistemic communities into collective decision making. For such a normative claim to be sustained, the epistemic community must share a common basis for validation of its understanding with the broader policy community. Conversely, I am much more skeptical about universal validity claims and am content to settle for the less ambitious internal truth tests. While in most cases, members outside the epistemic community may concur that validity claims exist, it is less clear that they would be able to identify or evaluate them.

15 According to A.M. Carr-Saunders, "What we now call a profession emerges when a number of persons are found to be practicing a definite technique founded upon specialized training." Subsequent sociologists have formulated a fuller definition that includes a reputation for authority, society's sanction, barriers to entry, a regulative code of conduct, and a service orientation.

16 This occurred in the context of efforts to control pollution in the Mediterranean, when several groups of natural scientists allied with the ecological epistemic community. While these scientists shared some of the causal beliefs and policy concerns of the epistemic community, they did not share its full array of normative and causal beliefs. See (Haas 1989).

17 Pitkin, following Wittgenstein, argues that language is a socially created artifact. Others argue for the strong influence of external reality on language, based on evidence of the correlation between languages with respect to particular perceptual concepts.

18 Rouse also notes that "even the simplest concepts, such as 'yellow' or 'ball,' have been said to have far-reaching theoretical assumptions."

19 Gouldner argues that the social consequences of conferring steering authority on particular groups remain unclear.

20 Lakatos offers a sophisticated extension of this argument, with normative suggestions for how to do scientific research and positive suggestions for how to evaluate truth claims from contending research programs. For an alternative viewpoint see (Feyeraband 1978).

21 Two other examples are noteworthy. Although U.S. Surgeon General C. Everett Koop opposed abortion on personal grounds, he did not succumb to political pressure to publicly

oppose it on medical grounds. A Catholic priest trained in carbon dating techniques offered evidence against the claims that the shroud found in Turin was Christ's shroud.

22 The actual definition of institutions in this literature appears remarkably fluid. Institutions may be anything from formal organizations to social forces (including capitalism) to culture.

References

Aberbach, J., R. Putnam, et al. (1981). *Bureaucrats and Politicians in Western Democracies*. Cambridge, MA, Harvard University Press.

Adler, E. (1986). "Brazil's Domestic Computer Industry." *International Organization* 40: 673–705.

Aho, M. C. and M. Levinson (1988). *After Reagan: Confronting the Changed World Economy*. New York, Council of Foreign Relations.

Alchian, A. E. (1950). "Uncertainty, Evolution, and Economic Theory." *Journal of Political Economy* 58: 211–221.

Allison, G. (1971). *Essence of Decision*. Boston, Little, Brown and Company.

Archer, M. S. (1982). "Morphogenesis Versus Structuration: On Combining Structure and Action." *British Journal of Sociology* 33: 455–483.

Art, R. (1973). "Bureaucratic Politics and American Foreign Policy." *Policy Sciences* 4: 467–490.

Ascher, W. (1983). "New Development Approaches and the Adaptability of International Agencies: The Case of the World Bank." *International Organization* 37: 415–439.

Ashley, R. K. and R. B. J. Walker (1990). "Speaking the Language of Exile Dissidence in International Studies." *International Studies Quarterly* 34(3).

Axelrod, R. (1976). *Structure of Decision*. Princeton, NJ, Princeton University Press.

Barnes, B. and D. Edge (1982). "General Introduction". *Science in Context*. B. Barnes and D. Edge. Cambridge, MA, MIT Press.

Ben-David, J. (1984). *The Scientist's Role in Society*. Chicago, University of Chicago Press.

Bendor, J., S. Taylor, et al. (1987). "Stacking the Deck: Bureaucratic Missions and Policy Design." *American Political Science Review* 81: 873–896.

Benjamin, R. and R. Duvall (1985). "The Capitalistic State in Context". *The Democratic State*. S. Elkin and R. Benjamin. Lawrence, University Press of Kansas: 19–57.

Benveniste, G. (1977). *The Politics of Expertise*. San Francisco, Boyd & Fraser.

Berger, P. L. and T. Luckmann (1967). *The Social Construction of Reality*. New York, Doubleday Anchor.

Blau, P. (1967). *Exchange and Power in Social Life*. New York, Wiley.

Blondell, J. (1982a). *Government Ministers in the Contemporary World*. Beverly Hills, CA, Sage.

Blondell, J. (1982b). *The Organization of Governments*. Beverly Hills, CA, Sage.

Boulding, K. (1956). *The Image: Knowledge in Life and Society*. Ann Arbor, University of Michigan Press.

Boulding, K. E. (1962). *Conflit and Defense*. New York, Harper.

Brennan, G. and J. M. Buchanan (1985). *The Reason of Rules*. Cambridge, Cambridge University Press.

Brooks, H. (1965). "Scientific Concepts and Cultural Change." *Daedalus* 94: 68.

Campbell, D. (1993). "Evolutionary Epistemiology". *Evolutionary Epistemology, Rationality, and the Sociology of Knowledge*. G. Radnitzky, W. W. Bartlet and K. Popper, Chicago, Open Court Publishing Company.

Campbell, D. and B. T. Paller (1989). "Extending Evolutionary Epistemology to 'Justifying Scientific Beliefs'". *Issues in Evolutionary Epistemology*. K. Halweg and C. A. Hooker. Albany, State University of New York Press.

Carbonell, J. (1981). *Subjective Understanding: Computer Models of Belief Systems.* Ann Arbor, University of Michigan Press.

Cardoso, F. H. and E. Faletto (1979). *Dependency and Development in Latin America.* Berkeley, University of California Press.

Cernea, M. M. (1988). Nongovernmental Organizations and Local Development. *Discussion Paper 40.* Washington, D.C., World Bank.

Chandler, A. D. (1966). *Strategy and Structure: Chapters in the History of the American Industrial Enterprise.* Garden City, NY, Doubleday.

Chandler, A. D. (1977). *The Visible Hand: The Managerial Revolution in American Business.* Cambidge, MA, Harvard University Press.

Clark, W. C. and R. E. Munn, Eds. (1986). *Sustainable Development of the Biosphere.* New York, Cambridge University Press.

Cole, M. and S. Scribner (1974). *Culture and Thought.* New York, Wiley.

Cooper, R. (1968). *The Economics of Interdependence.* New York, McGraw-Hill.

Crane, B. B. (1984). "Policy Coordination by Major Western Powers in Bargaining with the Third World: Debt Relief and the Common Fund." *International Organization* 38: 426.

Crane, B. B. and J. L. Finkle (1988). Population Policy and World Politics. *Fourteenth World Conference of the International Political Science Association.* Washington, D.C.

Crane, D. (1981). "Alternative Model of ISPAs". *Knowledge and Power in a Global Society.* W. M. Evan. Beverly Hills, CA, Sage.

DeLeon, P. (1988). *Advice and Consent: The Development of the Policy Sciences.* New York, Russell Sage.

Derber, C., W. A. Schwartz, et al. (1990). *Power in the Highest Degree.* New York, Oxford University Press.

Derian, J. D. and M. J. Shapiro (1989). *International/Intertextual Relations.* Lexington, MA, Lexington Books.

Dessler, D. (1989). "What's at Stake in the Agent-Structure Debate." *International Organization* 43: 441–473.

Dickson, D. (1981). "Limiting Democracy: Technocrats and the Liberal State." *Democracy* 1: 61–79.

Dietz, T. M. and R. Rycroft (1987). *The Risk Professionals.* New York, Russell Sage.

Dror, Y. (1986). *Policy Making Under Adversity.* New Brunswick, NJ, Transaction Books.

Dupree, H. (1963). *Science and the Emergence of Modern America, 1865–1915.* Chicago, Rand McNally.

Elster, J. (1983). *Explaining Technical Change Cambridge.* Cambridge, Cambridge University Press.

Evan, W. M. (1981). *Knowledge and Power in a Global Society.* Beverly Hills, CA, Sage.

Evans, P. (1979). *Dependent Development.* Princeton, NJ, Princeton University Press.

Evans, P. (1989). "Declining Hegemony and Assertive Industrialization." *International Organization* 43: 207–238.

Evans, P. B., D. Rueschemeyer, et al. (1985). *Bringing the State Back In.* Cambridge, Cambridge University Press.

Evera, S. V. (1984). "Cult of the Offensive and the Origins of the First World War." *International Security* 9: 58–107.

Ezhrahi, Y. (1980). "Utopian and Pragmatic Rationalism: The Political Context of Scientific Advance." *Minerva* 18: 111–131.

Feld, W. (1971). "Nongovernmental Entities and the International System." *Orbis* 15: 879–922.

Feyeraband, P. (1978). *Against Method.* New York, Schocken.

Fleck, L. (1979). *Genesis and Development of a Scientific Fact.* Chicago, Chicago University Press.

Ford, F. L. (1953). *Robe and Sword*. New York, Harper.

Foucault, M. (1973). *The Birth of the Clinic*. New York, Pantheon Books.

Foucault, M. (1978). *The History of Sexuality*. New York, Pantheon Books.

Foucault, M. (1979). *Discipline and Punishment: The Birth of the Prison*. New York, Vintage Books.

Fromuth, P. and R. Raymond (1987). "U.N. Personnel Policy Issues". *United Nations Management and Decision-Making Project*. New York, United Nations: 13.

Gais, T. L., M. A. Peterson, et al. (1984). "Interest Groups, Iron Triangles, and Representative Institutions in American National Government." *British Journal of Political Science* 14: 161–185.

Galambos, L. (1987). *The New American State: Bureaucracies and Policies Since World War II*. Baltimore, MD, Johns Hopkins University Press.

George, A. (1976). *The Causal Nexus Between Cognitive Beliefs and Decision-Making Behavior: The "Operational Code" Belief System*. Boulder, CO, Westview Press.

George, A. (1979). "Case Study and Theory Development: The Method of Structured, Focused Comparison". *Diplomacy: New Approaches in History, Theory, and Policy*. P. Gordon. New York, Free Press: 43–68.

George, A. (1980). *Presidential Decisionmaking in Foreign Policy: The Effective Use of Information and Advice*. Boulder, CO, Westview Press.

George, A. and T. J. McKeown (1985). "Case Studies and Theories of Organizational Decision Making." *Advances in Information Processing in Organizations* 2: 21–58.

Gillespie, B., D. Eva, et al. (1979). "Carcinogenic Risk Assessment in the United States and Great Britain: The Case of Aldrin/Dieldrin." *Social Studies of Science* 9: 265–301.

Gilpin, R. (1981). *War and Change in World Politics*. Cambridge, Cambridge University Press.

Gilpin, R. and C. Wright (1984). *Scientists and National Policy Making*. New York, Columbia University Press.

Goffman, E. (1974). *Frame Analysis*. Cambridge, MA, Harvard University Press.

Goldstein, J. (1988). "Ideas, Institutions, and Trade Policy." *International Organization* 32: 179–217.

Goldstein, J. (1989). "The Impact of Ideas on Trade Policy." *International Organization* 43: 31–71.

Gormley, W. T. J. (1987). *Professionalism Within Environmental Bureaucracies: The Policy Implications of Personnel Choices*. Madison, WI, La Follette Institute of Public Affairs.

Gouldner, A. W. (1979). *The Future of Intellectuals and the Rise of the New Class*. New York, Seabury Press.

Gourevitch, P. A. (1978). "The Second Image Reversed: International Sources of Domestic Politics." *International Organization* 32(4): 881–912.

Grindle, M. S. (1980). *Politics and Policy Implementation in the Third World*. Princeton, NJ, Princeton University Press.

Haas, E. B. (1975). "Knowledge, Technology, Interdependence." *International Organization* 29: 858–859.

Haas, E. B. (1980). "Why Collaborate? Issue Linkage and International Regimes." *World Politics* 32(3): 357–405.

Haas, E. B. (1990). *When Knowledge is Power: Three Models of Change in International Organizations*. Berkeley, University of California Press.

Haas, E. B., M. P. Williams, et al. (1977). *Scientists and World Order*. Berkeley, University of California Press.

Haas, P. (1989). "Do Regimes Matter? Epistemic Communities and Mediterranean Pollution Control." *International Organization* 43(3): 377–403.

Haas, P. M. (1990). *Saving the Mediterranean: The Politics of International Environmental Cooperation*. New York, Columbia University Press.

Habermas, J. (1971). *Knowledge and Human Interests*. Boston, Beacon Press.

Hall, P. A. (1986). *Governing the Economy*. Cambridge, Polity Press.

Hansen, J. M. (1987). Creating a New Politics: The Evolution of Agricultural Policy Networks in Congress, 1918–1980, Yale University. Ph.D. Dissertation.

Hays, S. P. (1987). *Beauty, Health and Permanence: Environmental Politics in the United States, 1955–1985*. Cambridge, Cambridge University Press.

Hechter, M. (1987). *The Principles of Group Solidarity*. Berkeley, University of California Press.

Heclo, H. (1978). "Issue Networks and the Executive Establishment". *The New American Political System*. A. King. Washington, D.C., American Enterprise Institute: 87–124.

Hirshleifer, J. and J. G. Riley (1979). "The Analytics of Uncertainty and Information: An Expository Survey." *Journal of Economic Literature* 17: 1375–1412.

Hoberg, G. J. (1990). "Risk, Science and Politics: Alachlor Regulation in Canada and the United States." *Canadian Journal of Political Science* 23: 257–277.

Hodgson, D. (1988). "Orthodoxy and Revisionism in America Demography." *Population and Development Review* 14: 541–569.

Hollis, M. and S. Luke (1982). *Rationality and Relativism*. Cambridge, MIT Press.

Holsti, O. (1977). *The "Operational Code" as an Approach to the Analysis of Belief Systems*, National Science Foundation.

Holsti, O. (1989). "Crisis Decision Making". *Behavior, Society, and Nuclear War*. P. E. Tetlock, J. L. Husbands, R. Jervis, P. C. Stern and C. Tilly. New York, Oxford University Press: 8–84.

Holzner, B. and J. H. Marx (1979). *Knowledge Application: The Knowledge System in Society*. Boston, Allyn & Bacon.

Hopkins, R. F. (1978). "Global Management Networks: The Internationalization of Domestic Bureaucracies." *International Social Science Journal* 30: 31–46.

Inkeles, A. (1975). "The Emerging Social Structure of the World." *World Politics* 27(4): 467–495.

Jackson, S. (1986). *Risk Management and Political Culture*. New York, Russell Sage.

Jacobson, H. K. (1974). "WHO: Medicine, Regionalism, and Managed Politics". *The Anatomy of Influence*. R. W. Cox and H. K. Jacobson. New Haven, CT, Yale University Press: 214.

Jervis, R. (1976). *Perception and Misperception in International Politics*. Princeton, NJ, Princeton University Press.

Jervis, R. (1988). "Realism, Game Theory, and Cooperation." *World Politics* 40: 317–349.

Johnson, B. (1975). "Technocrats and the Management of International Fisheries." *International Organization* 29: 745–770.

Jonsson, C. (1986). "Integration Theory and International Organization." *International Studies Quarterly* 30: 39–57.

Jonsson, C. and S. Bolin (1988). "IAEA's Role in the International Politics of Atomic Energy". *Politics in the United Nations System*. L. S. Finkelstein. Durham, NC, Duke University Press: 303–323.

Kahneman, D., P. Slovic, et al. (1982). *Judgement Under Uncertainty: Heuristics and Biases*: Cambridge, Cambridge University Press.

Katzenstein, P., Ed. (1977). *Between Power and Plenty: Foreign Economic Policies of Advanced Industrial States*. Madison, University of Wisconsin Press.

Katzenstein, P. J. (1986). *Small States in World Markets*. Ithaca, NY, Cornell University Press.

Keohane, R. O. (1984). *After Hegemony: Cooperation and Discord in the World Political Economy*. Princeton, NJ, Princeton University Press.

Keohane, R. O. (1988). "International Institutions: Two Approaches." *International Studies Quarterly* 32: 379–396.

Keohane, R. O. and J. P. Nye (2001). *Power and Interdependence*. New York, Longman.

Keohane, R. O. and J. S. Nye (1974). "Transgovernmental Relations and International Organizations." *World Politics* 27(1).

Keohane, R. O. N., Joseph S. and J. S. Nye, Eds. (1971). *Transnational Relations and World Politics*. Cambridge, Harvard University Press.

Keynes, J. M. (1936). *The General Theory of Employment, Interest and Money*. New York, Harcourt, Brace & World.

Kindleberger, C. (1975). "The Rise of Free Trade in Western Europe." *The Journal of Economic History* 35(1).

King, A. (1973). "Ideas, Institutions, and Policies of Governments: A Comparative Analysis." *British Journal of Political Science* 3: 291–313 and 409–203.

King, L. R. and P. H. Melanson (1972). "Knowledge and Politics: Some Experiences from the 1960s." *Public Policy* 20: 84.

Kingdon, J. W. (1984). *Agendas, Alternatives, and Public Policies*. Boston, Little, Brown.

Kistiakowsky, G. (1976). *A Scientist at the White House* Cambridge, MA, Harvard University Press.

Kornhauser, W. (1962). *Scientists in Industry*. Berkeley, University of California Press.

Krasner, S. D. (1983). "Regimes and Limits of Realism: Regimes as Autonomous Variables". *International Regimes*. S. D. Krasner. Ithaca, NY, Cornell University Press.

Krasner, S. D. (1984). "Approaches to the State." *Comparative Politics* 16: 223–246.

Krasner, S. D. (1985). *Structural Conflict*. Berkeley, University of California Press.

Kuhn, T. S. (1970). *The Structure of Scientific Revolutions*. Chicago, Chicago University Press.

La Porte, T. R. (1975). *Organized Social Complexity: Challenge to Politics and Policy*. Princeton, NJ, Princeton University Press.

Laird, F. (1985). Limiting Democracy: Participation, Competence, and Energy Policy, Massachusetts Institute of Technology. Ph.D. Dissertation.

Laitin, D. (1977). *Politics, Language and Thought*. Chicago, University of Chicago Press.

Lakatos, I. (1970). *Falsification and the Methodology of Scientific Research Programs*. Cambridge, Cambridge University Press.

Larson, D. W. (1988). "The Psychology of Reciprocity in International Relations." *Negotiation Journal*: 281–301.

Laudin, L. (1977). *Progress and Its Problems*. Berkeley, University of California Press.

Laudin, L. (1984). *Science and Values*. Berkeley, University of California Press.

Lave, J. (1988). *Cognition in Practice*. Cambridge, Cambridge University Press.

Lentner, H. A. (1984). "The Concept of a State: A Response to Krasner." *Comparative Politics* 16: 367–377.

Levy, M. J. (1966). *Modernization and the Structure of Societies: A Setting for International Affairs*. Princeton, NJ, Princeton University Press.

Lukes, S. (1982). "Relativism in Its Place". *Rationality and Relativism*. Hollis and Lukes. Cambridge, MA, MIT Press.

Macrae, D. (1976). "Technical Communities and Political Choice." *Minerva* 15: 169–190.

Maier, C. (1987). *Changing Boundaries of the Political*. Cambridge, Cambridge University Press.

Majone, G. (1984). "Regulatory Policies in Transition". *Jahrbuch fur neue politische Okonomie*. Tubingen, J. C. B. Mohr.

Mango, A. (1986). "The Role of Secretariats of International Institutions". *International Institutions at Work*. A. J. R. Grooms. New York, St. Martin's Press.

Mannheim, K. (1936). *Ideology and Utopia*. London, Routledge.

March, J. G. (1981). "Footnotes to Organizational Change." *Administrative Science Quarterly* 26: 563–577.
March, J. G. and J. P. Olsen (1984). "The New Institutionalism: Organizational Factors in Political Life." *American Political Science Review* 78: 734–749.
March, J. G. and J. P. Olsen (1989). *Rediscovering Institutions: The Organizational Basis of Politics*. New York, Macmillan.
Mefford, D. (1987). "Analogical Reasoning and the Definition of the Situations: Back to Snyder for Concepts and Forward to Artificial Intelligence for Method". *New Directions in the Study of Foreign Policy*. C. F. Hermann, C. W. J. Kegley and J. N. Rosenau. Boston, Allen & Unwin.
Moe, T. M. (1984). "The New Economics of Organization." *American Journal of Political Science* 28(4): 739–777.
Moe, T. M. (1989). "The Politics of the Bureaucratic Structure". *Can the Government Govern?* J. E. Chubb and P. E. Patterson. Washington, D.C., Brookings Institution Press: 267–328.
Morse, E. (1976). *Modernization and the Transformation of International Relations*. New York, Basic Books.
Mowery, D. C. and N. Rosenberg (1989). *Technology and the Pursuit of Economic Growth*. Cambridge, Cambridge University Press.
Mulkay, M. and K. D. Knorr-Cetina (1983). *Science Observed: Perspectives on the Social Study of Science*. London, Sage.
Mulkay, M. J. (1979). *Science and the Sociology of Knowledge*. London, Allen & Unwin.
Munz, P. (1985). *Our Knowledge of the Growth of Knowledge*. London, Routledge & Kegan Paul.
National Science Foundation (1985). *Federal Scientific and Technical Workers: Numbers and Characteristics, 1973 and 1983*. Washington, D.C., National Science Foundation.
Nau, H. R. (1974). *National Politics and International Technology*. Baltimore, Johns Hopkins University Press.
Nehru, J. (1989). "Rationale for Less Developed Countries to Turn to Science." *Is Science Necessary?* New York, E.P. Dutton: vii.
Nelkin, D. (1975). "The Political Impact of Technical Expertise." *Social Studies of Science* 5: 35–54.
Nelkin, D., Ed. (1979a). *Controversy: Politics of Technical Decisions*. London, Sage Publications.
Nelkin, D. (1979b). "Scientific Knowledge, Public Policy, and Democracy." *Knowledge Creation, Diffusion, Utilization* 1: 107.
Nisbett, R. and L. Ross (1980). *Human Inference*. Englewood Cliffs, NJ, Prentice-Hall.
Noble, D. F. (1977). *America by Design: Science, Technology, and the Rise of Corporate Capitalism*. Oxford, Oxford University Press.
Nordlinger, E. A. (1987). "Taking the State Seriously". *Understanding Political Development*. M. Weiner and S. P. Huntington. Boston, Little, Brown.
Odell, J. S. (1982). *U.S. International Monetary Policy*. Cambridge, MA, Harvard University Press.
OECD (1982). *Economic and Ecological Interdependence*. Paris, OECD.
Olson, M. (1965). *The Logic of Collective Action*. Cambridge, MA, Harvard University Press.
Perl, M. L. (1971). "The Scientific Advisory System: Some Obeservations." *Science* 173: 1211–1215.
Perutz, M. (1989). *Is Science Necessary?* New York, Dutton.
Pitkin, H. (1972). *Wittgenstein and Justice*. Berkeley, University of California Press.
Polanyi, M. (1962). "The Republic of Science." *Minerva* 1: 54–72.
Polyani, K. (1944). *The Great Transformation*. Boston, Beacon Press.
Popper, K. (1972). *Objective Knowledge*. Oxford, Oxford University Press.
Primack, J. and F.V. Hippel (1974). *Advice and Dissent*. New York, Basic Books.
Putnam, R. D. (1988). "Diplomacy and Domestic Politics: The Logic of Two-Level Games." *International Organization* 42(3): 427–560.

Rich, R. F. (1979). "The Pursuit of Knowledge." *Knowledge Creation, Diffusion, Utilization* 1: 6–30.
Rorty, R. (1979). *Philosophy and the Mirror of Nature.* Princeton, NJ, Princeton University Press.
Rothstein, R. (1984). "Consensual Knowledge and International Collaboration." *International Organization* 38(4): 733–762.
Rouse, J. (1987). *Knowledge and Power.* Ithaca, NY, Cornell University Press.
Ruggie, J. G. (1972). "Collective Goods and Future International Collaboration." *American Political Science Review* LXVI(3): 888–892.
Ruggie, J. G. (1983). "Continuity and Transformation in World Polity." *World Politics* 35: 261–285.
Ruggie, J. G. and E. B. Haas (1975). "International Responses to Technology." *International Organization* 29(3): 557–584.
Russell, R. W. (1974). "Transgovernmental Interaction in the International Monetary System, 1960–1972." *International Organization* 27: 431–464.
Sabatier, P. A. (1987). "Knowledge, Policy-Oriented Learning, and Policy Change." *Knowledge Creation, Diffusion, Utilization* 8: 649–692.
Schattschneider, E. E. (1975). *The Semisovereign People.* Hinsdale, IL, Dryden Press.
Schluchter, W. (1987). "Modes of Authority and Democratic Control." *Modem German Sociology.* V. Meja, D. Misgeld and N. Stehr. New York, Columbia University Press: 297.
Sclove, R. E. (1990). *Technology and Freedom.* Chicago, University of Chicago Press.
Segall, M., D. Campbell, et al. (1966). *The influence of Culture on Visual Perception.* Indianapolis, IN, Bobbs-Merrill.
Simon, H. (1978). "Rationality as a Process and as a Product of Thought." *American Economic Review* 68: 1–16.
Simon, H. (1983). *Reason in Human Affairs.* Stanford, CA, Stanford University Press.
Simon, H. A. (1985). "Human Nature in Politics: The Dialogue of Psychology with Political Science." *The American Political Science Review* 79: 293–304.
Skowronek, S. (1982). *Building a New American State: The Expansion of National Administrative Capacities, 1870–1922.* Cambridge, Cambridge University Press.
Slouka, Z. (1982). "International Law Making: A View From Technology". *Law Making in the Global Community.* N. G. Onuf. Durham, Carolina Academic Press.
Smith, B. L. R. (1990). *American Science Policy Since World War II.* Washington D.C., The Brookings Institution Press.
Snyder, G. and P. Diesing (1977). *Conflict Among Nations: Bargaining, Decision Making, and System Structure in International Crises.* Princeton, NJ, Princeton University Press.
Socolow, R. (1976). "Failures in Discourse". *Boundaries of Analysis.* H. Feiveson, F. Sinden and R. Socolow. Cambridge, Ballinger.
Spiegel-Rosing, I. and D. J. De Solla Price (1977). *Science Technology and Society.* Beverly Hills, CA, Sage.
Stein, A. A. (1990). *Why Nations Cooperate: Circumstance and Choice in International Relations.* Ithaca, NY, Cornell University Press.
Stein, J. G. (1988). "International Negotiation: A Multidisciplinary Perspective." *Negotiation Journal*: 221–231.
Steinbruner, J. D. (1974). *The Cybernetic Theory of Decision.* Princeton, NJ, Princeton University Press.
Suleiman, E. N. (1984). *Bureaucrats and Policy Making: A Comparative Overview.* New York Holmes and Meier.
Sundquist, J. L. (1978). "Research Brokerage: The Weak Link". *Knowledge and Policy: The Uncertain Connection.* L. E. Lynn. Washington, D.C., National Academy of Science. 5: 126–144.

Tesh, S. N. (1988). *Hidden Arguments: Political Ideology and Disease Prevention Policy*. New Brunswick, NJ, Rutgers University Press.

Toulmin, S. (1970). *Physical Reality*. New York, Harper & Row.

Toulmin, S. (1972). *Human Understanding*. Princeton, NJ, Princeton University Press.

Toulmin, S. (1990). *Cosmopolis: The Hidden Agenda of Modernization*. New York, Free Press.

Vertzberger, Y. (1990). *The World in Their Minds*. Stanford, CA, Stanford University Press.

Vollmer, H. and D. L. Mills (1966). *Professionalization*. Englewood Cliffs, NJ, Princeton-Hall.

Walker, J. L. (1981). "The Diffusion of Knowledge, Policy Communities, and Agenda Setting: The Relationship of Knowledge and Power". *New Strategic Perspectives on Social Policy*. J. E. Tropmen, M. J. Dluhy and R. M. Lind. New York, Pergamon: 75–86.

Wallerstein, I. (1979). *The Capitalist World-Economy*. Cambridge, Cambridge University Press.

Waltz, K. (1979). *Theory of International Politics*. Reading, MA, Addison Wesley.

Weinberg, A. (1972). "Science and Trans-Science." *Minerva* 10: 2009–2222.

Weingart, P. (1982). "The Scientific Power Elite: A Chimera". *Scientific Establishments and Hierarchies*. N. Elias, H. Martin and R. Whitley. Dordrecht, Netherlands, Reidel.

Weir, M. and T. Skocpol (1985). "State Structures and the Possibilities for 'Keynesian' Responses to the Great Depression in Sweden, Britain, and the United States". *Bringing the State Back*. Rueschemeyer, Evans and Skocpol. Cambridge, Cambridge University Press.

Wendt, A. (1987). "The Agent-Structure Problem in International Relations Theory." *IO* 41: 335–370.

Whorf, B. (1956). *Language, Thought and Reality*. Cambridge, MA, MIT Press.

Wilensky, H. L. (1964). "The Professionalization of Everyone." *American Journal of Sociology* 70: 137–158.

Wilensky, H. L. (1967). *Organizational Intelligence; Knowledge and Policy in Government and Industry*. New York, Basic Books.

Willets, P. (1982). *Pressure Groups in the Global System: The Transnational Relations of Issue-Orientated Non-Governmental Organizations*. New York, St. Martin's Press.

Williams, D. (1987). *The Specialized Agencies and the United Nations*. London, C. Hurst.

Williams, R. (1974). *European Technology: The Politics of Collaboration*. New York, Wiley.

Williamson, O. (1975). *Markets and Hierarchies*. New York, Free Press.

Wilson, J. Q. (1966). "Innovation in Organization: Notes Toward a Theory". *Approaches to Organizational Design*. J. D. Thompson. Pittsburgh, PA, University of Pittsburgh Press: 197–218.

Wilson, J. Q. (1980). *The Politics of Regulation*. New York, Basic Books.

Winner, L. (1975). "Complexity and the Limits of Human Understanding". *Organized Social Complexity*. T. R. La Porte. Princeton, NJ, Princeton University Press: 40–76.

Woolgar, S. (1988). *Science: The Very Idea*. London, Tavistock.

Wooster, W. (1973). "Interactions Between Intergovernmental and Scientific Organizations in Marine Affairs." *International Organization* 27: 103–113.

Yates, J. (1990). *Control Through Communication*. Baltimore, MD, Johns Hopkins University Press.

5

CONSTRUCTING ENVIRONMENTAL SECURITY FROM RESOURCE SCARCITY*

Peter M. Haas

Resource scarcity and environmental security are long-standing themes in international relations and security studies. These are important ideas because they inform or justify US and other countries' foreign policies. The commitment of resources, and possibly intervention, is warranted against these justifications. Claims that human societies are exhausting scarce resources, and that this process may contribute to armed conflict with and between states, has been part of the economics and politics literatures for at least 130 years. Yet an analytic focus on material scarcity as a provocation for conflict is but a meager part of much richer approaches to understanding conflict and cooperation more generally.

The question is, can these doctrines of resource scarcity and environmental security bear the weight of justifying US interventions? My argument is that these doctrines are flawed, and are selectively invoked by policy makers and inattentive academics in order to justify preexisting state goals. My interest here is why they enjoy such resilience, despite the fact that each is widely contested within the relevant policy communities, and that there are many other ways of formulating interesting questions about resource scarcity.

The attractiveness of these doctrines is not solely the nature of the argument they make, but rather their affinity to the values and beliefs invoked by the person making the argument. We have a lot of names for such things, including myths, mental models, worldviews, frames, narratives, ideologies, paradigms, epistemes, referentielles, or worldviews, all of which can be encompassed in a broad view of "discourses."

Discourses impart meaning to an ambiguous policy domain. Discourses are important because they institutionalize cognitive frames. They identify issues as problems, set agendas, and define the salient aspects of issues as problems for decision makers. Each discourse or perspective rests on different assumptions, goals and values, or ontologies and epistemologies as we now call them, and suggests different policy solutions. They have the effect of defining provocations or crises. One could wonder why the US construes Iraqi potential control over oil as a security threat but Saudi or Venezuelan control is not. The answer lies not in the nature of the resource, but with the presumptive motivations of the party with potential leverage over a resource that the US needs.

Why do these exaggerated beliefs about resource scarcity and their possible threats to environmental security persist? There are two reasons. The first reason is the absence of any consensual mechanism for reconciling interdiscourse (or inter-paradigm) disputes in either the arena of environmental issues or international politics. This absence of consensus distinguishes debates about environmental security, which are cast at the level of ritualized debates between highly polarized positions, from debates about multilateral environmental policy, around which consensual agreement has been reached about the utility of sustainable development approaches to environmental policy.

The second reason is the elective affinity between environmental and security discourses on the one hand, and other dominant discourses in social discussions such as neoclassical economics and realism in international relations and security studies on the other hand. Consequently self-interested political actors can borrow from discourses that are similar in their ontology and structure and that justify preexisting political ambitions. In a sense, none of these discourses is "right." A good approach to developing foreign policy, then, is procedural. It should be based on a public discursive dialogue between proponents of each. However, because each rests on its own assumptions, and is upheld by a different political constituency, it is difficult to imagine a committed debate, much less achieving closure.

Honest social scientists can do two things: identify myths that aren't widely appreciated, thus illuminating the effect of tacit assumptions on people's thinking, and help to dispel or debunk pernicious myths. I hope to do both.

Dominant discourses about environment and security

Cultural anthropologists who look at environmental politics have identified four overarching discourses or myths of Nature (Timmerman 1986; Schwartz and Thompson 1990; Thompson and Rayner 1998). Myths organize our thinking by presenting a tacit set of cause and effect relations that we can apply when considering an issue at hand, and also by providing a screen to select out factors that are not germane to the question at hand.

Cultural anthropologists organize the four dominant voices in environmental discourses about risk and the implications of environmental change on human affairs into the following categories: Cornucopian, Malthusian, Sustainable Development, and Radical/Postmodern [I am relying only on their taxonomy, and not on the grid-group scales that are sometimes associated with them]. In general authors fall into just one of the categories above, although some, such as Norman Myers, draw from Malthusian and Sustainable Development Discourses (Myers 1989).

Cornucopian writers see Nature as boundless, and thus unlikely to exercise significant constraints over human action (Simon 1981; Beckerman 1982; Simon and Kahn 1984; Lomborg 2001). Nature is seen to be essentially robust and benign. Summarizing this view, Jordan and O'Riordan write that "the management style associated with this view is relaxed, non interventionist and laissez-faire. It is associated with a market perspective on institutional functioning, and a belief in the prominence of the individual in coping with challenges" (Jordan and O'Riordan 1997). Economic growth encourages technological innovation and technical solutions to resource scarcity.

Malthusians see Nature as "precarious and unforgiving, vulnerable and constraining on human agency" (Myers 1989; Homer-Dixon 1991; Jordan and O'Riordan 1997; Klare 2000). Resources are seen as finite and limited and thus exercise constraints on human action. Population grows more rapidly than the resource base, and creates inexorable scarcities. "Human agency can "tip it over" to the point where human well-being itself is threatened" (Jordan and O'Riordan 1997). Malthusian policy responses tend to be hierarchically designed, with a strong state based managerial bent aimed at reducing population growth and resource usage.

Sustainable Development sees Nature as resilient within a context of some range of parameters that are themselves dynamic (Holling 1978; Clark and Munn 1986; Holdren 1991). "Nature is manageable so long as its limits are taken into account, either by conscious 'holding back' or by the application of ecological principles to human affairs" (Jordan and O'Riordan 1997). This perspective calls for deliberative procedural responses to environmental risks, stressing the need for involving multiple voices in formulating and assessing policy responses. Economic development and social growth is not seen as inconsistent with environmental preservation.

Radical views and Postmodern views see Nature as capricious, and unrelated to discourses about it or to social efforts to regulate it. Such authors dispute the underlying utility of discussing resource availability in the abstract. They argue that the true questions have to do with inequitable resource access and distribution (Gatlung 1973; Redclift 1987; Dalby 1999). They focus on redistributive policies to address questions of resource scarcity, and thus look at the social context in which resource decisions are taken, rather than taking resource availability as an exogenous factor.

To date, the principal debates have been between Malthusians and Cornucopians.

Intellectual history of resource scarcity

The intellectual history of concern about resource scarcity probably goes back to the Greeks. Early environmental determinism, and vulgar environmental analysis, were based on univariate causality in which climate or natural resources determine a country's development potential and political position in the international system. In its most vulgar form, environmental determinism claims that climate dictates national potential and individual psychological character. In its more sophisticated form it means that physical geography and access to and control over material resources influence national power capabilities in the international realm. The precise nature of these resources varied at historical moments: control over the European heartland and its agricultural production and transportation routes, control over naval routes, and control over enough physical territory to control food and all natural resources for *lebensraum* in interwar Germany. Most of these statements were just doctrinal justifications for expansion: We now know that these state objectives of access to resources can be more easily and efficiently achieved through trade, with less possibility of military provocation (Rosecrance 1986; Keohane and Nye 2001).

Thomas Malthus in 1798 argued, based on fairly primitive British demographic data, that population grows geometrically but food production only grows arithmetically. This simple observation was extrapolated into an iron law of universal resource scarcity. But he neglected family planning and birth control, as well as technological innovation.

The 19th century industrial revolution in England prompted concerns about scarcity of non-renewable resources. In 1865 William Jevons predicted coal shortages and an energy crisis (Jevons 1865). Materials consumption in England accelerated dramatically between 1880 and 1913. The 1954 Paley Commission in the United States expressed concern about scarcity and foreign control over natural resources necessary for the US economy.

In the 1970s similar concerns were expressed again, in a series of reports to the Club of Rome. The most famous one was the 1972 *Limits to Growth,* followed by a 1992 sequel (Meadows, Meadows et al. 1971; Meadows, Meadows et al. 1992). The *Limits to Growth* report predicted widespread collapse by the early 21st century of modern economic systems, from pollution and resource scarcity. The assumptions driving these extrapolations were profoundly Malthusian (Freeman and Jahoda 1978; Hughes 1985).

OPEC generated a burst of writings about whether resources and commodities could be controlled through cartels for political ends. Likewise, concerns were expressed about the manipulation or control of imported "strategic minerals" that were considered essential to the US economy and to the US military establishment because they were found in just a few countries and because no substitutes existed for them.

All these predictions, so far, have been wrong. There was no energy crisis in 19th century England; in fact the very foundations for the Industrial Revolution were based on a series of contingent historical events based on resource scarcity. The widespread deforestation of England in the middle ages as a consequence of burning wood for fuel and cutting down large trees for the Royal Navy raised the cost of fuel wood and encouraged the development of coal mining in Wales. By the late 1780s the most accessible coal had been extracted, and pumps had to be devised to extract the water from mines to facilitate deeper mining. In turn, these pumps provided the technological innovations for steam power that motivated the Industrial Revolution (Ashby and Anderson 1981).

Predictions of oil and food scarcity have also failed to materialize. Estimates of oil reserves, since 1952, have actually increased, and the years remaining before current reserves are exhausted have actually increased (Lynch). Food production has exceeded population growth for the last 20 years. FAO estimates indicate that food production per person worldwide has increased, outside of Africa, although maintaining these achievements rests on continued provision of chemical inputs and irrigation (Smil 2000).

Strategic Minerals shortfalls have also failed to materialize, due to stockpiling (Russett 1984). The Center for Defense Information concludes that "military intervention or posturing is more likely to threaten than protect important foreign raw materials, military force cannot guarantee the uninterrupted flow of raw materials, and the US leans too heavily upon its military might to protect access to foreign raw materials to the exclusion of more effective and less costly economic measures" (The Defense Monitor 1985).

Why have Malthusian estimates been consistently wrong? The discourse is generated by its assumptions, and the assumptions presume that scarcity will occur. In particular, Malthusian analysis typically neglects two vital social phenomena. One is that analysis is based on projections of consumption and production, but does not take account of prices or markets (Kaysen 1972). Consequently, there is no price signal feedback mechanism provided in Malthusian analysis, and thus no potential for social accommodation to resource pressures. Second, Malthusians do not take account of the potential for technological innovation or substitution for resources, as they become scarce. For instance, copper, which was widely feared in the 1970s as a future scarce resource, is now in ample supply as the principal use of copper, telephone cable, has been replaced by fiber optics. These discursive blinders in Malthusian thought are in large part the consequence of the disciplinary orientation of most Malthusians, who are natural scientists rather than economists, and thus rest their analysis on measurements of resource assessments in the physical environment rather than the price for those resources.

The irony is that it presently appears that the social, political and economic consequences of scarcity are now more appropriate for renewable resources (such as timber) rather than non-renewable resources. But this is because many governments, particularly in developing countries, pursue policies that encourage unsustainable use of renewable resources (such as trees) to generate foreign exchange or to benefit specialized interests within the country.

Environmental security

A shift in discourse occurred in the 1970s and 1980s from discussing resource scarcity to talking of environmental security. With the end of the cold war this discourse was picked up by traditional security policy networks in Washington as a way to think about new security threats, but their prior dominant realist worldview shaped their acceptance of the arguments to only their Malthusian kernel. The environmental security discourse was closely tied to the Malthusian geopolitical approach to resource scarcity.

Marvin Soroos distinguishes between two alternative views of environmental security (Choucri, Brown et al. 1990; Soroos 1995; Elliott 1996). A broad definition sees environmental security in terms of challenges to governability: literally governments become incapable of satisfying the needs of their population because of environmental threats, or figuratively because of a legitimacy crisis facing governability at all, in a Habermasian sense (Habermas 1973). This broad approach was popular in Scandinavia under the rubric of "comprehensive security," and as a positive image of peace by peace researchers, rather than as a provocation to war as it is viewed in the Malthusian discourse.

A narrower view sees environmental scarcity as a catalyst to violence. This argument is most forcefully advanced by Homer-Dixon, who presented the nuanced argument that resource scarcity is a contingent trigger to domestic and international violence. Homer-Dixon provides a progressive step to the Malthusian discourse through his attention to domestic/international interactions, and his probabilistic argument about the direct and indirect causal pathways to conflict potentially triggered by resource scarcity or degradation, including a focus on population dislocation and migration stimulated by environmental disruptions (Homer-Dixon 1991).

A full list of environmental issues cited in the environmental security literature includes climate change, stratospheric ozone depletion, acid deposition, deforestation, degradation of agricultural land, overuse and pollution of fresh water supplies, fishery decline, biodiversity loss, atmospheric pollution, and access to strategic minerals (Wieder and Brecke 1998).

General assessment of environmental security

Empirically none of these claims is upheld. One of the most frequently cited environmental triggers to conflict is that of access to fresh water (Cooley 1984; Gleick 1993). Egyptian posturing about the importance of the Nile is widely cited as an example of the strategic importance of water. In 1980 Egyptian President Anwar Sadat declared "If Ethiopia takes any action to block our right to the Nile waters, there will be no alternative for us but to use force." In 1985 then Foreign Minister Boutros Boutros-Ghali reiterated "The next war in our region will be over the waters of the Nile not over politics. . . ." (Myers 1989). Earlier Egypt had threatened to send troops to the southern border if Sudan had proceeded with its plans for a water diversion project. Yet, with water there are still no cases of interstate water wars, and most domestic water conflicts are prevented through state interventions or efforts to create markets for water and thus encourage conservation (Wolf 1997; Postel and Wolf 2001).

The empirical record of wars fought over resources is quite meager. There are few direct or indirect cases of violence, in Homer-Dixon's work or others'. In fact, virtually no one has been killed in direct international conflicts over water or any other resource. The standard argument in international relations is that war is rationally chosen as the extension of policy by other means. Therefore we would expect resource scarcity to galvanize armed conflict only under highly polarized circumstances in which states are looking for a provocation to fight.

Nor does the literature on state failure support the proposition that resource degradation and scarcity undermines state authority domestically. Gurr's extensive literature review suggests that resource inequality may be a determinant of economic growth, and its influence on violence is a matter of speed with which the scarcity is imposed (Gurr 1985). The important manipulable variable is economic equality and access to resources. The 1998 CIA State Failure Task Force Report, which relied in part on Gurr's database, found that "environmental change does not appear to be directly linked to state failure" (Esty, Goldstone et al. 1998).

Theoretical shortcomings with environmental security arguments

The environmental security argument suffers from a number of theoretical problems. Methodologically, authors are seizing on the independent variable; thus any examples of violence are attributed to the resource variable, rather than other possible events. They neglect the vast number of cases in which resource scarcity did not generate conflict.

More generally the argument suffers from the same set of theoretical blinders as the Malthusian approach to resource scarcity: a neglect of markets and policy interventions that could influence distribution, poverty reduction, and poor people's ability to pay for access to resources, or invest in producing more resources.

Daniel Deudney summarizes the major theoretical challenges to Homer-Dixon's framing of environmental scarcity as a security issue (Deudney 1990; Deudney 1997). First, he reiterates the problem of causal inference mentioned above: resource scarcity is not a cause but an intervening variable that influences violence. Here it is not the nature of resource scarcity that makes war likely; it is the nature of the political climate in which scarcity occurs. Similar demand for a resource in other regions is unlikely to provoke war. Scarcity may amplify pressures when they are already acute.

Second, Deudney says that the environmental security concept reflects a materialistic notion of security associated with realism, and has the effect of appropriating the political agenda at the expense of other, more comprehensive notions of ecological security. It provides too narrow a formulation of security.

Third, Deudney cites a rhetorical error in using the term security. The term "national security" has a rhetorical play to it, as Wolfers noted long ago, and that which is national security is whatever a state can successfully convince its citizens to embrace. To the majority of the foreign policy community, security invokes national security and the protection of borders and citizens from foreign threat, generally cast at the level of the state. (While the comprehensive definition of environmental security is cast at the individual level, this aspect of the argument is usually absent from discussions about environmental security.)

Yet few environmental issues share these properties of security as defined by the mainstream foreign policy community. While human survival may be at risk from environmental degradation, the risk doesn't stem from armed conflict. It is from starvation, cancer, and the long run degradation of the global ecosystems on which human survival depends. And it is from cumulative individual actions, at an unknown point in the future.

Deudney argues, correctly I believe, that the environmental security discourse is counterproductive. Efforts to improve environmental protection are tarnished by shoddy conceptualizations of environmental security. The environmental security discourse uses language that privileges the military, when the point is that military responses are inappropriate and the wrong perspective in which to view things.

There are two consequences of this error. The first is that it discredits efforts at environmental protection by associating it with a flawed causal argument that environmental scarcity contributes to violence. Environmental protection should be pursued because environmental protection is a valued goal, not because it can be instrumentally (and erroneously) linked to a related popular discourse. Once the flaws are made manifest any support for the goal of environmental protection is potentially jeopardized. The second consequence is that the ownership of security then resides with the wrong policy community. One of the initial tactical motivations in the 1970s behind the environmental security literature was to get the environment onto the US foreign policy agenda and to get positions in the administration associated with those issues. But this was counterproductive, as the issue was seized by the Pentagon and the majority of money in the US spent on environmental security went to cleaning up military bases, rather than any other items on the environmental agenda that may have posed a greater threat to human survival.

Conclusion

Resource scarcity debates have been around in various forms for a long time, and remain contested. Environmental threats are important and influence the quality of human life. On the other hand, they are seldom issues over which nations will go to war, either directly for resources, or even indirectly because of the stresses generated from resource scarcity. Violent conflict over resources at the national level is possible, but it seems unlikely to spill across borders. We shouldn't call such threats a matter of international security, because that nomenclature interferes with understanding and improving policy.

These discourses persist not because they are accurate, but because they are politically embedded discourses. Constructivists explain their persistence because they acquire a taken-for-granted quality as bureaucrats are socialized into accepting the worldview within the administration through standard operating procedures, and because the arguments are often expedient for achieving other goals. In addition lazy thinkers acquire the ideas in graduate school.

But discourses are also subject to challenge through deliberative debate in public about their foundations, and about alternatives. As self-conscious scholars of IR we can contribute to this process. Three options are available.

One is to debunk the arguments, and identify flawed use of theory to justify standard foreign policy goals.

A second is to discuss and promote alternative formulations to the environmental security and resource scarcity discourses. While not explicated at length in this essay, I prefer the Sustainable Development voice, because I feel that the core elements of Sustainable Development have greater technical consensus, political support and provide a better mechanism for achieving substantive and tactical linkages on international policy agendas (Haas 1996; Bernstein 2000).

If we remember that each of these four formulations is a discourse about the world, and consider the possible consequences of error (that is that the world behaves in a different way than presumed by the discourse), then the most tolerant policy to pursue is that of sustainable development. Even if wrong it promises some benefits through economic development and improved energy efficiency, as well as being attractive on political grounds insofar as it contributes to democratic processes of governance. The social costs of error for the other three are all severe. Assuming a Malthusian future reduces liberties. Assuming a radical future precludes efforts to redress environmental threats. Assuming a Cornucopian future risks environmental collapse.

Finally, there is a need for further research on institutional design that can improve environmental quality, and prevent violent responses to environmental scarcity or degradation. Amartya Sen noted that famines do not occur in democracies, where food allocation decisions are taken publicly, and that hunger is largely a consequence of ability to buy food rather than the absolute availability of food. What is often obscured in most studies is that environmental threats have actually been prevented through international governance. Further research can improve our understanding of effective responses.

Note

References

Ashby, E. and M. Anderson (1981). *The Politics of Clean Air.* Oxford, Clarendon Press.

Beckerman, W. (1982). "Global Warming and International Action". *The International Politics of the Environment.* A. Hurrell and B. Kingsbury. Oxford, Oxford University Press.

Bernstein, S. (2000). "Ideas, Social Structure, and the Compromise of Liberal Environmentalism." *European Journal of International Relations* 6(4): 464–512.

Choucri, N., J. W. Brown, et al. (1990). "Dimensions of National Security". *In the US Interest.* J. W. Brown. Boulder, CO, Westview.

Clark, W. C. and R. E. Munn, Eds. (1986). *Sustainable Development of the Biosphere.* New York, Cambridge University Press.

Cooley, J. K. (1984). "The War Over Water." *Foreign Policy* Spring.

Dalby, S. (1999). "Threats from the South". *Contested Grounds.* D. Duedney and R. Meyer. Albany, NY, SUNY Press.

Deudney, D. (1990). "The Case Against Linking Environmental Degradation and National Security." *Millennium* 19(3): 461–476.

Deudney, D. (1997). "The Limits of Environmental Security". *Flashpoints in Environmental Policymaking*. S. Kamieniecki, G. A. Gonzales and R. O. Vos. Albany, NY, SUNY Press: 281–310.

Elliott, L. (1996). "Environmental Conflict: Reviewing the Arguments." *Journal of Environment & Development* 5(2): 149–167.

Esty, D. C., J. A. Goldstone, et al. (1998). *State Failure Task Force Report: Phase II Findings*. Science Applications International Corporation, McLean, VA.

Freeman, C. and M. Jahoda (1978). *World Futures*. New York, Universe.

Gatlung, J. (1973). "The 'Limits to Growth' and Class Politics." *Journal of Peace Research* 10.

Gleick, P. H. (1993). "Water and Conflict." *International Security* 18(1).

Gurr, T. (1985). "On the Political Consequences of Scarcity and Economic Decline." *International Studies Quarterly*.

Haas, P. (1996). "Is Sustainable Development Politically Sustainable?" *Brown Journal of World Affairs* 3(2): 239–248.

Habermas, J. (1973). *Legitimation Crisis*. Boston, Beacon Press.

Holdren, J. (1991). "Population and the Energy Problem." *Population and Environment* 12(3): 231–255.

Holling, C. S., Ed. (1978). *Adaptive Environmental Assessment and Management*. New York, Wiley.

Homer-Dixon, T. F. (1991). "On the Threshold." *International Security* 16(2): 76–116.

Hughes, B. (1985). *World Futures*. Baltimore, MD, Johns Hopkins Press.

Jevons, W. S. (1865). *The Coal Question*. London, Macmillan.

Jordan, A. and T. O'Riordan (1997). Social Institutions and Climate Change. CSERGE Working Paper GEC 97–15.

Kaysen, C. (1972). "The Computer that Printed out W*O*L*F*." *Foreign Affairs* 50(3): 660–668.

Keohane, R. O. and J. P. Nye (2001). *Complex Interdependence*. New York, Longman.

Klare, M. T. (2000). *Resource Wars*. New York, Metropolitan Books.

Lomborg, B. (2001). *The Skeptical Environmentalist*. Cambridge, Cambridge University Press.

Lynch, M. "Data Provided by Michael Lynch, WEFA, Inc.".

Meadows, D. H., D. L. Meadows, et al. (1971). *The Limits to Growth: A Report for the Club of Rome's Project on the Predicament of Mankind*. New York, Universe Books.

Meadows, D. H., D. L. Meadows, et al. (1992). *Beyond the Limits*. Post Mills, VT, Chelsea Green Publishing Company.

Myers, N. (1989). "Environment and Security." *Foreign Policy* 74.

Postel, S. and A. T. Wolf (2001). "Dehydration Conflict." *Foreign Policy* September/October.

Redclift, M. (1987). *Sustainable Development Exploring the Contradictions*. London, Methuen.

Rosecrance, R. (1986). *The Rise of the Trading State*. New York, Basic Books.

Russett, B. (1984). "Resource Dependence and Analysis." *International Organization* 18(3).

Schwartz, M. and M. Thompson (1990). *Divided We Stand: Redefining Politics, Technology and Social Change*. New York, Harvester Wheatsheaf.

Simon, J. (1981). *The Ultimate Resource*. Princeton, NJ, Princeton University Press.

Simon, J. and H. Kahn (1984). *The Resourceful Earth*. Oxford, Basil Blackwell.

Smil, V. (2000). *Feeding the World*. Boston, MIT Press.

Soroos, M. S. (1995). "Environmental Security." *Phi Kappa Phi Journal* Winter: 20–24.

The Defense Monitor (1985). *Resource Wars: The Myth of American Mineral Vulnerability* 19(9).

Thompson, M. and S. Rayner (1998). "Cultural Discourses". *Climate Change Vol. 1*. S. Rayner and E. L. Malone. Columbus, OH, Batelle Press.

Timmerman, P. (1986). "Myths and Paradigms of Interactions Between Development and Environment". *Sustainable Development of the Biosphere.* W. C. Clark and R. E. Munn. Cambridge, Cambridge University Press.

Wieder, J. and P. Brecke (1998). The Pathways from Environmental Change to Violent Conflict: Connecting the Physical and Social Sciences. *International Studies Association Annual Convention.* Minneapolis.

Wolf, A. (1997). Water Wars and Water Reality. *NATO Advanced Research Workshop on Environmental Change, Adaptation and Human Security.* Budapest, Hungary.

PART II

Regimes and governance patterns

I don't care who write a nation's laws if I can write its economics textbooks.

(Silvia Nasar, interview with Paul Samuelson "Hard Act to Follow" The New York Times, *March 14, 1995, cited in Wapshott 2011, 232)*

International governance patterns vary with the presence or absence of epistemic communities. The pieces in this section analyze the discrete outcomes associate with the presence and absence of epistemic communities. Epistemic communities contribute to more effective and long-lasting environmental governance.

"Social constructivism" (Chapter 6) presents the broad history of collective environmental governance through the creation of international regimes. The chapter reviews the way in which epistemic communities become more institutionalized into world politics, and governments are socialized to the comprehensive ecological ideas promoted by epistemic communities active in international environmental governance. Multilateral negotiations and regimes are seen as potentially transformative social interchanges which provide information and empower new knowledge-based actors. Negotiations and regimes constitute governance by establishing new roles and new representatives as a consequence of institutionalized ideas and actors.

"Prospects for effective marine governance" and the skeletal presentation of the argument in the subsequent letter to the editor (Chapter 7) present the stylized framework of cooperation patterns shaped by different configurations of international institutions and epistemic communities. This framework is elaborated in "Epistemic communities and international environmental law" (Chapter 8).

Governance issues with strong international institutions and epistemic communities have distinct features. Collective governance occurs through a process of social learning. Treaties are comprehensive in form, and reflect the ecological understanding of the epistemic community about the ability of ecosystems to sustain stress from individual contaminants. In conservation regimes, environmental standards are adopted to protect ecosystems. In pollution regimes, environmental standards are adopted that reflect an understanding of the variety of environmental contaminants and their interactive effects. The regimes tend to be environmentally effective, because they reflect an accurate understanding of the ability of the environmental system to sustain stress. The governance arrangements tend to persist over time, and to respond quickly and accurately (minimizing type 1 and type 2 errors) to new environmental crises, because of the involvement of epistemic communities.

In the absence of epistemic communities governance is still possible, but will not involve any learning. There will be no discontinuities from past patterns of behavior, and state interests will remain unchanged in the absence of exposure to new ideas.

Reference

Wapshott, N. (2011). *Keynes Hayek: The Clash that Defined Modern Economics*. New York, W.W. Norton.

6

SOCIAL CONSTRUCTIVISM AND THE EVOLUTION OF MULTILATERAL ENVIRONMENTAL GOVERNANCE[1]

Peter M. Haas

Economic globalization creates transboundary and global environmental externalities. A system of multilateral international environmental governance has evolved over the last 25 years as the international community has attempted to address the ecological externalities of economic globalization. Most observers of the 1972 United Nations Conference on the Human Environment (UNCHE), when the international community first acknowledged the urgency of dealing with global environmental threats, were skeptical that effective governance could be established for such issues. Yet governments are now held accountable to new standards of environmental behavior by carriers of values who had no standing in 1945. Not only has a stable set of expectations about reciprocal state practice been established, its form has evolved over time to become more comprehensive, reflecting growing scientific understanding about the behavior of ecosystems and the sensitivity of human societies to such dynamics. The norms, rules and strategies for environmental governance are no longer widely contested (Kratochwil and Ruggie 1986; Haas 1992; Crawford and Ostrom 1995).[2] They are now enforced within a multilateral governance structure which systematically limits the role of the state. While states remain the putative sources of authoritative choice, officials are aware of two novel features of international environmental governance: that science is essential for the understanding of global environmental problems, thus shifting the determination of the scope of allocative decisions to the international institutions for science; and that states are increasingly accountable to domestic and transnational constituencies, thus shifting the locus of enforcement from states toward international institutions and NGOs.

In this chapter I describe the political character of international environmental problems, describe the major features of international environmental governance, analyze their development over the last 25 years, and discuss current prospects for this emergent governance structure. Against a backdrop of virtually no systematic governance existing in 1972 the current level is striking. It is far more extensive than in the past, and has been growing increasingly comprehensive (or ecological) in form. This evolution exceeds the theoretical ability of most conventional schools of thought in international relations—neorealism, neoliberal institutionalism, and political economy—to explain in a satisfactory manner on its own terms.

Two complementary definitions of governance underlie my discussion. Michel Foucault uses governance to mean a system of institutional arrangements which "structure the possible fields of actions of others" (Foucault 1982; Holsti 1992). Similarly, Douglass North writes:

> Institutions are the rules of the game in a society or, more formally, are the humanly devised constraints that shape human interaction. In consequence they structure incentives in human exchange, whether political, social, or economic. Institutional change shapes the way societies evolve through time and hence is the key to understanding historical change.
>
> *(North 1990)*

Governance thus consists of formal institutions designed to obtain collective goals generated from intersubjective beliefs and aspirations. International environmental governance is a process, but one which is principally impelled by changes in formal and informal institutions.

A constructivist approach helps to explain the evolution of this international governance system. The system's structural form—suffrage and decision-making rules—co-evolved with specific norms, rules and strategies for governments' environmental behavior. More than the norms, rules, and strategies which the new institutionalists discuss, this system also entails a set of interlocking social relationships through which such institutional features are developed; enforced; and diffused and reproduced over time. Here I argue that the institutionalization of a new ecological perspective, embodying norms, rules and strategies, was articulated by ecological epistemic communities and disseminated through formal governance institutions. This institutionalization of new beliefs developed concurrently with a general political decentralization at the international level. International decentralization means moving from an international system centered solely on states to a more decentralized arrangement in which states share responsibilities with international institutions and a variety of non-state actors. The widespread adoption of new ecologically informed state practices, based on new consensual understandings of the operation of physical ecosystems, is a central element of the explanation offered here, while other elements of the broader picture of international governance—the global decentralization of authority and the emergent influence of NGOs—are exogenous factors, possibly related to economic globalization, democratization, and the end of the Cold War. Once institutionalized within the formal structure of international governance arrangements, new ecological beliefs have acquired a strong role in shaping the policies and practices of member units, and of actors' expectations about those policies and practices. Over time, state practices informed by these ideas have become habitual and internalized, while attendant state regulations seem to be leading to the creation of a green international political economy.

Environmental markets finally emerged in the 1990s; they had not existed before. Donald Connors accurately states that "there is no natural market for environmental technologies, like there is for food or clothing" (Johnson 1995). It seems clear that their creation emerged against a backdrop of a more general evolution of more vigorous domestic environmental management driven by an internationally institutionalized recognition of the scale of new public threats and of convergent beliefs about appropriate unilateral and collective responses. In the absence of a strong international pressure on governments to launch new, national environmental policies, domestic practices would not have moved as quickly or as vigorously, and firms would have lacked incentives to invest and commercialize environmental innovations as rapidly as they did.

Ontologically I hesitate to privilege either international systems or state units, or ideas or units. Given the complexity of the international political system as it has developed since 1648, each factor now plays a role. New ecological beliefs did not constitute new units involved in international environmental governance, although they may have privileged the role of non-state actors and thus helped to reinforce other international centrifugal forces that have led to their increasing participation in international politics. These beliefs did constitute institutional regularities: the environmental norms, rules and strategies now widely pursued by governments and firms. In this regard, the new beliefs preceded changes in state practices, making it clear that the system of governance was constituted by ideas associated with a discrete and identifiable group of individuals. This interpretation contrasts with the oft-cited view that practices precede norms, and that norms are merely expedient justifications or codifications of prior practices and the preferences associated with the conduct of those practices.

The constructivist methodology pursued here consists of deductively identifying the possible consequences of ideas on practices, specifying a credible mechanism linking new ideas to changes in practices, combined with process tracing to determine if the hypothesized changes occurred through the proposed mechanisms. Counterfactuals and process tracing help to dispel alternate hypotheses including the null hypothesis about the source of change in environmental governance. Without the involvement of epistemic community members with their ecological perspective, national measures and international regimes would have been much more disjointed. They also would have been weaker, as they would have reflected the overarching desire of many firms and industries to avoid additional production costs and would have reduced the desire of politicians to appear green to domestic voters. Such forces, untempered by an awareness of ecological realities, would have yielded least common denominator agreements at best, agreements which would have been insufficient to drive market change.

Observed changes cannot be explained by technological change alone, as much that technological change has occurred as a consequence of states' efforts to enforce international commitments at home. Even the control of CFCs, which has been widely hailed as a triumph for technological innovation, occurred as a result of government stimulus (Haas 1992; Levy, Keohane et al. 1993). Widespread domestic environmental concern did not appear until the late 1980s and 1990s, and thus occurred too late to explain much of the change discussed below. Even now the depth of public commitment and its political impact remains open to doubt. State power has not been concentrated in the major areas in which dramatic changes have occurred, and, to the extent to which US hegemonic influence was available in environmental issues, it was not systematically deployed. Thus, neither technological change nor the distribution of power in the international system explains observed changes in international environmental regimes.

As I discuss at the end, comprehensive notions of ecological environmental management have increasingly informed international environmental regimes over the last 25 years, and many governments have adopted new national practices in accordance with their new international obligations. Less demanding environmental regimes have also been developed during this period, raising the problem of causal inference for analysts regarding the question as to whether the observed change in governance is due to ideas or more diffuse multilateral action. In the absence of the application of more comprehensive ecological policy approaches—that is if all environmental regimes were merely the consequence of compromise between governments, as analyzed by neoliberal institutionalists—the overall tenor of environmental regulation, including the major markets of Germany, Japan and the USA, would be much weaker, and thus the incentives and pressure for adopting new technologies and products would be weaker. Firms in this counterfactual scenario would be responding to the policies introduced by governments in order to achieve the more modest goals negotiated internationally, and thus would not have strong incentives for innovation. Consequently, the role of ideas demands particular analytic attention in order to explain the governance revolution in the area of the international environment over the last 25 years. Goals, and a more general theoretical account, should focus on the broad process by which new notions of governance emerge and diffuse at the international level, and subsequently become adopted and embedded in national political settings.

Thus my chapter complements those of Lake, Sandholtz and Cerny. But, as I have just argued, environmental governance does not rest solely or even predominantly on contracting. Irreconcilable differences in principled interests about transboundary and global environmental threats continue to inhibit effective North–South cooperation for climate change, among other issues. Before international institutions could be deployed to achieve cooperation, potentially aggregated interests had to be established. Moreover, empirically (see below) very few international legal institutions appear to fulfill the contracting functions specified by Lake. More common are institutional arrangements that provide information about the nature of threats and potential solutions, rather than providing information about compliance. Learning takes precedence over assurance, then, in terms of the functional performance of international institutions in the environmental realm, although some degree of assurance occurs as well. Unlike Sandholtz, who focuses primarily on the overarching embedded ideational arrangements within different formal institutional structures, I try to focus on the process by which such institutions become embedded, and the potential variation between formal institutional structures and the potential for embedding dramatically new ideational orientations. Cerny, lastly, focuses on the material underpinnings of global governance. I argue, however, that current environmental technology markets would not have existed without prior concerted international efforts at applying ideas and creating formal institutional structures for environmental governance. As Polanyi argued for markets in the early 19th century, and North thereafter, effective markets rest on shared economic beliefs and formal social institutions of information and enforcement.

The nature and scope of international environmental problems

While a sense of urgency has accompanied efforts for the management of environmental risks over the last 25 years, due to a fear of nonlinear environmental effects of widespread environmental degradation and its possibly irreversible consequences, international environmental problems are singularly difficult to manage owing to both technical and political reasons.

Technically, efforts to cope must be comprehensive if they are to effectively confront the complex array of casual factors associated with environmental threats. Yet this is organizationally difficult to achieve as few governments or international institutions are organized to cope effectively with the multiple dimensions of environmental problems, and many governments lack the technical resources to develop and apply comprehensive efforts. Effective management is politically difficult because solutions must be negotiated, and there are few strong proponents for vigorous or comprehensive environmental management, either domestically or internationally. Domestically the costs of action are generally concentrated and short-term, while the benefits are diffuse and long-term, creating an Olsonian social choice problem in which the collective good suffers. The management of transboundary and global environmental threats requires countries to mutually adjust their policies, but governments are reluctant to participate unless they can be sure that other significant polluters will actively participate in the regime as well. Thus, the difficulty of obtaining international policy coordination also inhibits the pursuit of environmental protection domestically.

Vigorous international action is also difficult because governments often have different policies which must be coordinated, and because national preferences are often at odds. Many developing countries (LDCs) still have not yet established environmental policies, and those adopted are generally narrower in scope than in the advanced industrialized countries. LDC environmental policies tend to focus on the management of specific resources, rather than integrating environmental considerations in all development planning, such as is emerging in the industrialized countries. Developing country delegates to UNCHE and the 1992 UN Conference on Environment and Development (UNCED) expressed concerns that a focus on industrial pollution and waste management may distract attention from their environmental priorities of deforestation, including unsustainable agricultural practices, population pressures, water quality, and soil erosion. Many LDCs also fear that environmental policy coordination with industrialized states can threaten to retard short-term economic growth by imposing new production and operating costs (Latin American and Caribbean Commission on Development and Environment 1991; South Centre 1991; Soto 1992).

The architectonics of international environmental governance

The current governance system can be described in constitutionalist terms. Governance entails a set of commitments which command obedience (rights and responsibilities, conventions, principles and norms), actual rules and procedures by which these commitments are to be realized, a set of authoritative actors who may participate in decision making (rules of suffrage), a set of formal institutions through which activities are coordinated, and a procedure for adjudicating disagreements and challenges (Dahl 1971; Dahl 1989; Wirth 1994).

No less than in domestic society, governance is exercised through political and administrative processes performed by a legislature, executive and judiciary. Internationally, these functions are performed by international institutions, although the judiciary role is seldom exercised as vigorously as at the national level and is not as equally respected by governments. Authoritative values are developed in international treaty law, through UN General Assembly decisions, and under the auspices of UN agencies. The discussion below of the international environmental governance norms and rules are based on the coding of the texts of 133 of the 142 multilateral environmental treaties signed from 1920 through 1992 as collected by UNEP. A body of 'soft law' has also been developed which international lawyers regard as less binding than the formal treaties and customary law developed for the environment. Executive branch and administrative functions are performed by UN specialized agencies. The enforcement and arbitration function is performed only through informal and decentralized channels. Unlike domestic society, these decisions always rest on choice. However, as in domestic society, a lasting and legitimate exercise of governance rests on its legitimacy in the eyes of member governments.

Norms

International environmental norms establish the expectations against which states are willing to be judged. While such instruments of international law may lack the formal requirements of 'law' in a more legalistic sense, they do express shared purposes and expectations when voluntarily developed.

Legal scholars concur that a set of principles have been established in international treaties with the effective claim of normative injunctions which constrain the conduct of states by creating new norms and obligations to restore ecosystems' health. States now share the obligations of preventing and minimizing harm and appreciable risk from environmental contamination (Lang, Nuehold et al. 1991; Schachter 1991; Birnie and Boyle 1992; Iwama 1992; Weiss, Magraw et al. 1992; Kiss and Shelton 1993; Sands 1993; Weiss 1993). Toru Iwama writes that "...natural or cultural resources are regarded as international or global commons whose uses are limited by a common interest of mankind. The states are regarded as guardians or custodians of the international community"(Iwama 1992).

More broadly, international treaties concluded since UNCHE reveal an appreciation of the saliency of the issue as well as an acceptance that an increasingly comprehensive approach should be taken toward environmental management.

Environmental governance arrangements have become increasingly 'ecological' in form, heeding the ecological laws espoused by scientific ecologists and focusing on the sustainable management of ecosystems rather than containing threats to environmental quality. Species management is cast in terms of a habitat's ability to support multiple species rather than in terms of protecting individual populations living in the area. Environmental impact assessments are now widely required by governments and international organizations in order to weight the environmental consequences of economic or development decisions. International debates now regularly consider new concepts such as "ecological sensitivity values" to bound the rates of economic growth.

As written, environmental norms complement norms of sovereign rights. In practice, these commitments make governments abnegate claims of sovereignty. While the doctrine of sovereignty maintains an ambiguous standing in UNCHE Principle 21, subsequent international laws establish a clear set of constraints on the actual exercise of sovereignty while leaving the doctrine intact. Governments have become increasingly willing to sacrifice their operational sovereignty in these treaties. Before 1973, only 56% of multilateral environmental treaties regulated domestic activities. Since 1973, 85% of treaties have covered domestic activities. Industrialized countries have been much more willing to sign treaties which circumscribe their sovereignty than have developing countries.

These norms command widespread support from the industrialized countries, and, more recently, from the former centrally planned economies that often lack the domestic resources needed for coping with the ecological devastation caused by central planning. Almost every country is party to at least one multilateral environmental treaty, and most countries are parties to many treaties regulating activities in many different environmental media, be they marine, terrestrial, or air. Yet the norms are not universal in their claims to legitimacy. Forty percent of 85 international environmental treaties lack the ratification of at least one country that could play a significant role in resolving the problem at hand. Developing countries have been less willing to cede domestic authority than industrialized countries: China, Brazil, India, and Indonesia have each signed fewer environmental treaties which constrain domestic authority than have the USA, Japan, and Germany (Choucri, Sundgren et al. 1994). Developing countries are also more likely to ratify treaties with general declarative statements than those which deal with operational aspects of environmental management (Sand 1992).

Rules

Multilateral environmental treaties and regimes contain a number of concrete rules for national practices. Governments accept the duty to inform each other of risks, to consult in order to assess risks, to notify neighbors in cases of emergency situations and to cooperate in scientific research and systematic environmental monitoring.

Most common are declaratory treaties, which establish collective intent but do not entail concrete domestic commitments for signatories. Thirty-one percent of all multilateral environmental treaties have been declaratory.

Environmental commitments generally take the form of regulatory injunctions administered through a combination of numerical targets, procedural permits, and reporting systems. Twenty-nine percent of all multilateral environmental treaties entailed some form of operational standards or obligations for management. Air and marine pollution control regimes apply emission and ambient standards for individual and groups of contaminants, typically in the form of "black" (banned) and "gray" (controlled) lists. These lists, too, are becoming increasingly comprehensive in light of the growing ecological sophistication of the regimes' principles and an awareness of possible interactive effects of individual contaminants. Efforts to protect species typically have a set of appendices listing categories of species according to the degree to which their survival is endangered. Rules which stipulate what technologies can be used for various applications have been applied to the management of fisheries (bans on drift-nets) as well as to the control of operational oil pollution from ships.

Ecosystems-based management models have been applied in:

- the 1971 RAMSAR Convention on Wetlands of International Importance;
- the 1973 Agreement on the Conservation of Polar Bears;
- the 1979 Bonn Convention on the Conservation of Migratory Species;
- the 1980 Convention on the Conservation of Antarctic Marine Living Resources;
- as well as for managing the North American Great Lakes (1978).

Management activities based on scientific assessments of the environmental carrying capacity have been developed for protecting stratospheric ozone, where the chlorine-loading potential dictates the volume of CFCs which countries may emit under the Montreal Protocol and its subsequent amendments, and limiting European acid rain, where critical loads appraisals provide guidelines for national sulfur emissions under the 1994 revised sulfur protocol.

Other rule-making techniques are less common, and are declining in frequency of use. Twenty percent of multilateral treaties contain actual bans on prescribed activities. Fewer treaties rely on the use of insurance or liability mechanisms. Insurance funds have been established to pay for oil spill cleanups and compensate victims. Clear provisions for liability for environmental damage have only been developed for: nuclear damage, civil liability for pollution damage caused by maritime transport of oil, civil liability for pollution damage caused by offshore operations, and civil liability for damage caused by inland transport of dangerous substances in Europe (Doeker and Gehring 1992).

Decision-making procedures typically take the form of annual or biannual intergovernmental meetings to review the activities of international secretariats, authorize new projects, and hash out differences in national approaches to environmental policy. Increasingly diplomats apply a set of legal innovations which have been developed since UNCHE by a small cadre of ecologically sensitive international lawyers for international environmental treaties. These lawyers participated in a wide variety of negotiations and tended to reapply their earlier successful legal experiments as precedents when drafting each new regime. Examples of such legal innovations are: a framework convention developed in 1976 for the Mediterranean which is devoid of concrete obligations and specific protocols dealing with environmental protection; technical black and grey lists introduced in the 1972 London Dumping Convention which specify the substances regulated by the treaty, yet which can be modified by expert agreement without having to reconvene political parties; an iterated negotiating process first elaborated by a UNEP committee of legal experts in 1981 and developed for stratospheric ozone protection in which states first adopt a framework convention signaling common aspirations and goals followed by more technical and binding protocols; the establishment of dedicated voluntary trust funds so that regimes may be self-supporting which has the effect of making the conventions financially self-supporting as well as creating more resources for programmatic activities; and the creation of committees for monitoring treaty compliance, first developed for the Montreal ozone protocol, and now used for the climate change treaty and LRTAP.

Monitoring efforts remain largely the domain of governments, although most treaties require the provision of periodic reports. National reporting to secretariats about their environmental protection activities is often poor, and many secretariats lack the resources or authority to check data submitted by governments. Many NGOs are now capable of monitoring environmental quality, as well as national compliance, and could help compensate for the dearth of effective environmental quality data, as well as providing an independent quality check on data collected through other sources. Much of the environment can be monitored remotely from satellites, and does not require the active collection and submission of data by governments. Remote sensing and satellite monitoring would also enhance verification of trends in natural resource use, marine pollution from organic sources and from oil, as well as in monitoring levels and production of greenhouse gases, although ground truthing is still necessary to confirm remote sensing data. Satellite and airplane based monitoring is less effective at monitoring inorganic marine contamination and urban air quality, for instance, which requires localized sampling and monitoring.

Suffrage

The primary figures in international environmental governance are, of course, representatives of governments who claim the legal authority to take authoritative decisions in international institutions. Governance through the late 1980s primarily occurred through dialogues between environmental ministries and environmental scientists, and with other governmental bodies.

Over the last 20 years, participation in environmental governance has spread beyond the society of states. The scientific community participates widely in international discussions as well as engaging in preliminary discussions at the national level. Environmental issues were initially unfamiliar to policy makers, and they often had to solicit technical advice from environmental scientists. Those exercising authoritative control over knowledge command great potential influence because technical information is a critical resource for effective environmental governance. As environmental scientists became systematically involved in decision making at the national level for domestic and international issues, environmental governance came to increasingly reflect their improved understanding of ecosystems. Many regimes have established independent advisory panels, with the effect of institutionalizing the participation of independent scientists in environmental governance. Developing country governments that lack their own indigenous capacity rely on such panels for independent scientific appraisals of environmental risks.

While originally established in the industrialized countries, environmental NGOs are no longer restricted to the North. Many have flourished in Malaysia, the Philippines, India, and Brazil, and umbrella groups exist on every continent (Barnes 1984; Clark 1990; Dubash and Oppenheimer 1992; Livernash 1992; Fisher 1993; Princen and Finger 1994; Weiss and Gordenker 1995). At UNCED, 70% of the registered NGOs came from industrialized countries, with the heaviest representation from the USA, Canada, and the UK. The most heavily represented developing countries were India, Philippines, Nigeria, Kenya, Sri Lanka, and Pakistan, with over ten NGOs accredited from each. Many of the developing country NGOs are still politically and organizationally weak, lacking large staffs, stable financing, capability to assess science, and experience with dealing with governments.

Multinational corporations kept a low profile in international environmental politics until recently. Many firms decided that environmental regulations did not constitute significant threats to their competitiveness and were content to lobby their governments and temper policies they found onerous. More recently they have responded to growing government intrusiveness and greater interest in the environment on the part of voters and consumers to become more vigorous participants. MNCs could make a positive contribution to environmental governance because they control the products, technologies, and knowledge about markets that are essential for efficient environmental protection.

International administration by international organizations

Many executive branch functions are performed by international organizations. Like executive branches in national governments, the UN specialized agencies and the Bretton Woods institutions can compel or induce action by reluctant members of society as well as legitimating the political participation of members of society. International organizations are engaged primarily in national capacity building, generating and diffusing information, collecting information about actions taken at the national level, evaluating national performance, and building domestic constituencies of sympathetic actors such as scientific networks, environmental NGOs, and, increasingly, private firms.

A number of international intergovernmental organizations have come to design and apply economic development projects and conservation projects which integrate environmental considerations with traditional responsibilities (Skolnikoff and Kay 1972; Haas and Haas 1995). In the United Nations, the Commission on Sustainable Development (CSD) is now responsible for overseeing and coordinating the environment and development activities of the UN specialized agencies. UNEP has been the environmental catalyst as well as the environmental conscience of the UN system. The World Bank's development style has been transformed to incorporate environmental considerations into project development. The World Health Organization (WHO) publishes environmental standards that are widely accepted in national laws, particularly in developing countries where governmental agencies are unable on their own to set appropriate domestic standards. Other IOs are also expanding their activities to integrate environmental considerations into regular projects

Enforcement

The judiciary function is weakly performed in international environmental governance. Very few international environmental regimes contain strong provisions for adjudication. Seventeen (including the Montreal Protocol, CITES, the Basel Treaty, and 10 species conservation treaties) contain provisions for trade sanctions against violators. Such provisions have rarely been invoked and a few that have are being challenged in the World Trade Court of the World Trade Organization.

Enforcement in practice depends largely upon shame (Chayes and Chayes 1991; Chayes and Chayes 1993; Mitchell 1993). International institutions provide a key function by collecting and disseminating data and information, receiving reports on treaty implementation by states, facilitating independent monitoring and inspection, and acting as fora for reviewing the performance of individual states (Boyle 1991; Lykke 1992; French 1994).

Actual enforcement is exercised by political oversight of parties and NGOs at the domestic level in elections. NGOs perform useful functions in international governance by providing information and counterbalancing the claims of both governments and private actors. Some information about governmental performance that influences election outcomes comes from the information provided by international institutions. In addition, a country's reputation in environmental matters is increasingly being considered by multilateral development banks and foreign aid agencies and even private sources of foreign investment. Consequently, enforcement is voluntary, subject to fear of opportunity costs of not appearing green. The effectiveness of such decentralized and informal techniques depends heavily on the widespread availability of credible information, the ecological literacy of publics, and the political efficacy of policies.

Many NGOs are now capable of evaluating their government's environmental performance, and could submit their own reports or confirm official submissions to secretariats. Greenpeace now regularly monitors trade in hazardous wastes and in flora and fauna, and publicizes shipments that are in violation of international treaties. The publicity generated by the NGO monitoring is often sufficient to inform recipient governments of activities of which they may have been unaware, as well as pressuring them to enforce their international commitments and to refuse entry of such products. Many NGOs have become virtual watchdogs over private activities in the field as well, replacing or supplementing the monitoring activities of national enforcement agencies. Because governments are often unwilling to cede the semblance of authority to NGOs, private monitoring of governments' actions and of the environment may best be accomplished through independent scientific panels, which have access to a variety of sources of information. Surprise visits by independent inspectors are used in some regimes as a means of verification. Long a part of the Antarctic Treaty System, CCAMLR provides for such visits, and they have been considered by the Helsinki Commission in the management of the Baltic Sea area. The concept is accepted by Eastern European and OECD countries, but not by LDCs.

Explaining the evolution of international environmental governance

The evolution of environmental governance has occurred within a matrix of scientific understanding, international institutional guidance, market forces, national leadership, and mass concern. While individual regimes and treaties were developed through different configurations of forces, over the last 20 years these efforts have accumulated to form a coherent set of expectations about national behavior and to a lesser degree, a common set of actual practices exercised by governments and by companies. The combination of knowledge and institutional influence and, more recently, public concern and market pressures have interacted like an isometric set of vector forces to maintain a resilient international regime. The norms, rules, and strategies for ecological preservation gain in standing as the domestic and international political costs associated with their violation grow stronger for governments and firms, due to public expectations.

An evolutionary institutional model of environmental governance

The interplay of these factors can be best understood within an evolutionary framework. Ideas provide the genetic material of international governance. This framework accounts for both the frequency with which ideas are adopted and the functional and geographic variation of their adoption.

The loose framework I propose is evolutionary because it consists of independent mechanisms by which new ideas are generated (a variation mechanism, although not purely random since knowledge accretes over time) and adopted by authoritative actors (a selection mechanism); and because past decisions affect present circumstances and choices in ways not apparent to most actors (Nelson and Winter 1982; Elster 1983; Krasner 1988; Adler 1991; Adler and Hass 1992). Ideas become embedded in institutions through standard operating procedures, organizational cultures, and codified routines, and shape subsequent perceptions and decisions taken under their auspices. Ideas condition practical policy choices by actors influenced by institutions, in turn leading to convergent action between those actors, be they governments, firms, or individuals. The impact of ideas persists beyond the initial constellation of factors responsible for their institutionalization as they are internalized to become new organizational routines and institutional incentives to which individuals respond. By creating established roles for all the multiple actors involved in elaborating and enforcing practices which are suggested by these ideas, they create new informal international institutions of shared beliefs and expectations. They inform collective decision making and contribute to the creation of a shared community of understanding. Finally, such ideas are converted to laws and regulatory policy at the national level.

An evolutionary model offers a probabilistic view of trajectories or paths of behavior, with a politically specifiable set of variables likely to explain which idea will be chosen at particular branch points in the trajectories. Branch points follow major crises or systemic shocks because shocks alert publics and leaders to undesirable circumstances and create demands for decision makers to undertake new efforts, and also undermine the authority of policy experts associated with prior policy orientations. Once institutionalized, the new ideas imported at the point of crisis shape perceptions and choices until subsequent crises arise, while also helping to define what would constitute a crisis, because frameworks also help identify plausible anomalies. In the absence of new ideas at such moments, old ideas and perspectives will continue. This model provides essentially a sticky, path-dependent explanation of human behavior, in which many choices or decisions are largely irreversible and the array of possible choices at time (t) is causally related to decisions taken at time (t-1). Ideas may exercise a causal impact later in time regardless of the reasons for their initial selection, as they continue to guide decisions even in the absence of original conditions which contributed to their widespread acceptance. Overall and over time frequency of recourse to policy and practices based on the new ideas increases.

The primary actors are states suffering from uncertainty. Leaders and decision makers are driven by a need to survive politically and institutionally. I assume that because of the complex nature of contemporary interdependent international activities, decision makers operate with incomplete and even scanty information about the nature of the physical and social environment they seek to influence. Consequently they are also unsure of their preferences, about their policy alternatives, and about the preferences of other actors with whom they will have to interact strategically. Under such circumstances neither conventional descriptive nor prescriptive decision-making models based on subjective expected utility or prospect theory apply, because actors are fundamentally unable to identify or choose between the array of choices. Moreover, decision making operates according to a satisficing model in which actors express procedural rationality, at best, subject to the time constraints imposed by bounded rationality, and do not engage in continuous information searches about the state of the policy environment or the efficacy of their own efforts (Simon 1983; Simon 1985; March 1994). Lacking experience with environmental problems, few policy makers have sufficient experience or knowledge to draw policy inferences from direct environmental observations. Policy makers are often unclear about how their wealth and power are likely to be affected by new policies, or what measures will promote wealth and power acquisition in the new issue-area.

Thus, changes in information processing are likely to occur following a well-publicized shock or crisis, which imparts a sense of urgency. With tightly interconnected media and communications, the crisis need not be domestic to precipitate national and international calls for prompt state action. At such times, decision makers search for new ways of organizing experience, and they defer to groups with recognized authority. During subsequent, less revolutionary periods these new doctrines or orthodoxies assume the status of taken-for-granted assumptions, or dogma, which persist until called into question again by external anomalies. In the absence of available consensual knowledge they will remain bound by existing policy paradigms and resort to available techniques based on power and compromise.

The ability to articulate new ideas and control their distribution are critical sources of power because they confer control over meaning and inference. Since the Second Industrial Revolution, experts in science and technology have been regarded as authoritative sources of advice under many circumstances. They monopolize the legitimate use of certain types of knowledge. *Epistemic communities* (networks of professionals who share common normative and causal beliefs, accept common truth-tests and are engaged in a common policy enterprise) serve as the cognitive baggage handlers for knowledge-based epistemes and for less abstract policy-relevant concepts. They transmit new ideas between actors, and articulate ideas' programmatic implications for the formal institutions in which they successfully establish influence.

Once new ideas are established, they diffuse more widely, subject to the leverage commanded by the institution in which the epistemic community has acquired and consolidated its influence. States disseminate the ideas of epistemic communities to other states and to international regimes according to their ability to compel or induce others to apply such practices. Weaker states may adopt new ideas by emulating lessons observed abroad.

Not all states or international institutions are equally likely to consult members of an epistemic community or defer to their advice. Institutions with strong science and technology capabilities, and those representing pluralistic societies, will be most likely to be in the first wave of those embracing new ideas. It is there that the scientific culture of epistemic communities will have a close affinity, and new groups will be able to quickly articulate new policy initiatives.

There are several alternatives regarding the likelihood of which ideas will be selected at a moment of crisis. If there is only one new candidate, there is little problem. If there are many, choices may be conditioned by calculations of relative political gain from each idea. In either case, new instrumental ideas will only endure if they are loosely commensurate with deeper seated beliefs and do not threaten strategic political alliances.

Table 6.1 presents a fuller variety of mechanisms by which ideas diffuse internationally.

Institutional learning occurs as groups apply evolving consensual knowledge to institutional practices. Learning may occur directly, through interpersonal persuasion, communication, exchange, and reflection, leading to the recognition or appreciation of new causal models and shared values. Leaders are socialized to accept new views and to empower their expositors. Alternatively, learning may occur through administrative recruitment, as epistemic community members or their confidantes replace officials informed with alternative perspectives. It may also occur indirectly, as actors alter behavior subject to the influence wielded by institutions that embody the ideas. Different patterns may occur along different time frames, as well. While some effects may be felt immediately through new recruitment decisions, and persuasion, broader shifts in public opinion and societal effects may occur on the order of decades.

Once in place, ideas are likely to persist once they acquire a taken-for-granted element and future generations of policy makers are likely to justify their actions through reference to established legitimate principles. It also becomes politically costly to reverse such practices as new interest groups mobilize around them after recognizing that material gains are possible from the application of the new ideas.

Application of the model to explain the evolution of environmental governance

The development of international environmental governance, particularly its increasingly comprehensive form, can best be understood in light of the evolutionary model presented above. In 1972, most environmental practices by governments were narrow and unconnected to broader social and economic policies. By 1995, international regimes and most national practices were increasingly comprehensive. Both the increasing frequency with which such new ideas about environmental policy are applied and the national and functional distribution of where such policies are applied can both be understood by tracing the development of new ecological management beliefs and their international institutionalization.

The dominant epistemic community in international environmental issues, until the late 1960s, was composed of neoclassical economists and resource managers. They were widely discredited by broadly publicized environmental disasters

TABLE 6.1 Diffusion mechanisms for environmental ideas

From/to	*IO*	*State*	*NGO*	*Epistemic community*	*Firm*	*Society*
IO	• interagency coordination • jointly administered programs • cofinancing	• training • demonstration effect • project funding • anticipation of project funding • leadership by IO officials • sponsor meetings	• training	• convene conferences • support research and monitoring	• publications • study groups, workshops, • technical panels • information clearinghouse • contracting arrangements	• public education/ consciousness raising
State	• coercion • recruitment choices • leadership • funding	• coercion/sanctions • persuasion • market impacts • compromise • demonstration effect • imitation/mimesis • leadership • rewards (foreign aid, bargaining linkages)			• tax/investment policy • regulation • contracting arrangement	• education
NGO	• publicity • mobilizing international pressure	• grants • publicity • mobilizing international pressure	• common campaigns • communication		• demonstration effect	• demonstration effect • education • advertising • publicize and translate epicom findings • education
Epistemic community	• recruitment • persuasion • advisory panels	• recruitment • persuasion • advisory panels	• persuasion • communication through overlapping networks	• communication		
Firms					• joint venture	• advertising • demonstration effect
Society		• elections • polls • civil unrest	• contributions • memberships		• purchasing choices	

and the international energy crisis of the 1970s, which they had been unable to predict, and attendant popular fears of widespread resource depletion.

New advice came from an emergent research program with an associated epistemic community; scientific ecology. It had flourished in the United States and Europe in the late 1960s, and is generally regarded as being relatively uncompromised by political and institutional influence (McIntosh 1985; Worster 1987; Bowler 1992; Hagen 1992; Sachs 1992). Paul Sears called the new approach "subversive" because it challenged the public and private right to contaminate the environment, regardless of rights of usufruct or other social conventions, and was based on a holistic approach to international policy.

Ecological epistemic communities offered an alternate formulation of environmental problems, replacing a focus on managing discrete resources with a focus on preserving ecosystems through the management of their multiple uses. They shared a perspective that treats environmental contamination as part of a broader set of interconnected problems subject to conditions of high uncertainty, with high potential for surprise and nonlinear response from environmental stress.

Galvanized by a sense of urgency and environmental crisis accompanying UNCHE, most governments had to create formal institutions to deal with their new environmental responsibility. From 1972 to 1982 the number of industrialized countries with environmental agencies grew from 15 to 34. Many were staffed by members of the ecological epistemic community, as they were the only legitimate group which commanded a reputation of technical environmental expertise. UNEP was created in 1973 and was staffed principally by young epistemic community members with firm ecological convictions. UNEP elaborated a political process that Maurice Strong had initiated for UNCHE, which pressed states to concurrently engage in environmental research and monitoring while also drafting collective policy measures. By urging states to negotiate before they were confident of their interests he managed to avert some of the more acute Olsonian threats to regime creation.

LDCs were later to appeal to the ecological epistemic community for policy advice. The number of developing countries with environmental agencies grew from 11 in 1972 to 110 in 1982. Developing countries were slower to introduce new environment protection measures than were industrialized countries. Initial attention was directed toward solving domestic environmental problems associated with poverty—water quality, public health, and natural resource conservation. Gradually measures were adopted to cope with transboundary and global environmental threats as well. By the mid 1980s many developing countries had seriously begun to develop pollution control legislation, although environmental investments and implementation lagged well behind the North. By the end of the 1980s most Latin American and Caribbean governments had adopted legislation protecting water, forests, wildlife, soils, coasts, natural resources and sanitation (Tolba and El-Kholy 1992; Weale 1992; Brenton 1994). Southeast Asian governments started introducing legislation and investing in sewage treatment and pollution control in the 1990s. Few African governments have done so.

While unilateral efforts were taken to deal with domestic environmental threats (some of which occur ubiquitously around the world and thus were subject to international influence from USAID, UNEP, and the World Bank), most transboundary and global environmental problems were not seriously addressed unilaterally. This was because the full costs of environmental degradation were not encountered by any single government, and unilateral efforts would not lead to changes in environmental quality unless they were widely reciprocated by other states.

From international governance to national governance

National environmental policy styles became more comprehensive over time, as they were caught up in an increasing number of ecological regimes and were subject to influence from a growing number of ecologically informed international institutions, primarily USAID, UNEP, and the World Bank (after 1987). National efforts tended to become more vigorous and stringent over time under such institutional pressures. This shift is indicated in part by the increasing use of environmental impact assessments as an obligatory component of environmental planning. Table 6.2 presents the evolution of use of such instruments by different countries over time.

The use of integrated coastal zone management policies (ICZM), integrative practices which reflect ecosystems management perspectives, has also increased over the last 25 years. The planning approach for the management of coastal resources and environments was first developed in the early 1970s and was first converted to national regulation in the 1972 US Coastal Zone Management Act. By 1984, 13 coastal states had ICZMs and 42 states had them by 1990. In each instance the numbers are split fairly evenly between industrialized and developing countries (Sorensen 1993).

There are interactive effects between international and national level changes over time. Many of the countries which developed more comprehensive environmental planning approaches—both for transboundary and global environmental resources and also for domestic ones—were directly affected by epistemic communities or were part of a regime which was informed by an epistemic community.

TABLE 6.2 Introduction of formal EIA requirements

	–1972	*1973–1982*	*1983–1992*	*1993–*	*Total*
Advanced industrialized countries	2	4	14	3	23
Former centrally planned economies	0	0	2	0	2
Developing countries	2	5	2	3	12
NIC	2	2	2	0	6
Non-NIC	0	3	0	3	6

Source: calculated by the author from (Gilpin 1995).

While a comparative study of domestic and foreign environmental policy changes in ozone, European acid rain and marine protection indicates occasions of changes guided by shifts in public opinion, domestic political alignments, and foreign pressure, these factors do not dominate. They appear in only roughly half of the cases. Individual policy changes are equally influenced by inputs from domestic and transnational epistemic communities, and from international institutional forces which themselves are epistemically informed.

Towards a green political economy

International markets for pollution control technology finally came into existence in the 1990s, following the establishment of international environmental governance arrangements. Green markets were not conceived immaculately. Governmental regulations and emergent popular concern created the demand for new products and processes. Regulations created legal requirements for pollution reduction, creating niches for new products and procedures to reduce pollution. European and US industries report that corporate environmental protection efforts were taken in response to legal requirements (Porter 1991; Vaughn and Mickel 1993; United States Congress 1994; Christie and Rolfe 1995; The Economist 1995).

Estimates vary on the size of these markets. The OECD estimates that the current global market for environmental products and services is $200 billion. It is projected to grow to $300 billion by the end of the century (Ayres and Simonis 1994; National Academy of Sciences 1994; Socolow 1994). The largest and most technically advanced environment markets developed in the US (with 45% of the environmental technology market in 1990), Germany, and Japan (with 7% of the market), countries with the most comprehensive and effective environmental regulations and also the most competitive environment industries. Environmental technology markets grew most rapidly from 1990 to 1992 in Eastern Europe, Canada, Latin America, Mexico, China, Taiwan, and South Korea (Environmental Business Research 1993; The Economist 1995). The Worldwatch Institute finds that "some 80–90 percent of this burgeoning environmental industry, which includes an estimated 30,000 U.S., 20,000 European, and 9,000 Japanese firms, is in industrial nations, though it is now growing rapidly in the Third World as well" (Flavin and Young 1993).

With the combination of market opportunities, public concern, and national regulation, there are now incentives for global firms to produce with greenest standards which may be in excess of actual legal requirements in countries of operation. Frances Cairncross, former environmental editor for *The Economist*, observes (Cairncross 1992):

> Without a good environmental record it will be harder to find new sites for expansion, because local communities will be less likely to trust a company with a record of being dirty. Good environmental practice reduces the risks of unwittingly committing an environmental offense—breaking one

> of those regulations and so incurring a large fine or a worse punishment. . . . Reducing toxic emissions means less risk of nasty accidents; using fewer dangerous materials may keep down the bill for insurance premiums; managing to a high standard may be a defense in court for a company charged with some petty environmental transgression.

Some of the largest MNCs have called for global uniform environmental standards based on some of the most stringent measures (Schmidheiny 1992). Many MNCs have endorsed a variety of voluntary industrial guidelines and codes of conduct for environmental practices, ecological accounting procedures, and public environmental accounting. The major dynamics of international environmental diplomacy began to change in the late 1980s with the onset of popular environmental concern. Earlier, green parties had failed to exercise widespread influence over national environmental policy in Western Europe, and, while the public expressed widespread concern, it was generally unwilling to commit resources or act based on its concerns. By the early 1990s a growing sense of environmental concern became evident worldwide. A 1992 survey of public opinion in 24 countries conducted by the Gallup International Institute demonstrated increasing widespread concern about environmental contamination, in industrialized, developing, and formally centrally planned economies, combined with growing demands for international action. Most striking is the universal doubling in the percentage of respondents who felt that their own health was more seriously affected by environmental contamination than ten years ago (Dunlap, Gallup et al. 1992).[3] Such widespread concern has the effect of reinforcing the institutional and constructivist forces which contributed to the environmental gains of the 1970s and 1980s; making it extremely difficult to reverse decisions which have already been taken.

The future of international environmental governance

It is now politically inconceivable that there can be a return to pre-UNCED levels of environmental governance, given the isometric interplay between public opinion, economic opportunity, and government regulation. A number of potentially limiting factors to future effective international environmental governance remain.

Compliance

National compliance with these prescribed (and nominally accepted) international norms, rules, and practices still varies widely. National compliance is often a function of national will, capacity, and institutional design. The correct assignment of enforcement responsibilities may avoid conflicts of interests and thus expedite compliance, although will and capacity are probably the most important factors affecting compliance. Clifford Russell notes that commitments are most efficiently enforced when there is public oversight of private obligations, and private oversight of public commitments (Russell 1992).

While states may wish to comply, not all are capable. Many developing countries and formally centrally planned economies have greater difficulties in complying with international obligations than do industrialized countries, due to less developed administrative systems and fewer monitoring and financial resources which can be devoted to enforcement. International institutions may help build national capacity through training, resource transfers, and technology transfer.

Some states lack the will to enforce commitments. International institutions may help build state will. In pluralistic societies they may amplify and refract domestic forces onto governments through publicizing the state of environmental quality. Elsewhere they help to establish domestic demand by supporting incipient NGOs and grassroots movements, through public education, by disseminating information, and by including scientists in international discussions. Institutions may also directly offer incentives for state enforcement through green conditionality.

Other states are relatively impervious to the shaming elements in the governance structure. Without pluralistic societies and without the need to seek credit or finance from international institutions, governments of many Newly Industrializing Countries (NICs) are relatively insensitive to the array of international and domestic political influences on governments to protect the environment. Future environmental governance for these countries may be a matter of host-country MNC relations. While MNCs, particularly the largest ones engaged in manufacturing, are now part of the new green IPE and will be prone to apply universally the practices they are now pursuing in the North; their domestic counterparts may be resistant to what they see as unnecessary and additional short-term expenses (Pearson 1987; Leonard 1988).

The NICs pose a big environmental threat which may not be easily met through international governance mechanisms. Most of the NICs have weak environmental legislation and economic growth is driven by some of the world's most pollution-intensive sectors: electronic components, basic metals, vehicle parts, plastics, chemicals, leather products, printing, glass, paper and cellulose, cements, and fertilizers (Harris 1992; The Economist 1993a, b). Loss of biodiversity, toxic wastes, heavy metals, sulfur dioxide, nitrogen oxide, and carbon dioxide emissions are likely to increase dramatically with increased economic growth in China, Taiwan, South Korea, Malaysia, India, Indonesia, and Thailand (The Economist 1990; Lenssen 1993; The Economist 1993a, b; Zich 1993). While some recent increases in environmental regulation and investment are evident, they are unlikely to be adequate (Howard 1993; O'Connor 1994). India, South Korea, and China, whose ambitious national industrialization plans offer a potentially devastating impact on the global climate, biodiversity, and on regional seas, remain ambivalent about many global and international commitments.

Emergent environmental threats

Potential environmental threats are numerous. Acid rain in South East Asia, West Africa, and parts of Latin America has been recorded, and will increase in severity as industrialization continues to accelerate. Water quality in the Third World is an overriding national environmental threat to public health because diarrhea remains the major source of infant death in the developing world. Few systematic efforts have been developed to regulate offshore drilling and uses of the seabed, or to collectively manage river basins in sustainable ways. Nuclear decommissioning in the former USSR, and counteracting environmental destruction in Eastern Europe, are major items on the future agenda. There is insufficient information for informed public policy for the over 1,500 chemicals annually which are widely used globally (Tolba and El-Kholy 1992).

Substantive gaps and inconsistencies

The successes to date in environmental governance raise a new set of problems as well. The multitude of pollution controls need to be harmonized. Treaties and regimes were developed to cope with discrete problems, or pollutants. The result has been a patchwork of standards for the use of different ecosystems. Industries are presented with different and inconsistent rules for the same substance for different bodies of water, or for emissions into the air or disposal on land. For instance, the Mediterranean and South East Pacific have identical coverage regarding land-based sources of pollution, while they differ substantially from the North Sea and Baltic (Meng 1987; Davis 1993). Elsewhere, wastes may be incinerated on land although they may not be directly discharged into oceans. This is ecologically problematic as once in the air they reach oceans and thus undermine purely marine-based efforts for environmental protection. Government operations, particularly those connected with national security and defense, remain immune from international governance.

Organizational overload and competition

The proliferation of environmental treaties raises the specter of organizational overload and competition. Few environmental or foreign ministries are adequately staffed to be able to effectively participate in all of the meetings organized under the auspices of the dozens of environmental regimes. Many international secretariats are too small and poorly financed to support the many activities necessary for the maintenance of their regimes. The current crisis of multilateralism further threatens the funding for a number of isolated secretariats, and UNEP has been seeking to consolidate many of them.

Some regimes call for harmonization, as their environmental problems have elements which are consigned to different international organizations. In marine dumping, for example, some activities fall under the jurisdiction of IMO, while others fall under IAEA or secretariats organized to administer regional regimes. While overlapping institutional jurisdictions may actually improve the strength of international environmental governance because of the possibility of ratcheting up national commitments through linkages and nesting, such a complicated agenda is frustrating for government ministries and inhibits effectiveness.

Financing environmental governance

The costs of global environmental protection are enormous. The cost of implementing Agenda 21's proposals for sustainable development was estimated by the UNCED secretariat at roughly 600 billion dollars a year for 1993–2000. Estimates of the annual capital costs to reduce greenhouse gas emissions by 5% by volume by 2005 are $27 billion for Eastern European states, $18 billion for the OECD states, and $22 billion for the rest of the world. Providing adequate water supply and sanitation services to 90% of the rural and urban populations by the end of the century is estimated to require an average annual investment of $28.2 billion. Estimates of containing tropical deforestation are of the order of $40 billion over five years. Containing desertification by the year 2000 may require $2.4 billion per year, and protecting biodiversity may cost $1 billion per year (Kimball 1992). The World Bank estimates that programs to address the major environmental problems in the developing countries will require an additional annual investment of 66.5–77.5 billion dollars a year in the year 2000. Expenditures for environmental protection would account for between 1.29–1.49% of anticipated aggregate developing country GDP by 2000 (at a projected growth rate of 4.7% from the present), or 3.15–3.65% of actual GDP growth during the period.

Significant distributional problems are associated with foreign investment. From 1991 to 1994, 89% of total private capital flows to developing countries were concentrated in 12 countries: China (29%), Mexico (13%), Argentina (8%), South Korea (6%), Malaysia (6%), Portugal (6%), Brazil (5%), Thailand (4%), India (4%), Turkey (3%), Hungary (3%), and Indonesia (2%) (World Bank 1996). Other developing countries will be hard-pressed to attract sufficient financing for investment in clean technology. This is potentially damaging to the environment, since some of the major potential sources of industrial and consumer greenhouse gases and sulfur dioxide are India and China, which have failed to attract such large volumes of international capital.

Political support by many of the least developed of the developing countries for many environmental regimes may be threatened by fears of having to compete with former centrally planned economies for financial resources from the World Bank and bilateral aid agencies.

Information gaps

Information constraints on many public domain goods inhibit ready recognition of their existence by either sellers or buyers (Report of the Secretary-General 1994). Many potential users outside the OECD lack information about environmentally benign technologies and few providers are familiar with these markets. Information clearinghouses are needed to collect information about available environmental technologies and to unite potential buyers and sellers.

Doctrinal backlash

Success breeds backlash. The international environmental principles and rules which have been assembled over the last 25 years are encountering two forms of ideational backlash. In international regimes, they are encountering challenges from actors who are involved in trade regimes that have been developed on entirely different doctrinal foundations. Aspirations for environmental protection now conflict with regime rules and doctrines developed for other international activities. While environmental governance is fundamentally regulatory, international trade governance has been liberal and free market based. International institutions have varied in their positions on these matters, depending on the institutionalized beliefs within them. The WTO, charged essentially to promote liberal trade internationally, has questioned the environmental effects of trade liberalization, whether trade policies should be used to enforce environmental standards, and whether environmental standards constitute barriers to trade in two major decisions. Conversely, the EU, with an eye towards economic integration and political integration, has been less prone to interpret environmental regulations as threatening free trade.

Summary and conclusion

New forms of environmental governance have emerged with the entry of new authoritative actors onto the international diplomatic arena. The international environmental accords concluded since UNCHE establish normative and procedural benchmarks against which governments are held accountable by their citizens, by other governments, and by influential international organizations. A broad base of environmental governance now exists.

Substantively, governments are increasingly modifying economic policies in ways that are believed to be less environmentally destructive. Through the growing influence of ecological epistemic communities in international environmental institutions and regimes new ideas about ecologically sustainable development have disseminated internationally, and become increasingly embedded in national regulatory structures; through SOPs of foreign ministries and environmental ministries; through domestic administrative procedures and policies; through patterns of enforcement; and through broader patterns of expectations about governmental behavior. Articulated by epistemic community members, these ideas were first institutionalized internationally in regimes and in programmatic activities by international institutions, followed by national lock in through policies, recruitment of like-minded people to serve in key bureaucratic positions, and enforcement. In turn these efforts have given rise to new economic markets for environmental goods, locking in the governance structure at the national level, and contributing to a better potential for mitigating environmental harm and contributing to clean ups. Further effectiveness depends on easily accessible information about environmental quality and environmental policies, transparent actions by international organizations and governments, the continued participation in environmental diplomacy by non-state actors, sustained levels of public concern in major countries, and improved national capacity for environmental protection.

Such environmental governance is likely to be reasonably robust as the new regulations propagated for environmental protection and the emergent markets for pollution control technology and clean technologies mobilize domestic constituencies to support the international regimes and the national policies which enforce them. As MNCs and firms recognize the potential for financial profits from environmental protection a new transnational political economy coalition emerges behind sustainable development.

Its complete continuation depends of course on the maintenance of the structural underpinnings of the governance structure. Without continued commitment to multilateral institutions and the existence of domestic liberal political structures it would be increasingly difficult for such patterns of governance to persist. While existing commitments might well remain locked in, future environmental threats would not be addressed subject to the rules and values of the current environmental governance framework.

More profoundly, recent patterns of international environmental governance signal a shift in fundamental international governance principles. While units and authority remain territorially grounded, the officials responsible for those units and the effects for which they are responsible are increasingly thinking in non-territorial terms. Decision making occurs primarily in territorially defined and grounded political units, but others whose identities are not defined by political territorial location also make important choices. Ecological epistemic community members' overarching solidarity is with the ecological systems they seek to protect; their identities are expressed functionally at different geographic scales in different issues. Consequently, a shift is underway from ego-based national interests to eco-based ones, as states are informed with such new perspectives and increasingly appreciate that they can no longer regard themselves as discrete entities capable of providing their own wherewithal.

International environmental governance is thus increasingly grounded on notions of connectedness. Problems are no longer easily decomposable. To the extent that such beliefs diffuse more broadly from environmental governance, substantive linkages based on the causal connections between problems may increasingly characterize international negotiations and relations. Future research can profitably address the ways in which regimes initially developed to manage discrete areas of activity are modified to cope with growing appreciation of the increasingly global interconnections between such activities. Under the influence of such a new overarching policy vision, environmental and economic policies may shift from a focus on proximate causes of problems of concern to address more fundamental issues which are believed to be causally implicated within the broader vision of a thick causal tapestry of international politics associated with an deeper recognition and appreciation of complex systems (Social Learning Group 2001).

Notes

1 Copyright © 1999 Routledge, the Taylor & Francis Group. This chapter was first published in Aseem Prakash and Jeffrey A. Hart (eds) *Globalization and Governance*, London: Routledge, 1999, pp. 103–133. Reprinted with permission of the Taylor & Francis Group.

2 Norms are principled statements about what states (in this instance) *should* do. Rules are the arrangements by which states intend to fulfill these obligations. Strategies are the policies or practices which states apply domestically to fulfill their international commitments.
3 The countries are Germany, USA, Portugal, Great Britain, Canada, Ireland, Switzerland, Norway, Netherlands, Japan, Denmark, Finland, Russia, Poland, Philippines, Nigeria, India, Turkey, Uruguay, Hungary, Mexico, Brazil, Chile, and South Korea.

References

Adler, E. (1991). "Cognitive Evolution". *Progress in Postwar International Relations*. E. Adler and B. Crawford. New York, Columbia University Press.
Adler, E. and P. M. Hass (1992). "Conclusion." *International Organization* 46(1): 367–390.
Ayres, R. U. and U. E. Simonis (1994). *Industrial Metabolism: Restructuring for Sustainable Development*. Tokyo, UNU Press.
Barnes, J. (1984). "Non-Governmental Organizations." *Marine Policy*.
Birnie, P. W. and A. E. Boyle (1992). *International Law and the Environment*. Oxford, Oxford University Press.
Bowler, P. J. (1992). *The Environmental Sciences*. New York, W.W. Norton & Co.
Boyle, A. E. (1991). "Saving the World? Implementation and Enforcement of International Environmental Law Through International Institutions." *Journal of Environmental Law* 3(2): 229–245.
Brenton, T. (1994). *The 'Greening of Machiavelli: The Evolution of International Environmental Politics*. London, Royal Institute of International Affairs Energy and Environmental Programme.
Cairncross, F. (1992). "UNCED, Environmentalism and Beyond." *Columbia Journal of World Business* 27(3): 15–16.
Chayes, A. and A. H. Chayes (1991). "Compliance Without Enforcement: State Behavior Under Regulatory Regimes." *Negotiation Journal*: 311–330.
Chayes, A. and A. H. Chayes (1993). "On Compliance." *International Organization* 47(2): 175–206.
Choucri, N., J. Sundgren, et al. (1994). "More Global Treaties." *Nature* 367(3): 405.
Christie, I. and H. Rolfe (1995). *Cleaner Production in Industry*. London, Policy Studies Institute.
Clark, J. (1990). *Democraticizing Development*. West Hartford, CT, Kumarian Press.
Crawford, S. E. S. and E. Ostrom (1995). "A Grammar of Institutions." *American Political Science Review* 89(3): 582–600.
Dahl, R. A. (1971). *Polyarchy: Participation and Opposition*. New Haven, CT, Yale University Press.
Dahl, R. A. (1989). *Democracy and Its Critics*. New Haven, CT, Yale University Press.
Davis, J. W. (1993). "The Need for a New Global Ocean Governance System". *Freedom of the Seas in the Twenty-First Century*. Z. Durwood. Washington, D.C., Island Press.
Doeker, G. and T. Gehring (1992). "Liability for Environmental Damage". *The Effectiveness of International Environmental Agreements*. P. H. Sand. Cambridge, Grotius Publications.
Dubash, N. K. and M. Oppenheimer (1992). "Modifying the Mandate of Existing Institutions: NGOs". *Confronting Climate Change*. I. M. Mintzer. Cambridge, Cambridge University Press.
Dunlap, R., G. Gallup, et al. (1992). *Health of the Planet Survey*. Princeton, NJ, Gallup International Institute.

Elster, J. (1983). *Explaining Technical Change Cambridge*. Cambridge, Cambridge University Press.

Environmental Business Research. (1993). *Research Assessment of U.S. Environmental Technology Strengths and Applications*. U.S. Agency for International Development.

Fisher, J. (1993). *The Road from Rio: Sustainable Development and the Nongovernmental Movement in the Third World*. Westport, CT, Praeger.

Flavin, C. and J. E. Young (1993). "Shaping the Next Industrial Revolution." *State of the World*. Worldwatch Institute. New York, W.W. Norton: 182.

Foucault, M. (1982). "The Subject and Power". *Beyond Structuralism and Hermeneutics*. H. Dreyfus and P. Rabinow. Chicago, University of Chicago Press.

French, H. F. (1994). "Strengthening International Environmental Governance." *Journal of Environment and Development* 3(1): 59–69.

Gilpin, A. (1995). *Environmental Impact Assessment*. Cambridge, Cambridge University Press.

Haas, P. (1992). "Introduction: Epistemic Communities and International Policy Coordination." *International Organization* 46(1): 1–37.

Haas, P. and E. B. Haas (1995). "Learning to Learn: Improving Global Governance." *Global Governance* 1(3): 255–284.

Haas, P. M. (1992). "Banning Chlorofluorocarbons." *International Organization* 46(1): 187–224.

Hagen, J. B. (1992). *The Entangled Bank: Origins of Ecosystem Ecology*. Brunswick, NJ, Rutgers University Press.

Harris, N. (1992). "Wastes, the Environment and the International Economy." *Cities*: 179.

Holsti, K. J. (1992). "Polyarchy in Nineteenth-Century Europe". *Governance Without Government*. J. N. Rosenau and E.-O. Czempiel. Cambridge, Cambridge University Press: 32.

Howard, M. C. (1993). *Asia's Environmental Crisis*. Paris, OECD.

Iwama, T. (1992). "Emerging Principles and Rules for the Prevention and Mitigation of Environmental Harm". *Environmental Change and International Law*. E. B. Weiss. Tokyo, UNU Press.

Johnson, J. (1995). "Selling Blue Skies, Clean Water." *Environmental Science and Technology* 29(6): 262A–267A.

Kimball, L. A. (1992). *Forging International Agreement: Strengthening Inter-Governmental Institutions for Environment and Development*. Washington, D.C., World Resources Institute.

Kiss, A. and D. Shelton (1993). *Manual of European Environmental Law*. Cambridge, Grotius.

Krasner, S. D. (1988). "Sovereignty: An Institutional Perspective." *Comparative Political Studies* 21(1): 66–94.

Kratochwil, F. and J. G. Ruggie (1986). "International Organization: A State of the Art on the Art of the State." *International Organization* 40(4): 753–776.

Lang, W., H. Nuehold, et al. (1991). *Environmental Protection and International Law*. Boston, M. Nijhoff.

Latin American and Caribbean Commission on Development and Environment (1991). *Our Own Agenda*, Washington, D.C., Inter-American Development Bank, United Nations Development Programme.

Lenssen, N. (1993). "All the Coal in China." *Worldwatch*: 22–30.

Leonard, H. J. (1988). *Pollution and the Struggle for the World Product*. Cambridge, Cambridge University Press.

Levy, M. A., R. O. Keohane, et al. (1993). "Improving the Effectiveness of International Environmental Institutions". *Institutions for the Earth*. M. A. Levy, R. O. Keohane and P. M. Haas. Cambridge, MA, MIT Press: 420–421.

Livernash, R. (1992). "The Growing Influence of NGOs in the Developing World." *Environment* 34(5).

Lykke, E. (1992). *Achieving Environmental Goals.* London, Belhaven Press.
March, J. G. (1994). *A Primer on Decision Making.* New York, The Free Press.
McIntosh, R. P. (1985). *The Background of Ecology.* Cambridge, Cambridge University Press.
Meng, Q.-N. (1987). *Land-Based Marine Pollution.* London, Graham & Totman.
Mitchell, R. B. (1993). "Compliance Theory: A Synthesis." *Review of European Community and International Environmental Law* 2(4): 327–334.
National Academy of Sciences (1994). *The Greening of Industrial Ecosystems.* Washington, D.C., National Academy Press.
Nelson, R. R. and S. G. Winter (1982). *An Evolutionary Theory of Economic Change.* Cambridge, MA, Harvard University Press.
North, D. C. (1990). *Institutions, Institutional Change and Economic Performance.* Cambridge, Cambridge University Press.
O'Connor, D. (1994). *Managing the Environment with Rapid Industrialization.* Paris, OECD.
Pearson, C. S., Ed. (1987). *Multinational Corporations, Environment, and the Third World.* Durham, NC, Duke University Press.
Porter, M. E. (1991). "America's Green Strategy." *Scientific American.*
Princen, T. and M. Finger (1994). *Environmental NGOs in World Politics.* London, Routledge.
Report of the Secretary-General (1994). *Report of the Secretary-General on Technology Transfer, Cooperation and Capacity-Building* E/CN.17/WG.I/1994/2.
Russell, C. S. (1992). "Environmental Enforcement". *Innovation in Environmental Policy.* T. Tietenberg. Aldershot, Edward Elgar: 219.
Sachs, W. (1992). "Environment". *The Development Dictionary.* London, Zed Books.
Sand, P., Ed. (1992). *The Effectiveness of International Environmental Agreements.* Cambridge, Grotius.
Sands, P. (1993). *Greening International Law.* London, Earthscan.
Schachter, O. (1991). "International Environmental Law." *Journal of International Affairs* 44(2): 457–493.
Schmidheiny, S. (1992). *Changing Course: A Global Business Perspective on Development and the Environment.* Cambridge, MA, The MIT Press.
Simon, H. (1983). *Reason in Human Affairs.* Stanford, CA, Stanford University Press.
Simon, H. A. (1985). "Human Nature in Politics: The Dialogue of Psychology with Political Science." *The American Political Science Review* 79: 293–304.
Skolnikoff, E. and D. Kay, Eds. (1972). *World EcoCrisis.* Madison, University of Wisconsin Press.
Social Learning Group (2001). *Social Learning and the Management of Global Environmental Threats.* Cambridge, MA, MIT Press.
Socolow, R. (1994). *Industrial Ecology and Global Change.* Cambridge, Cambridge University Press.
Sorensen, J. (1993). "The International Proliferation of Integrated Coastal Zone Management Efforts." *Ocean & Coastal Management* 21: 45–80.
Soto, A. (1992). "The Global Environment: A Southern Perspective." *International Journal.*
South Centre (1991). *Environment and Development.* Geneva, South Centre.
The Economist (1990). "Pollution in Asia." *The Economist*: 19–22.
The Economist (1993a). "Pollution in Asia." *The Economist.*
The Economist (1993b). "World Chemicals: The Challenge of Asia." *The Economist*: 27–30.
The Economist (1995). "How to Make Lots of Money and Save the Planet too." *The Economist*: 57.
Tolba, M. K. and O. A. El-Kholy (1992). *The World Environment 1972–1992.* London, Chapman & Hall.

United States Congress (1994). Industry, Technology and the Environment: Competitive Challenges and Business Opportunities. Washington D.C., U.S. Government Printing Office. OTA-ITE-586.

Vaughn, D. and C. Mickel (1993). *Environmental Profiles of European Business*. London, Royal Institute of International Affairs.

Weale, A. (1992). *The New Politics of Pollution*. Manchester, Manchester University Press.

Weiss, E. B. (1993). "International Environmental Law: Contemporary Issues and the Emergence of a New world Order." *Georgetown Law Journal* 81(1): 675–693.

Weiss, E. B., D. B. Magraw, et al. (1992). *International Environmental Law: Basic Instruments and References*, Transnational Publishers.

Weiss, T. G. and L. Gordenker (1995). "NGOs, The UN, and Global Governance." *Third World Quarterly* 16(3): 248.

Wirth, D. D. (1994). "Reexaming Decision-Making Processes in International Environmental Law." *Iowa Law Review* 79(4): 769–802.

World Bank (1996). "World Bank Data." *Finance & Development*: 32.

Worster, D. (1987). *Nature's Economy*. Cambridge, Cambridge University Press.

Zich, A. (1993). "Taiwan." *National Geophraphic*: 10–32.

7

PROSPECTS FOR EFFECTIVE MARINE GOVERNANCE IN THE NW PACIFIC REGION[1]

Peter M. Haas

Introduction

This article looks at the prospects for effective regional action to protect the NW Pacific from environmental harm. In this article governance refers to patterns of state environmental practices; generally shaped by international institutions and shared beliefs and commitments. Effective governance arrangements are ones that are likely to influence states to adopt policies that are likely to improve environmental quality.

The region faces a wide variety of environmental threats, yet has not yet developed an adequate set of governance arrangements to significantly address the problems or reduce emissions into the shared sea from national sources. The authoritative *Random House Atlas of the Oceans* (produced in 1991 by the International Union for the Conservation of Nature) lists a full array of regional environmental degradation (International Union for the Conservation of Nature 1991). A wide range of land-based contaminants, including industrial pollution, domestic sewage and agricultural runoff, pollutes coastal waters. The Yellow Sea, East China Sea and Sea of Japan receive large amounts of wastes from the extensive coastal industrial activities, including manufacturing, mining, refining and municipal wastes, from China, Japan, South Korea, and North Korea. Specific contaminants range from chemicals such as PCBs and DDT, to tar balls and heavy metals. The area is also subject to extensive resource depletion. There is extensive fishing and mariculture (particularly for kelp). Threatened marine species include bowhead whales, northern right whales, humpback whales, fin whales, loggerhead turtles, green turtles, and short-tailed albatross. The coastal zone has heavy tanker traffic transmitting crude oil to China, Korea, Japan, and eastern Russia, so that there are ongoing problems of operational oil pollution, and risk of collisions and groundings. Low-level radioactive waste dumping occurs in the Sea of Japan. Potential offshore oil and gas deposits exist in the South China Sea, creating potential threats from offshore exploration and drilling. The World Resources Institute reports that 50% of marine catch in the region is at or above replacement limits (World Resources Institute 1997).

The region faces many of the generic political features that often inhibit effective collective action. A decade of rapid national economic growth has generated environmental externalities in the region, so that countries' economic growth may interfere with one another. Much of the waste and resources concentrate in the *res communis* outside the 200 mile limit and thus beyond the national control of any of the states. While concern about collective threats is recent, many of the countries have already been engaged in formulating and enforcing national environmental policies since the 1970s. The Japanese Environmental Agency was established in 1972, the Chinese Ministry of Urban and Rural Construction and Environmental Protection was established in 1974, the Russian State Committee for Hydrometeorology and Control of the Environment was established in 1973, and South Korea established an environmental administration within the Ministry of Health in 1980. Basic environmental protection laws were introduced in Japan in 1967, in South Korea in 1977, in China in 1979 (although promulgated in 1989), in North Korea in 1986, and a number were adopted in Russia between 1957 and 1963 (Kim 1999). Thus, a kaleidoscope of incommensurate national standards already exist in the region which must be coordinated in order to yield effective governance (Republic of Korea 1995; OECD 1997; Sorensen 1997); yet most states are reluctant to modify pre-existing standards when such actions may entail additional economic costs for their economies and potentially threaten their firms' competitiveness. Chinese enforcement of environmental regulations, for instance, is in general quite weak, and subordinated to state interests in promoting economic growth (Bachner 1996).

Regional environmental governance works best when it is multilateral. This article reviews the lessons and experience over the last 25 years from collective efforts at regional marine environmental governance from UNEP's Regional Seas Program, comprising 34 treaties (of which 29 are in force) concluded between 1976 and 1996, as well as efforts to protect the North Sea and the Baltic seas from pollution (Haas 1991; Woods Hole Oceanographic Institute Marine Policy Center 1993; Haas 1997). These cases span over a hundred countries across levels of economic development, political systems, and cultural backgrounds. They reflect experiences from regions with highly disparate levels of economic development within the region, such as the Mediterranean and Caribbean, as well as cases where the climate of political antagonism was potentially quite high (such as the Mediterranean and Gulf of Kuwait).

Existing governance arrangements for the NW Pacific

A limited number of global and regional environmental conventions exist with regional application, so they provide a fragile basis on which to build for effective regional environmental governance. Few of the arrangements command strong support in the region, compliance is meager, and no binding provisions yet exist for land-based sources of marine pollution.

Three broad environmental regimes apply to the NW Pacific Region. None apply directly to the array of specific problems encountered in the region. The regimes are listed in Table 7.1; along with an indication of which countries have ratified the relevant treaties.

The 1973 Convention for the Prevention of Pollution from Ships as modified by the Protocol of 1978 (collectively known as MARPOL 73/78) seeks to reduce oil in the oceans by establishing procedural elements for operational handling of oil on tankers, and also introduces design standards for new and existing tankers of specified sizes. Yet a 1991 US General Accounting Office survey found that only 30% of the members of the MARPOL convention submitted reports on operation oil pollution (Sand 1990; Sand 1992; USGAO 1992).

The Convention for the Regulation of Whaling and the International Whaling Commission set annual limits on the capture of whales. This regime lacks popular support in Japan, and is hardly likely to command support as a legitimate regime from which to build a broader governance framework.

The London Dumping Convention of 1972 requires permits for the offshore disposal of a wide range of industrial and municipal wastes, and bans the dumping of a "black list" of the most environmentally dangerous substances. Only about 60% of the parties to the 1972 London Dumping Convention complied with reporting obligations (Sand 1990; USGAO 1992); and widespread violation of the regime is assumed by most observers and analysts of the LDC. Russia has been identified as one of the most flagrant violators of the LDC (Stokke 1998).

No regimes exist for addressing land-based sources of pollution in the region. The 1995 Global Plan of Action for the Protection of the Marine Environment from Land-based Activities provides broad guidelines, but is not legally binding. The only formal institutional arrangement that includes Japan, China, South Korea, and Russia is the Northwest Pacific Action Plan (NOWPAP), adopted in 1994, after 3 years of negotiation under the auspices of the United Nations Environment Program (UNEP) and the Intergovernmental Maritime Organization (IMO) (UNEP 1997). However, NOWPAP has been lying dormant for lack of high-level

TABLE 7.1 Environmental treaties in force for the NW Pacific Region

Country/treaties in force	*MARPOL 73/78*	*Whaling*	*London Dumping Convention*
USA	X	X	X
Japan	X	X	X
Russian Federation	X	X	X
North Korea	X	X	
South Korea (ROK)	1978 but not 1973	X	X
China	X	X	X

Source: Environmental Treaties and Resource Indicators 1998.

commitment by the governments of the region. No governments have yet committed funds to the budget, so that the program has limped along from 1994 to 1996 with 208,000$ a year from the UNEP Environment Fund, and Japan will not recognize North Korea.

Patterns of regime creation

International experience with regional marine governance suggests that five factors are key in constructing regional environmental regimes: national leadership, international institutions, transnational scientific networks, NGOs, and public concern (Haas 1999; Young 2000; Social Learning Group 2001). Different combinations of factors generate different regulatory forms and patterns of regime compliance. The most effective regimes have been those with strong international institutions working in conjunction with transnational scientific networks, known as epistemic communities (Haas 1990; Haas 1997; Boesch 1999) (Mediterranean, SE Pacific, S Pacific, and possibly the Black Sea). The combination of strong institutions and transnational scientific involvement has led to robust and comprehensive arrangements that take an effects-based approach to environmental protection and have led to demonstrable reductions in many sources of pollution (Table 7.2).

Strong institutions alone have also contributed to effective regimes for the Baltic and North Sea, although the collective goals have been reached through political compromise and thus are less likely to generate technical results at an optimal economic cost than arrangements worked out in conjunction with regional experts. In the absence of any of these factors collective efforts are likely to be very weak and only generate commitments which are tolerable to the least interested party ("lowest common denominator" responses).

While potentially contributing to effective regional governance, domestic pressure and NGOs have not played a strong role in regional marine management to date. In Europe environmental concern was very modest until the late 1980s, and only took off in the rest of the world in the early 1990s (Dalton 1994). A Gallup poll prepared for UNCED demonstrated increased worldwide concern, but very

TABLE 7.2 Examples of regional marine environmental governance

	Institutionally strong (dense/thick)	*Institutionally weak (thin)*
Transnational scientific involvement	**Mediterranean** Persian Gulf South pacific SE Pacific	
No transnational scientific involvement	Red Sea Black Sea Baltic **North Sea (1987–)**	Caribbean West Africa East Africa North Sea (1972–1987)

Boldface indicates regimes with significant reductions in pollution attributable to the regime itself.

little emphasis was placed on marine issues. Public opinion remained highly issue specific, like siting individual factories rather than regional planning. In the North Sea regime moments of public concern about marine environmental quality offered opportunities for ambitious national environmental ministers to submit vigorous new proposals for collective action, but such a degree of public concern is rare, and few other regional regimes have shown much influence.

Surveys of public concern in Japan and South Korea, the only East Asian countries for which public opinion polls have been conducted on environmental matters, are consistent with these global observations. The surveys reveal only moderate levels of concern about marine pollution. While 66% of Japanese expressed "a great deal or a fair amount of concern of personal concern about the environment, and 80% in South Korea, focused concern about marine issues was much lower, with 43% of Japanese surveyed responding that they thought that pollution of rivers, lakes and oceans was "very serious," with 49% of South Koreans expressing such concern. In Japan only 12% volunteered environmental problems as the most important problem facing the nation, with 9% in South Korea (Dunlap, Gallup et al. 1992).

Surveys conducted recently found significant increases in public demand for more stringent national environmental policy. The poll reported that in Russia 73% of respondents now support major action, and that in China 56% now support strong measures, unlike 1997 when majorities in each country were undecided or unsure (1998).

NGOs have also been rather absent from regional marine governance. In the 1990s Greenpeace International has launched campaigns aimed at mobilizing public concern in the Mediterranean and North Sea. In the Mediterranean Greenpeace has publicized violations of the regime, and pushed successfully for the adoption of a protocol on the transport of hazardous wastes. In the North Sea Greenpeace's Brent Spar campaign successfully induced Shell Oil to dispose of obsolete oil-drilling platforms on shore, rather than at sea.

Some environmental NGOs are present in the region. More than two NGOs attended UNCED from each of the following countries: Japan, USA, South Korea, and Russia. But NGOs are unlikely to exercise a strong political role in regional marine governance for two reasons. First, few of the environmental NGOs focus on problems of the appropriate regional geographic scale. Most NGOs are organized locally or globally, focusing on global issues such as global warming, or very local issues of industrial siting (Kim 1999). Second, NGOs lack political access and legitimacy in many of the region's countries. Only in Japan and South Korea do environmental NGOs engage in the domestic political process.

These political circumstances are not unique to Northeast Asia. In 1989 Eastern Europe faced many similar conditions, and foreign-aid agencies quickly recognized that effective environmental protection efforts which would involve local participation required the creation of local social and political institutions ("civil society") by which local groups could be effectively involved in national policy discussions (that is the process by which interests could be clearly articulated and aggregated for democratic national judgment) and help to implement environmental clean-up efforts (Keohane and Levy 1996). In 1990 the USA, EU, and Hungary helped establish the Regional Environmental Center for Central and Eastern Europe (REC), in Szentendre, Hungary. The center helps to establish and support NGOs in Eastern Europe and to provide a network of NGOs organized around the REC node. The CGIAR arrangement for agricultural research, funded by private US philanthropic foundations is a similar model.

National leadership

The guidance of any one country capable and willing to exercise leadership is often regarded as a key ingredient of successful regional cooperation and governance. In the Mediterranean, France helped to unilaterally provide the collective good of diplomatic leadership in designing and guiding early negotiations. The United States has selectively chosen to lead talks in the Caribbean, but not in other regional seas programs in which the USA is a member. State leadership may be important for agenda setting, guiding negotiations, and for enforcing compliance. If the leader commands respect it may be able to exercise leadership without expending a great degree of political influence, and inducing support and compliance through positive inducements. However, if other states are suspicious of the leader, or fearful about the distribution of costs from compliance, then political leadership is much more challenging and enforcement may well rely on threats of sanctions, and the governance arrangements may well collapse once the leader's hegemony erodes or if it fails to pay close attention to the state's compliance.

Few governments appear to have a vision of regional environmental leadership at this time. Most of the governments in the region are distracted by economic issues and are still responding to financial crises. The USA has some geopolitical interest in being seen as a constructive world leader, and has immediate national interests in maintaining unhindered navigation in the region. China might be willing to exercise leadership out of interest of managing oil conflicts in the South China Sea. Markets for environmental protection technology resulting from regional environmental governance may be an attractive option for some governments, although most discussions to date have been cast in terms of energy-efficient technology that would reduce greenhouse gas emissions.

International institutions

International institutions, when permitted by their member states, can play an important role in promoting regional governance and sustainable development. International institutions can help build more comprehensive regimes and encourage compliance by providing a venue for international cooperation, by building national capacity, and by building political will. In particular this means providing politically tractable instruments to groups within countries who are interested in sustainable development and marine protection, and building stable political coalitions who are able to press their governments within and among countries around such issues.

The primary functions performed by international institutions that improve regional marine governance have been improving the contractual environmental, enhancing national capacity for addressing environmental threat, and building national concern. Each of these will be addressed in greater detail below.

Northeast Asia is one of the most under-institutionalized regions of the world. A study group composed of regional government officials and academics, convened in 1998 by the Nautilus Institute, identified the following minimal practical needs for institutional measures for protecting the NW Pacific (The Nautilus Institute 1998). Pursuing these would also have the effect of strengthening the political functions performed by international institutions, and expanding the array of pressures on governments and incentives for marine environmental protection.

1 cooperation in developing a regional marine monitoring network,
2 cooperation in developing systems of integrated coastal zone management in the Sea of Japan,
3 cooperation in developing a sea vessel traffic control scheme in the straits between Korea and Japan,
4 cooperation in standardizing port management practices for ship wastes,
5 cooperation in developing oil spill response mechanisms in the region.

Few if any international institutions are readily able to perform these functions for the region's countries. There are no regional environmental institutions; environmental campaigns have been only briefly waged by institutions with much larger memberships which are informed by institutional missions with which the environment does not closely fit, and regional efforts are swamped by broader concerns such as economic cooperation around the Pacific Rim or global warming, or where the membership is so large as to make regional efforts such as this ungainly (Valencia 1990; Davis 1996).These broader institutions include APEC, ASEAN, UN-ESCAP, UNEP, UNESCO IOC, the World Bank, IMO, UNDP, and the Asian Development Bank.

Regional governance with the support of an international institution has occurred in the past, though. UNEP served to focus regional discussions in 9 regional seas despite (or due to) the absence of a regional environmental institutional authority, and The World Bank successfully established regional governance for the Black Sea. The GEF does have a mandate for oceans.

What prospects are there for specific institutional functions in this region?

Contractual environment

Elsewhere, successful governance efforts improved the contractual environment within which environmental diplomacy was conducted largely through the provision of regular meeting facilities to small numbers of parties, providing environmental monitoring data, verifying national policies and compliance, conducting talks at an appropriate level or profile such that entrepreneurial national environmental ministers would have an opportunity to introduce innovative policy proposals (such as the periodic North Sea Ministerial Conferences), linking negotiations and discussions to other areas of collective concern and thus improving the possibility for compromise, and by nesting the discussions within an overarching shared framework.

The small number of countries to coordinate in the region—only 6 countries, including the USA—greatly facilitates the logistics of meetings, and of monitoring and verifying national actions. It can also facilitate horizontal linkages between marine issues and more salient issues which command concern by or for the region. There is no regularly available source of monitoring or policy information for the region. Limited options exist for linking talks about marine pollution to other sets of ongoing discussions in the region about navigation, and possibly technology transfer discussions under the rubric of the Kyoto Protocol in climate change. Zarsky and Hunter have argued for the potential benefits of simple policy reforms: linking macroeconomic policy reforms to marine pollution by stressing reforms which improve energy efficiency, reduce energy use and thus reduce production costs as well as reducing airborne emissions (Zarsky and Hunter 1997). Policy reforms which make economy more efficient would also generate resources for pollution control. Involving the World Bank in regional talks about marine pollution would enable such linking.

Similarly, regional marine pollution talks may be nested under prior collective projects that provide precedents and frameworks for national commitments. Valencia lists 40 IMO navigational and oil pollution treaties to which one combination of the 5 regional countries is party (Valencia 1990). At the moment the dominant policy frames for the region appear to be related to financial recovery, with little potential for environmental protection.

Enhancing national capacity

International institutions have successfully overcome national reluctance to engage in multilateral environmental protection efforts by enhancing national capacity for environmental protection. Such efforts have the effect of providing additional benefits or inducements for states that are indifferent to the environment, as well as enabling states to formulate environmental policy domestically, and to enforce measures. For countries, particularly developing countries, which are suspicious of international initiatives arising from the North, building domestic institutions makes these governments more confident negotiators and thus more willing to acquiesce to collective deals.

Capacity building has generally occurred by providing financing and technology transfer of monitoring equipment, or establishing clearinghouses with information about access to pollution control technology, and by providing training seminars for government officials in coastal zone management or in research and monitoring techniques for scientists and technicians. In the short term such capacity building activities often merely provide an expedient justification for appearing to consult science-based policy; but in the longer term they may transform states' willingness to rely on science-based policy by enhancing the domestic political standing of the community, and also alerting the state to possible benefits from its use.

European states trying to protect the North Sea and Baltic Sea have successfully experimented with task forces organized around the quest for best available technologies and best environmental practices with voluntary leadership by countries that already enjoy comparative advantage in particular environmental technologies. Through the use of lead-countries for technologies associated with different economic sectors, countries with an advantage are offered an opportunity for sharing that information with others and thus creating new markets for their firms. In the Mediterranean periodic oil spill conferences with parallel trade shows provided governments the opportunity to promote their country's technology, as well as to learn about others'.

Simply participating in international meetings and chairing panels or working groups confers prestige on governments that are seeking to elevate their international political stature.

A striking feature of the North Asian region is the relatively high state capacity for formulating environmental policy (Baker, Bassett et al. 1985; Kim 1999). Competent state environmental agencies exist in all countries—except North Korea—with adequate numbers of professional staffs, and with adequate budgets. Less is known about the autonomy or discretion of the agency with the broader government, and thus its ability to influence policies with marine consequences taken by other functional agencies. Consequently, few governments would be attracted to collective talks solely out of the hope for material capacity building. The Japan Environmental Agency was established in 1972. By 1982 it had a staff of 907, of which 374 were scientific professionals. In 1993 1.1% of the total national budget went for the environment, and 3.5% in 1998 (OECD 1997; Kim 1999). In China, the Ministry of Urban and Rural Construction and Environmental Protection was established in 1974. A bureau of Environmental Protection was established in 1982 with a staff of 1,358, out of which 413 were scientific professionals. Its 1982–83 budget was 2.15 million $US. In 1991 0.8% of GDP was spent on environmental protection, and around 1% in 1998. South Korea created in 1980 an environmental administration within the Ministry of Health, and in 1990 established an office of the environment. The environmental administration had a staff of 233 of which 88 were scientific professionals. In 1998 2.24% of the South Korean budget went for environmental protection. Russia inherited many national institutional arrangements from the USSR. The State Committee for Hydrometeorology and Control of the Environment was established in 1973. Of GNP 1.3% was spent on environmental protection in 1988/1989 (Pryde 1991). The United States established the Environmental Protection Agency in 1970. The EPA is the largest environmental agency in the world, with over 17,000 employees 1992–1997, and with a budget of roughly 6 billion dollars (OECD 1996; Davies and Mazurek 1998).

Building national concern

By building state concern about the environment international institutions can improve the prospects for multilateral efforts at environmental protection. Building national concern about environmental threats occurs in two ways. Directly, institutions may provide information to governments and elites that may increase their willingness to commit resources domestically and collectively to environmental protection. Indirectly institutions may provide information to the mass public, which will then be channeled up to governments as demand for environmental action.

Heightening governmental concern has occurred by providing information and timely and high-quality scientific data that is important to these states. Joint monitoring, research and policy efforts all have had the effect of building governmental concern directly. Environmental NGOs may be important vehicles or allies for environmental institutions hoping to affect states' environmental policies.

Indirectly, state concern can be built through amplifying state–society connections. International institutions have been able to magnify domestic pressures on governments through publicity and involvement of national NGOs and scientists The provision of public education programs, the creation and strengthening of NGOs, and promoting the findings and individual status of scientists involved in transnational scientific networks have enhanced national concern, as has publicizing environmental threats by providing items to the media or through public education and museum exhibits.

In Northeast Asia domestic social and political institutions capable of aggregating and articulating interests for the government are quite weak. Japan and South Korea are political democracies with a fair degree of pre-existing domestic environmental concern, but Russia, North Korea, and China, may all be relatively immune to such institutional efforts to mobilize organic political concern; with no independent media in China, and few NGOs organized around regional marine pollution issues. Freedom House ranks China among the countries where the press is the least free. International institutions may have a better chance at public education and mobilizing environmental concern in the "free" presses of South Korea and Japan, or the "partly free" press of Russia (Freedom House 1998).

Transnational scientific networks and communities of expertise

In addition to the institutional 3Cs discussed above must be added a 4th C: the community of expertise. The systemic inclusion of science into regional environmental governance provides a firm technical foundation for collective action. In its absence political action will be driven by compromise and anticipated potentials for short-term gain. It is only in cases where regional marine scientists have been involved in collective political efforts that treaties and efforts have contained long-range planning and research components, and have efforts been based on what scientists regard as ecologically justified policies. While institutionalized science leads to more comprehensive and more judicious efforts, they develop more slowly than those negotiated through institutions in which science is not allowed to play a significant role.

Reviews of UNEP's efforts to build a regional community of expertise in its regional seas program, and similar efforts elsewhere in international environmental diplomacy, offer the following lessons for building a community of expertise (Haas 1991; Kimball 1992):

1. Carefully survey the population of scientists. In the Mediterranean a UNEP consultant spent 9 months visiting national laboratories to inventory national capabilities and to personally build the scientific network.
2. Recruit carefully for national and regional institutions. Base judgments on professional credentials and networking ability.
3. Avoid relying on 1 national institution to provide research and training.
4. Provide professional outlets for members through conferences and publications in refereed professional journals. This also elevates the domestic profile of individual scientists in the community of expertise who may then be recruited to fill positions in national administrations.
5. Avoid government designation of scientists to international meetings.
6. Try to make use of joint international panels for environmental risk assessment rather than relying on national assessments.
7. Assure the timely submissions in advance of meetings avoid single state sponsorship of collective research.
8. Avoid capture by 1 scientific discipline or school of expert analysis.
9. Arrange for focused interactions between scientists and policy makers to discuss the technical substance of the issues.
10. Maintain momentum within the community by continuing to have projects and research opportunities so those members don't drift away. This avoids having to reconstitute the community each time a new problem emerges.

NOWPAP has accorded priority to applying many of these lessons in the region. NOWPAP intends to publish a directory of marine environmental institutions and leading researchers in the region; inventory the technical capabilities of regional research institutions; develop collaborative research projects; and establish a collaborative regional monitoring program.

The raw potential for a community of expertise appears high in Northeast Asia. Table 7.3 summarizes crude measures of national scientific capability, although such an inventory should surely be updated.

A second feature of communities of expertise is their independence from their governments, and ability to operate as a transnational network, thus serving as consultants and advisors, and synchronizing policy advice across countries.

In China most scientists are dependent upon the state for institutional support and funding, and thus operate subject to close governmental sponsorship. Governmental support for oceanography and marine research is guided by Chinese geopolitical

TABLE 7.3 International Directory of Marine Scientists

Country	*Number of marine scientists*	*Number of institutions/laboratories*	*Disciplinary strengths?*
China	No data	No data	No data
South Korea	121	22	No data
North Korea	No data	No data	No data
Russia (Vladivostok, Kamchatka, Sakhalin)	81	9	No data
Japan	Thousands	Many	Extensive

Source: UNESCO and FAO 1983.

interests in establishing territorial claims in the S China Sea, finding fishing resources, searching for offshore oil and gas, and promoting maritime defense (Song 1989; Economy 1997; Kim 1999). Very little appears to be known about the organization of marine scientists in other countries in the region.

Conclusion

Regional environmental governance works best when it is multilateral. Yet prospects are not promising for effective regional environmental governance in the NW Pacific. International institutions are weak; knowledge is only weakly organized, and appears difficult to mobilize, and no country appears willing to commit economic or diplomatic resources towards regional environmental leadership.

The absence of transnational channels and institutional functions is a consequence of deeper regional realities. Few governments are interested in establishing strong regional institutions. Moreover, domestic circumstances that were necessary for effective institutional influence elsewhere in the world are absent in the region. The most effective international institutions worked with civil society by amplifying pressures on governments, and by creating and reinforcing public concern. In this region these conditions are generally absent, calling for a deeper long-run strategy for effective governance of fostering public environmental consciousness, building transnational environmental networks of expertise, and encouraging democratization.

Note

1 Copyright © 2000 Elsevier. This chapter was first published in *Marine Policy*, Vol. 24, No. 4, July (2000) pp. 341–348. Reprinted with permission of Elsevier.

References

(1998). "Tougher Environmental Laws Wanted, World Poll Finds." Retrieved 6/8/98, from http://envirolink.lycos.com/news/newfeed/jun05-1998-02.html.

Bachner, B. (1996). "Regulating Pollution in the Peoples' Republic of China." *Colorado Journal of International Environmental Policy and Law* 7(2): 373–408.

Baker, M., L. Bassett, et al. (1985). *The World Environmental Handbook*. New York, World Environment Center.

Boesch, D. F. (1999). "The Role of Science in Ocean Governance." *Ecological Economics* 31(2).

Dalton, R. J. (1994). *The Green Rainbow*. New Haven, CT, Yale University Press.

Davies, C. and J. Mazurek (1998). *Pollution Control in the United States*. Washington D.C., Resources for the Future.

Davis, E. v. w. (1996). "Global Conflicts in Marine Pollution: The Asian Pacific." *The Journal of East Asian Affairs*: 192–222.

Dunlap, R., G. Gallup, et al. (1992). *Health of the Planet Survey*. Princeton, NJ, Gallup International Institute.

Economy, E. (1997). "Chinese Policy-Making and Global Climate Change". *The Internationalization of Environmental Protection*. M. A. Schreurs and E. C. Economy. Cambridge, Cambridge University Press: 94–95.

Environmental Treaties and Resource Indicators (1998). Environmental Treaties in Force for the NW Pacific Region.

Freedom House (1998). "Global Press Freedom Declines." *The Economist*: 98.

Haas, P. M. (1990). *Saving the Mediterranean: The Politics of International Environmental Cooperation*. New York, Columbia University Press.

Haas, P. M. (1991). "Save the Seas". *Ocean Yearbook 9*. E. M. Borgese, N. Ginsburg and J. R. Morgan. Chicago, University of Chicago Press: 188–212.

Haas, P. M. (1997). "Scientific Communities and Multiple Paths to Environmental Management". *Saving the Seas*. A. Brooks and S. VanDeveer. Maryland Sea Grant.

Haas, P. M. (1997). "Introduction: Epistemic Communities and International Policy Coordination". *Knowledge, Power and International Policy Coordination*. P. M. Haas. Columbia, University of South Carolina Press.

Haas, P. M. (1999). "Social Constructivism and the Evolution of Multilateral Environmental Governance". *Globalization and Governance*. A. Prakash and J. A. Hart. London, Routledge: 103–133.

International Union for the Conservation of Nature (1991). *Random House Atlas of the Oceans*. New York, Random House.

Keohane, R. O. and M. A. Levy, Eds. (1996). *Institutions for Environmental Aid: Pitfalls and Promise*. Cambridge, MA, MIT Press.

Kim, H.-J. (1999). International Environmental Cooperation in Northeast Asia. *40th International Studies Association annual convention*. Washington D.C.: 6.

Kimball, L. A. (1992). *Forging International Agreement: Strengthening Inter-Governmental Institutions for Environment and Development*. Washington D.C., World Resources Institute.

OECD (1996). *Environmental Performance Review: United States*, Organisation for Economic Co-Operation and Development.

OECD (1997). *Environmental Performance Review: Japan*, Organisation for Economic Co-Operation and Development.

Pryde, P. R. (1991). *Environmental Management in the Soviet Union*. Cambridge, Cambridge University Press.

Republic of Korea. (1995). *Korea's Green Vision 21*. Ministry of Environment.

Sand, P., Ed. (1992). *The Effectiveness of International Environmental Agreements*. Cambridge, Grotius.

Sand, P. H. (1990). *Lessons Learning in Global Environmental Governance*. Washington, World Resource Institute.

Social Learning Group (2001). *Social Learning and the Management of Global Environmental Threats*. Cambridge, MA, MIT Press.

Song, Y.-H. B. (1989). "Marine Scientific Research and Marine Pollution in China." *Ocean Development and International Law Journal* 20: 601–621.
Sorensen, J. (1997). "National and International Efforts at Integrated Coastal Management: Definitions, Achievements, and Lessons." *Coastal Management* 25(34): 3–41.
Stokke, O. S. (1998). "Nuclear Dumping in Arctic Seas". *The Implementation and Effectiveness of International Environmental Commitments*. D. Victor, K. Raustiala and E. Skolnikoff. Cambridge, MA, MIT Press.
The Nautilus Institute (1998). Energy, Security, Environment in Northeast Asia Project Energy-Related Marine Issues in the Sea of Japan. Tokyo.
UNEP (1997). *Action Plan for the Protection Management and Development of the Marine and Coastal Environment of the Northwest Pacific Region*, United Nations Environmental Program.
UNESCO and FAO (1983). *International Directory of Marine Scientists*.
USGAO (1992). *Strengthening the Implementation of Environmental Agreements.* United States General Accounting Office. GAO/RCED-92-0188.
Valencia, M. (1990). "Sea of Japan." *Marine Policy* November: 507–525.
Woods Hole Oceanographic Institute Marine Policy Center (1993). *Comparative Efforts for Regional International Programs to Control Land-Based Pollution*. Woods Hole Oceanographic Institution.
World Resources Institute (1997). *World Resources 1996–1997*. New York, Basic Books.
Young, O. R. (2000). *Governance in World Affairs*. Ithaca, NY, Cornell University Press.
Zarsky, L. and J. Hunter (1997). "Environmental Cooperation at APEC: The First Five Years." *Journal of Environment and Development* 6(3): 222–251.

Appendix: A letter to the editor[1]

I am writing in response to Mark Valencia's letter to the editor about my *Marine Policy* piece on "Prospects for Effective Marine Governance in the Northwest Pacific Region." I hope to clarify my argument and to respond to Mr. Valencia's critique. I am trying to identify the driving forces behind effective multilateral environmental cooperation.

My overall argument, that I develop more fully in other published pieces,[2] is that there exist several discrete patterns of international environmental cooperation, depending on the array of such factors as the existence and willingness of key states to exercise leadership, the existence of strong international institutions, the presence of a strong transnational scientific involvement (epistemic communities), and the involvement of NGOs.

The combination of strong institutions and transnational scientific involvement has led to robust and comprehensive arrangements that take an effects-based approach to environmental protection and have led to demonstrable reductions in many sources of pollution. When strong institutions are able to embed knowledge into standard national practices—through socializing governments, establishing norms, and rewarding compliance—then distinctive patterns of environmental governance occur via social learning, as states come to change their notions of the value of environmental protection, and the means by which to achieve it. Institutionalized science leads to regimes that are more comprehensive, judicious, and slower than regimes that are negotiated through institutions in which science is not allowed to play a significant role. Such regimes have goals that reflect scientific consensus about the ability of ecosystems to sustain stress, and apply differentiated national obligations.

Strong institutions alone have also contributed to effective regimes, although the collective goals have been reached through political compromise and thus are less likely to generate technical results at an optimal economic cost than arrangements worked out in conjunction with regional experts. Environmental regimes developed through institutional bargaining contain legal efforts that are uniform, and commitments that tend to entail across-the-board emission cuts, or numerical limits in species taking.

In the absence of any of these factors collective efforts are likely to be very weak and only generate modest and bland commitments that are acceptable to the least interested party ("lowest common denominator" responses). With the presence solely of epistemic communities in the absence of strong institutions, state practices will vary according to domestic factors. Negotiated arrangements will be weak and modest, but "over-compliance" may occur in countries with strong epistemic communities or vigorous NGOs.

Different configurations yield different distinct patterns of cooperation. Stacey VanDeveer, in a recent piece in *Environment* magazine, offers a similar argument about the systematic patterns of behavior associated with different international conditions.[3] The North Sea provides a good example. Prior to 1987, all factors were weak, and international cooperation remained largely ineffective. While a regime was in place that adopted a number of pieces of binding legislation, the substance of each one tended to reflect a least-common denominator political dynamic for its adoption, and thus did not impose significant burdens on any of the parties and did not significantly contribute to the improvement of the quality of the Sea.[4] After 1987 the institutional structure became stronger, in conjunction with shifts in domestic environmental concern in the region, and the collective pattern of cooperation became much more vigorous and actually yielded some discernible improvements in the quality of the sea.

Professor Valencia advances a number of critiques of my piece. I will address each individually.

First of all, my analysis does not derive solely form the Baltic and North Sea, as Professor Valencia suggests. They are areas that provide a poor comparison for the Northwest Pacific region, because of the relatively more homogeneous nature of the parties in those regions, and the higher level of domestic concern. Rather, my analysis is based, as I explicitly note in the piece, on a broader base that includes all of UNEP's Regional Seas Programmes, including many developing countries and regions much more comparable to that of the Northwest Pacific.[5]

Second, I argue only that scientific agreement is most important for yielding comprehensive forms of environmental cooperation, not for yielding any cooperation *at all.* It is only in regions with epistemic communities and scientific consensus that have sought to control land-based sources of pollution, or to set environmental standards that reflect scientific appreciation of the ecosystem's ability to withstand stress. In cases without epistemic communities, but driven by other political impulses, the cooperative arrangements have been quite different. For instance, the North Seas has uniform emission cuts of 30–50% for a variety of contaminants, but the justification for this amount of reduction or the selection of the specific contaminants was done by political rather than scientific judgement. My argument for the Northwest Pacific region is that the linkages are insufficiently developed, and not likely to be strengthened in the next 5 years or so, to enable the kind of political dynamics that have led to more effective efforts elsewhere in the world.

Third, as to the catalytic function of an environmental disaster or crisis. It is true that well publicized environmental threats have occurred in the region. But a crisis is a necessary but not sufficient condition for galvanizing international action. Surely, many environmental disasters occur that do not galvanize responses. Yet for a response to occur, and for governments to seek advice from international institutions or transnational networks of scientists, a crisis is necessary to shake them up.

As to regional leadership, I am unaware of any significant efforts on the part of South Korea to exercise regional environmental leadership, and to commit resources to inducing the other states to follow.

Finally, as to the existence of regional institutional structures. UNEP has indeed provided the institutional bulwark when regional institutions were absent. But UNEP has varied widely in its successes, as UNEP well appreciates. In the absence of domestic commitments and political capacity on the part of national governments, UNEP's efforts have been disappointing. For instance West African and East African regional seas efforts are largely moribund, with the regional governments reluctant to take over control of the regime, or to even ratify many of the agreements they have signed. Elsewhere action plans have been adopted, but have languished in the absence of binding pieces of international law, such as in the South Asian Seas, Eastern Atlantic (Brazil, Uruguay and Argentina),[6] and, indeed, I argue, in the Northwest Pacific, where the inability to move beyond NOWPAP demonstrates the absence of significant political wherewithal in the region.

While I remain modestly optimistic about the prospects for effective international cooperation for the protection of the seas overall, I do not feel that the foundations for cooperation are strong in this particular region. In this piece I was seeking to demonstrate the reasons for this pessimism.

Notes

1 Copyright © 2000 Elsevier. First published in *Marine Policy*, Vol. 24, No. 6, December (2000) pp. 499–500. Reprinted with permission of Elsevier.

2 Peter M. Haas "Institutionalized Knowledge and International Environmental Politics" in John Ikenberry and Vittorio Parsi, eds. *Handbook of International Relations*. Rome, forthcoming; Peter M. Haas "Social Constructivism and the Evolution of Multilateral Environmental Governance" in Aseem Prakash and Jeffry Hart, eds. *Globalization and Governance*. Routledge, 1999, pp. 103–133; Peter M. Haas "Scientific Communities and Multiple Paths to Environmental Management" in L. Anathea Brooks and Stacy D. VanDeveer, eds. *Saving the Sea*. Maryland Sea Grant College, 1997, pp. 193–228.

3 Stacey D. VanDeveer "Protecting Europe's Seas." *Environment*, Vol. 42 No. 6 July/August 2000.

4 Sunneva Saetevick *Environmental Cooperation between the North Sea States*. London, Belhaven Press, 1998 and Peter M. Haas "Protecting the Baltic and North Seas" in Haas, Keohane and Levy pp. 133–181.

5 Paragraph 4, and Peter M. Haas "Save the Seas: UNEP's Regional Seas Programme and the Coordination of Regional Pollution Control Efforts" in Elisabeth Mann Borgese, Norton Ginsburg, and Joseph R. Morgan eds. *Ocean Yearbook 9*. University of Chicago Press, 1991, pp. 188–212.

6 For summaries and assessments of most of these efforts, see http://www.ngo.grida.no/ggynet/ and http://www.unep.ch/seas/rshome.html

8

EPISTEMIC COMMUNITIES AND INTERNATIONAL ENVIRONMENTAL LAW[1]

Peter M. Haas

1.0 Introduction

As the world becomes more globalized, decision makers grow uncertain about what their interests are and how best to achieve them, and ideas become increasingly important as maps or frames for decision makers in an unfamiliar environment. With the end of the Cold War decision makers cannot rely on geopolitical doctrines as a guide for various areas of foreign policy and international practice.

The environment provides a telling issue area in which to address the role of ideas in an increasingly uncertain global policy environment. This chapter looks at the concept of 'epistemic communities,' the role of epistemic communities in institutionalizing ideas in international relations, and in particular the role played by ecological ideas in the development of international environmental law and the role played by sympathetic international environmental lawyers in converting such rules of nature to rules of man.

2.0 Epistemic communities

2.1 The concept of epistemic communities

Epistemic communities are a concept developed by "soft" constructivist scholars of international relations concerned with agency. Soft Constructivists in general focus on the role of various types of norms, principled beliefs, causal beliefs and discourses in establishing roles and rules in international relations: that is determining the identities, interests, and practices that shape the identification of actors in international relations.

The concept helps international relations (IR) scholars to describe and understand the actors associated with the formulation of causal beliefs and the circumstances, resources and mechanisms by which new ideas or policy doctrines get developed and influence the political process as well as to understand and explain the broader process by which their ideas shape politics. Epistemic communities are important actors who are responsible for developing and circulating causal ideas and some associated normative beliefs, and thus, help to identify state interests and preferences, as well as help to identify legitimate participants in the policy process and influence the form and content of negotiated outcomes by shaping how conflicts of interest will be resolved. Attention to epistemic communities provides a way to understand agency in politics and policy formation. Under conditions of complexity, decision makers are uncertain of their goals or how to achieve them. Ideas may play important roles in framing the policy debate, and in predisposing outcomes highlighted by each perspective.

Epistemic communities are networks—often transnational—of knowledge-based experts with an authoritative claim to policy relevant knowledge within their domain of expertise. Their members share knowledge about the causation of social or physical phenomena in an area for which they have a reputation for competence, and a common set of normative beliefs about what actions will benefit human welfare in such a domain. Members are experts with professional training who enjoy social authority based on their reputation for impartial expertise. In the environmental realm they are primarily scientists and engineers, and, in economic matters, they are economists. Such individuals may hold positions in multiple locations over their careers: academia, think tanks, governments, and even within firms. Few are drawn to non-governmental organizations (NGOs) because most NGOs value principled beliefs over causal beliefs in formulating and justifying their policy discourses. Their beliefs are an analytic level below those of Dryzek's "discourses," although their political involvement can contribute in a causal way to the institutionalization of such broader discourses.[2] In particular, they are a group of professionals, often from a number of different disciplines, who share all of the following characteristics:

- Shared consummatory values or principled beliefs. Such beliefs provide a value-based rationale for social action by the members of the community.
- Shared causal beliefs or professional judgment. Such beliefs provide analytic reasons and explanations of behavior, offering causal explanations for the multiple linkages between possible policy actions and desired outcomes. While they need not agree on every element—in fact they are likely to disagree vehemently about some elements—they do agree about the core assumptions and causal forces from which their models of the world derive and agree on the means by which such differences can potentially be reconciled.
- Common notions of validity: intersubjective, internally defined criteria for validating knowledge.
- A common policy enterprise: a set of practices associated with a central set of problems which have to be tackled, presumably out of a conviction that human welfare will be enhanced as a consequence.

This combination of factors—especially the socialized truth tests and common causal beliefs—distinguish epistemic communities from other types of policy networks and groups active in politics and policy making. Unlike other organized interest groups active in politics and policy making, epistemic communities are bound by the truth tests to which they were socialized. Thus, they are more likely to provide information that is politically untainted, and therefore, more likely to 'work,' in the political sense that it will be embraced and followed by political authorities concerned about the need for appearing impartial. Their advice is also more likely to be technically effective, in the sense of obtaining the desired goals, than would be policies derived by political compromise. Politically their reputation for expertise and impartiality provides a social provenance, in the sense of the pedigree imparted, for example, by antiques appraisers that confers confidence in their advice by potential consumers.

2.2 Intellectual history of scholarship about epistemic communities

Constructivism has achieved an intellectual status as one of the dominant approaches in IR (Danish 2007). The ideational focus was absorbed into the broader constructivist program developed in international relations and comparative politics that looked at the role of beliefs and ideas in shaping state interests and practices, with epistemic communities providing a mechanism by which new ideas are developed and circulated, and providing a means for focusing on agency when studying international cooperation and governance. Consequently, the role of causal ideas and epistemic communities in explaining international cooperation has received increased attention. Causal ideas and epistemic communities are now generally regarded as significant sources of state interests under conditions of complexity and uncertainty, as main source in the origins and forms of multilateral regimes, and as important actors in the broader processes of social learning, reflexive governance, and identity formation. In other words, Constructivists have demonstrated that globalization makes actors uncertain about their interests and policies, and that various forms of social constructs (ideas) play a significant role in helping actors plot their course in uncertain conditions.

The focus on causal ideas and their transmission mechanisms have been largely eclipsed since the late 1990s by more attention to the role of norms and principled beliefs. A variety of critiques have been offered of the epistemic communities' research program (Dunlop 2000).[3] Some of these challenges include the following:

- Epistemic communities' knowledge base is itself socially constructed, and represents the potential systematic bias of those responsible for articulating research agendas;
- It is difficult to isolate the influence of the causal beliefs of epistemic communities from the institutional influence of the formal institutions with which they are associated;

- Epistemic communities' scholarship lacked a compelling theory of the state and did not pay sufficient attention to the domestic politics of the countries in which epistemic communities exercised a potential influence; and
- Better appreciation is needed for which ideas are likely to prove attractive to decision makers in response to uncertainty.

These critiques have led to increased analytic clarification and to stronger theoretical foundations and refinements for the constructivist epistemic communities' research program. While states are key juridical actors and act willfully, they vary in their administrative and political ability to formulate environmental (and other) policy, to absorb ideas domestically and from abroad, and to enforce policies at home. States are functionally differentiated, and vary widely according to their state/society relations and their technical capacity to formulate and enforce public policies in technical domains. State capacity in this sense is issue specific: that is regulatory potential varies by issue because the nature of the state's institutions, their resources, and policy networks vary by issue. Consequently, governance varies depending on the issue area.

Epistemic communities are most likely to emerge initially under circumstances affected both by issue and state characteristics. They are likely to be found working on substantive issues where scientific disciplines have been applied to policy-oriented work such as environmental protection, economic development, and energy policy. Thus, epistemic communities arise in disciplines associated with natural science, engineering, ecology and even economics. They usually emerge around specific topics whose research support is relatively inexpensive and not directly related to the explicit mission of the sponsoring research body. Consequently, the research and policy guidance of the epistemic community will be relatively independent of the explicit agenda of the research sponsor. Epistemic communities thus will enjoy broader political influence on issues such as the environment and ecology, rather than nuclear energy, because their findings are seen as being impartial or independent of explicit political influence. Environmental research is relatively inexpensive, compared to that of nuclear energy, and is conducted through networks of researchers who are not directly associated with the government or government laboratories. Epistemic communities are also most likely to emerge in countries with well established research capacity and where scientists enjoy some autonomy from the state. In such social settings the knowledge base will be relatively untainted and as a result enjoy a stronger political authority and legitimacy. States will look for new ideas after highly publicized crises. Such crises, potential or actual environmental disasters, or economic crises will prompt decision makers to take action to ameliorate pressure from domestic constituencies and from abroad.

The empirical experience with the global reception of environmental ideas does not directly engage the question of which ideas will matter at such points, other than saying that new ideas must have some affinity with existing beliefs or discourses, and that there is a strong path-dependent element to the dynamics of causal ideas once they are institutionalized. In the early 1970s, when international institutions were founded to address global environmental threats under the influence of the United Nations Conference on the Human Environment (UNCHE), there were few contending epistemic communities. In fact, there were only three major intellectual approaches to international environmental management. Traditional Resource managerial approaches were widely discredited by the Limits to Growth study and the 1973 Oil Crisis. Market-based approaches to environmental management were weakly advanced by economists, the World Bank and the General Agreement on Tariffs and Trade at UNCHE, but they and their institutional backers generally paid little heed to the conference. Market-based approaches did receive more attention in the 1990s, particularly in the drafting of the Kyoto Protocol. Access to information—a form of economic instrument that confers information and improves the prospects for informed consent and contracting—also became more prominent, especially in the 1998 Aarhus Convention on Access to Public Information, Public Participation in Decision-Making and Access to Justice in Environmental Matters and the 1998 Convention on Prior Informed Consent Procedure for Certain Hazardous Chemicals and Pesticides in International Trade. What is striking analytically, however, is the relative infrequency with which the negotiated outcomes that correspond to the dominant expertise paradigm—economics—have been adopted. Even in the Climate Change regime, for example, market mechanisms are instrumentally applied to achieve the substantive goals, identified by scientists, of ensuring that anthropogenic greenhouse gas emissions do not exceed the carrying capacity of the atmosphere. Market mechanisms as such, however, are primarily concerned with means rather than ends. They seek to provide mechanisms that improve environmental quality, but they are agnostic about the desirable level of environmental quality to be pursued. Conversely, epistemic communities specify the desirable environmental target and work backwards to develop the policy instruments that meet the desired ends. Even with the popularity of market-based discourses in the 1990s, negotiated outcomes—including the 2001 Stockholm Convention on Persistent Organic Pollutants—continue to reflect the command and control approach of the ecological epistemic community which derived environmental standards from the causal beliefs of the community. The ecological epistemic community's ideas persist.

Capturing the ideational high ground after the UNCHE left the nascent ecological epistemic community as the only game in town to assert a technical claim to expertise for addressing the new agenda of issues. Subsequently, this approach has retained its dominance as the presumptive approach to environmental management. Still, the ecological beliefs of the epistemic community at least corresponded loosely to the overarching policy paradigm, and political imperatives of the time, by continuing to assert the role of the state in promoting environmental protection, even while specifying the legitimate domain to which the state should exercise sovereignty.

More generally, some hypotheses were offered about which ideas are likely to be favored in response to uncertainty and crises. Thus, some analysts suggested that it isn't the accuracy of objective truth that matters but, rather, the political function that potential ideas serve. In either case truth alone is insufficient since politicians must believe that these ideas serve some political utility in order to apply them. Judith Goldstein has suggested that ideas that have been regarded as successful in the past are likely to enjoy favor (Goldstein 1993), such as trade liberalization, while John Odell suggested that simple ideas will enjoy an edge over more complicated ones (Odell 1990). Moreover, Jeffrey Garrett and Barry Weingast suggest that ideas that are favorable to cementing political coalitions, either internationally or domestically, will prevail (Garrett and Weingast 1993). In general, if there is only one candidate it will be likely to prevail. The political factors associated with selection are only salient if there exist a large number of competing ideas. In practice, the universe of available causal policy ideas is not very large, and there may not be a vast opportunity for political competition at moments of selection following crises.

More attention also has been paid to the broader dynamics by which ideas are transferred internationally. While they may initially be developed and consolidated in advanced democracies, they will be emulated and borrowed in other countries with sufficient domestic scientific capacities. International institutions are also likely to play an influential role in the broader transmission pattern. Major formal international institutions (IOs) that can play a role are those that enjoy strong organizational resources, and some autonomy from their member governments, through a combination of secretariat skills, porous organizational boundaries, and executive leadership. Such organizations are most likely to be immediately responsive to recognizing changes in the policy environment, and to absorbing new ideas about their underlying missions. In the environmental realm, these institutions included the UN Environment Programme (UNEP), the World Bank, and the UN Economic Commission for Europe (UNECE), and the UNCHE. Subsequently organizations may transmit the new ideas to member states through training programs and the demonstration effect, as well as to other international organizations through joint activities.

This makes it possible to analytically disentangle the causal effects of ideas from the institutions associated with them. It is possible to control the influence of the institution by looking at the same institution's behavior over time, subject to the ideas that enjoy popularity within it. For instance, the World Bank encouraged much stronger environmental policies after the 1987 reforms that had the effect of lodging more ecological ideas within the bank. Thus, epistemic communities and causal ideas have played a determinative role in influencing the ends to which institutional resources will be deployed in practice.

More generally, studies have separated two distinct mechanisms by which ideas disseminate internationally. One is through active social learning, through which individual decision makers are persuaded of the virtue of new ideas through direct exposure to epistemic community members or international institutions acting on their behalf. Such social learning is most likely to lead to robust changes in national practices as a consequence of profound long-term shifts in the appreciation of the national interest and how countries' prospects for material welfare are influenced by ecological stability. An example of this pattern is the European acid rain regime (Long-Range Transboundary Air Pollution Convention (LRTAP)), in which the concept of critical loads was developed by UNECE and the Institute for Applied Systems Analysis and disseminated through key countries to inform international treaties on sulfur and nitrogen oxide emissions, as well as national policies and legislation adopted to implement treaty obligations.

A second pattern of dissemination lacks the social element of direct contact that social learning entails. This second pattern is one of policy borrowing—of emulation of patterns of behavior and policies from salient countries and organizations. While policy borrowing is a major pattern by which ideas diffuse internationally and lead to policy harmonization, it would not be possible without the causal operation of social learning that institutionalizes ideas in the major countries and institutions from which others borrow. Instances of this have to do with adoptions of economic incentives for environmental protection in Europe, and the widespread application of toxic chemical standards developed by the United States or the World Health Organization.

Social learning is better understood now as a path-dependent process of collective social change. New ideas are solicited by uncertain policy makers in the aftermath of well publicized disasters or shocks, which encourage decision makers to seek guidance for responding to new and uncertain issues. Such conditions are more frequent under conditions of globalization and complexity. Through the steps described earlier in this chapter, new ideas imparted to decision makers by epistemic community members have the effect of changing state understandings of the policy environment and of their own national interests, and eventually to changes in state practices as well. These beliefs become institutionalized through the redeployment of state resources and through the consolidation of administrative practices and laws.

2.3 Ecological epistemic communities and multilateral environmental governance

A new environmental management doctrine based on ecological principles emerged in the 1960s. Since 1972, the ecological epistemic community has increasingly institutionalized this new management doctrine in state policies and practices, in the programmatic activities of international institutions, and in international regimes. As a consequence of the international institutionalization of ideas held by the ecological epistemic community, states have undertaken more comprehensive styles of environmental management for transboundary and global environmental threats, leading to selective improvements in environmental quality. Members of the ecological epistemic community subscribed to holistic ecological beliefs about the need for policy coordination subject to ecosystemic laws (see (Tarlock 2007)). Their ideas about ecological management were based on a systems perspective of environmental and social systems. Ecological management proposals favor setting comprehensive environmental standards based on conservative estimates of the ability of ecosystems to sustain stress, subject to the epistemic community's technical understanding of the behavior of particular ecosystems. They promoted international environmental regimes that are grounded on policies that offered coherent plans for the management of entire ecosystems, sensitive to interactions between environmental media (such as air and water), sources of pollution, and contending uses of common property resources. Rather than limiting themselves to more traditional policies for managing discrete activities or physical resources spaces within fairly short-term time horizons, they proposed treaties, in which bans and emissions limits were set for multiple contaminants, with environmental standards for each contaminant set according to scientific understanding about its environmental impact and its interactive effects with other contaminants.

Over time, the epistemic community added a broader social component regarding the need for widespread stakeholder participation in research and negotiations, out of the belief that such non-state actors, and particularly indigenous peoples, would have access to vital ecological knowledge outside the traditional domain of ecological sciences (Brash 2007). Moreover, such non-state actors, it was felt, would prove powerful political forces for compliance once management efforts were designed.

The ecological epistemic community has been associated with distinctive patterns of environmental cooperation. Treaties concluded with their input reflect their causal beliefs about environmental management. These treaties are based on environmental standards that reflect the epistemic community's technical understanding of a specific environment's abilities to sustain stress. An example is the 'critical loads' approach of the 1994 Sulphur Protocol to the LRTAP Convention, which differs from more common patterns of institutional bargaining where negotiated outcomes are solely based on compromise. Patterns of institutional bargaining embrace universal standards that are based on political compromise but have no clear relationship to environmental quality, such as the arbitrary, across-the-board cuts in pollution contained in early European air pollution agreements or North Sea emission controls. In the absence of technical consensus on the behavior of the North Sea and the impact of individual pollutants on the overall health of the ecosystem, North Sea diplomats adopted 30 percent and 50 percent reductions in the emissions of a wide array of chemicals. Such standards, while politically attractive, lack any clear impacts on overall environmental quality, and thus may impose economic costs without any offsetting environmental benefit.

In contrast, when regimes are negotiated with the involvement of epistemic communities and strong international institutions, they develop through a process of 'social learning.' Negotiations occur with a scientific discourse, in which political debate and compromise reflect expert consensus on the behavior of ecosystems and their ability to sustain stress. The substance of regimes reflects scientific consensus about the most important environmental threats, and negotiated standards reflect consensus about the degree of environmental stress the target environment can sustain. Social learning generates treaties with differentiated national obligations and substantive commitments, based on expert consensus on causes and environmental effects. Agenda setting is responsive to scientific advice and publicized crises. States interests may be altered as a consequence of being exposed to new ideas, and negotiated treaties will reflect an increased willingness to sacrifice short-term economic gains for longer term collective environmental protection.

Since the 1970s, an increasing number of regimes have been developed through the process of social learning. Only regimes negotiated in the shadow of epistemic communities and strong international institutions have yielded patterns of comprehensive environmental management through a process of social learning. Ecological epistemic communities, often working with UNEP, have helped draft comprehensive international environmental regimes governing marine pollution, acid rain, stratospheric ozone protection, wetlands protection, protecting migratory species, polar bears protection, and the preservation of Antarctica. An increasing proportion of all environmental regimes are now based on the comprehensive ecological approach promoted by the ecological epistemic community as the epistemic community has increased in vigor and influence, and countries have institutionalized their ideas. In 1973, three out of 11 international environmental regimes (or clusters of MEAs) were based on ecological management styles; in 1985, seven out of 22; and in 1995, 11 or 12 out of 25, if the climate change regime's commitment to stabilizing atmospheric concentrations at safe levels is interpreted as an ecological management frame. The application of ecological management ideas to environmental regimes spanning a number of geographic areas and functional activities means that most states have accepted ecological obligations for governing a wide variety of human activities.

Conversely, treaties and regimes concluded without epistemic communities yielded political compromises that were based on across-the-board reductions, or least common denominator type negotiated outcomes. Instances of the non-social learning regimes include fisheries management, whaling, marine dumping, and the Baltic and North Seas. Table 8.1 provides examples of some of the variety of negotiated environmental outcomes that characterize the environmental regimes that were established since the UNCHE. Social learning has yielded comprehensive ecological efforts with the combination of epistemic community involvement and strong international institutions. Such comprehensive efforts entail differentiated national obligations and substantive commitments, based on experts' consensus on the causes of environmental degradation. Institutional bargaining results from strong institutions encouraging states to compromise in the absence of epistemic consensus. Such treaties and regimes tend to involve disjointed treaties with uniform national regulatory obligations chosen for their political and emotional appeal. Finally, least common denominator results of weak collective obligations result from negotiations undertaken in the absence of strong institutions or strong scientific consensus, such as with most fisheries agreements where the institutional context for treaty negotiation and national enforcement is extremely weak, and while fisheries biologists can demonstrate that overfishing is occurring they cannot convincingly identify sustainable fishery yields.

The climate change and biodiversity regimes shed light on the nature of consensual knowledge shared by the ecological epistemic community. They, however,

TABLE 8.1 Major multilateral environmental agreements (in force) and their negotiated form (as of 2005)

	Strong institutions	*Weak institutions*
Epistemic communities	Social learning Ozone European acid rain Persian Gulf (pre-1980) South Pacific SE Pacific Antarctica Migratory species Polar bears Wetlands	
No epistemic communities	Institutional bargaining Red Sea Black Sea Baltic North Sea (1987–) London Dumping Convention/marine dumping MARPOL/operational oil pollution Whaling	Least common denominator Caribbean West Africa East Africa North Sea (1972–1987) Fisheries management Persian Gulf (1980–)

have not been included in Table 8.1 because, while in each case there exists a cohort of like-minded environmental specialists who believe that action is necessary in order to prevent calamitous climate change or loss of biodiversity, the knowledge base is insufficiently mature (or 'usable') in each instance to provide meaningful targets and deadlines for decision makers to apply (Andresen and Skjaerseth 2007).

3.0 Epistemic communities and international environmental lawyers

International environmental lawyers do not themselves constitute an epistemic community. However, they have played a key role in institutionalizing ecological epistemic communities in the formation of international environmental law and regimes.

3.1 International environmental lawyers are not an epistemic community

While there is now an active community of international environmental lawyers involved in the development and interpretation of international environmental law, they remain a policy community and not an epistemic community. Their shared beliefs take a different form from that of the ecological epistemic community discussed above, and they lack the social authority enjoyed by the ecological epistemic community.

International law as a field lacks the social authority or legitimacy of the technical authority commanded by epistemic communities. Legitimacy in the ecological epistemic community rests on the internally consistent substantive nature of the ideas put forth for policy makers by experts, and the transparent way in which such ideas are developed and gain consensus. International environmental law and international law more generally, is philosophically and epistemologically different from the social domain of ecological science. Whereas ecology and the technical domains in which epistemic communities operate are the realms of hybrid and brute facts, where claims about physical phenomena may be evaluated by recourse to mechanisms independent of the subject of study, law is a different activity. Law operates in the domain of social facts. The substantive domain of law is not subject to the type of shared tangible understandings that characterize the political domain in which ecological facts are identified for public policy. While international environmental law as a corpus may provide a presumptive backdrop against which negotiations and treaty drafting occurs (Johnston 2005), international lawyers lack the authority in environmental policy circles that scientists and epistemic communities enjoy directly.

International environmental law also lacks epistemic status for sociological reasons. International lawyers lack the public respect for impartial views about the world to which their advice is deployed, for the reasons mentioned above. In addition, the professional knowledge base of international environmental law is insufficiently institutionalized to generate common truth tests and a tight sociological network. Law school curricula tend to offer only one elective course on international environmental law, and there are few professional career path opportunities for international environmental lawyers in the absence of litigation opportunities in the realm of international environmental law. The array of professional niches for international environmental lawyers remain fairly narrow, limited to policy analysts, and law school professors.

International environmental law texts now exist, but they are largely collections of treaties rather than resting on a deductive set of propositions from which conclusions about institutional design and appropriate behavior may derive. Thus, international environmental law lacks the core causal beliefs and truth tests that define an epistemic community.

3.2 International environmental lawyers and the translation of epistemic communities' ideas into international environmental law

International environmental lawyers have played an active role in the development of international environmental law, and in the conversion of causal ideas of the ecological epistemic community into treaties for the management of environmental threats, by institutionalizing scientists' involvement in collective decision making and in drafting treaties based on the scientists' insights. Since the UNCHE, a core network of individuals has been involved in the formulation of much of the corpus of international environmental law. A small clique of lawyers initially worked for the Stockholm secretariat, and cut their teeth drafting the 1972 Convention on the Prevention of Marine Pollution by Dumping of Wastes and Other Matter. Subsequently, their rules of thumb were reproduced, and they acquired the status of precedent amongst the broader community of international lawyers. The cadre of sympathetic lawyers grew, with their foundations and institutional memories residing in UNEP's legal office (now called the Environmental Policy Development and Law unit), in the World Bank's legal office, and in some government offices. Some figures included Patrick Szell (United Kingdom), Johann Lammers (Netherlands), Winfried Lang (Austria), and Peter Sand (Food and Agriculture Organization, UNEP, the UNECE, and the World Bank).

These lawyers developed a kernel of procedural techniques by which ecological principles of environmental management would be applied to environmental law. These techniques include the following:

- The umbrella convention protocol sequence—this technique was developed in the 1972 Convention on the Prevention of Marine Pollution by Dumping Waste and Other Matter (London Convention) and the Protocols on Marine Dumping and Emergency Co-operation. Initially, in the regional seas treaties, this technique required states to commit not only to broad amorphous principles but also to more specific binding commitments as well, since states in most cases had to ratify at least one protocol. However, this no longer seems to be true for current multilateral environmental agreements (MEAs), which allow a state to become a party to the main treaty only, without ratifying the subsequently adopted protocol. The climate change and biodiversity regimes are both relevant examples (Gehring 2007);
- conference of the parties (COPs) and treaty amendments—COPs, through their decisions develop the MEA-based regimes, thereby circumventing the problem of having to amend the treaty. In addition, by provisions such as those developed under the auspices of the Montreal Protocol on Substances that Deplete the Ozone Layer, diplomats, based on expert advice, can approve environmental standards by amending an annex to the protocol without having to go through a time-consuming domestic ratification process (Hey 2007; Ulfstein 2007);
- lists of banned or regulated substances—the 1972 London Convention differentiates between banned substances (black list) and regulated substances (grey list). As of March 2006, for those states that are party to the 1996 protocol, this system will be replaced by a ban on all dumping of wastes at sea, except those substances listed (the so-called reverse listing approach). Both systems required identification of substances and the setting of appropriate standards and, thus, the involvement of environmental specialist;
- standing expert committees and panels—most international environmental regimes have been designed to have standing scientific panels associated with

them for the provision of timely environmental monitoring and policy information. Such arrangements provide the regularized mechanism by which individual scientists and epistemic community members may be recruited and involved in international environmental treaty making; and

- the establishment of dedicated voluntary trust funds so that regimes may be self-supporting—these funds have the effect of making the conventions financially self-supporting as well as creating more resources for programmatic activities.

More broadly, international environmental lawyers have come to appreciate international environmental negotiations as an exercise in communication and learning, where individual parties develop new ways of understanding the world, others, and themselves as a consequence of being exposed to the causal ideas articulated by epistemic communities. Such a reflexive approach also provides the means by which international law development can accelerate past a snail's pace dictated by the least enthusiastic party.

4.0 Conclusion

Over the last 30 years the environment has become firmly established on the international diplomatic agenda and, through regime formation, binding rules have been developed for most human activities affecting environmental quality. Substantively the focus on international environmental regimes has shifted from discrete environmental standards to ecosystem protection. Procedurally it has come to rely much more heavily on the expert advice of ecological epistemic communities.

Epistemic communities are one of the principal vehicles by which new ideas of ecological sustainability and governance have evolved. International environmental lawyers have helped provide the institutional mechanisms by which these ideas have reached and been disseminated to states and other actors. Constructivist theories of IR help to provide the explanatory parameters by which these processes occur and their effects.

Notes

1 Copyright © 2007 Oxford University Press. This chapter was first published in pp. 791-806 Ch.34: 'Epistemic Communities' by Peter Haas from "Oxford Handbook of International Environmental Law" edited by Bodansky, Daniel; Brunnée, Jutta & Hey, Ellen (2007). By permission of Oxford University Press.
2 See Dryzek chapter in this volume.
3 For an overview of these critiques and some trenchant rejoinders see (Dunlop, 2000).

Further reading list

Emanuel Adler 2001 "Constructivism in International Relations" in W. Carlsnaes, T. Risse, B Simmons eds. *Handbook of International Relations.* Beverly Hills, CA: Sage.

Steinar Andresen et al. 2000 *Science and Politics in International Environmental Regimes.* Manchester: Manchester University Press.

Paolo Contini and Peter H. Sand 1972 "Methods to Expedite Environmental Protection: International Ecostandards" *American Journal of International Law* 66 (1): 37–59.
Ernst B. Haas 1990 *When Knowledge is Power.* Berkeley: University of California Press.
Peter M. Haas ed. 1992/1997 "Knowledge, Power and International Policy Coordination" special issue *International Organization* 46(1) reprinted as Peter M. Haas ed. 1997 *Knowledge, Power and International Policy Coordination.* Columbia, SC: University of South Carolina Press.
Peter M. Haas 2001 "Environment: Pollution" in P.J. Simmons and Chantal de Jonge Oudraat, eds. *Managing a Globalized World*, Carnegie Foundation, pp. 310–353.
Peter M. Haas and Ernst B. Haas 1995 "Learning to Learn" *Global Governance* 1(3): 255–285.
Edward L. Miles, Arild Underdal, Steinar Andresen, Jorgen Wettestad, Jon Birger Skjaerseth, and Elaine M. Carlin 2002 *Environmental Regime Effectiveness.* Cambridge, MA: MIT Press.
John Gerard Ruggie 1998 What Makes the World Hang Together?" *International Organization* 52(4): 557–584.
John Searle 1995 *The Construction of Social Reality.* New York: The Free Press.

References

Andresen, S. and J. B. Skjaerseth (2007). "Science and Technology: From Agenda Setting to Implementation." *The Oxford Handbook of International Environmental Law.* D. Bodansky, J. Brunnee and E. Hey. Oxford, Oxford University Press.
Brash, R. L. (2007). "Indigenous People." *The Oxford Handbook of International Environmental Law.* D. Bodansky, J. Brunnee and E. Hey. Oxford, Oxford University Press.
Danish, K.W. (2007). "International Relations Theory." *The Oxford Handbook of International Environmental Law.* D. Bodansky, J. Brunnee and E. Hey. Oxford, Oxford University Press.
Dunlop, C. (2000). "Epistemic Communities: A Reply to Toke." *Politics* 20: 137–144.
Garrett, G. and B. Weingast (1993). "Ideas, Interests and Institutions." *Ideas & Foreign Policy.* J. Goldstein and R. O. Keohane. Ithaca, Cornell University Press.
Gehring, T. (2007). "Treaty-Making and Treaty Evolution." *The Oxford Handbook of International Environmental Law.* D. Bodansky, J. Brunnee and E. Hey. Oxford, Oxford University Press.
Goldstein, J. (1993). *Ideas, Interests and American Trade Policy.* Ithaca, Cornell University Press.
Hey, E. (2007). "International Institutions." *The Oxford Handbook of International Environmental Law.* D. Bodansky, J. Brunnee and E. Hey. Oxford, Oxford University Press.
Johnston, I. (2005). "The Power of Interpretive Communities." *Power in Global Governance.* M. Barnett and R. Duvall. Cambridge, Cambridge University Press: 185.
Odell, J. S. (1990). "Understanding International Trade Policies." *World Politics* 43: 139–167.
Tarlock, D. (2007). "Ecosystems." *The Oxford Handbook of International Environmental Law.* D. Bodansky, J. Brunnee and E. Hey. Oxford, Oxford University Press.
Ulfstein, G. (2007). "Treaty Bodies." *The Oxford Handbook of International Environmental Law.* D. Bodansky, J. Brunnee and E. Hey. Oxford, Oxford University Press.

PART III

Institutions and learning

The pieces in this section look at various mechanisms by which international institutions induce social change. "Institutions" include global conferences, regimes, and international organizations. In all cases I look for instances of influence and the conditions where international institutions are able to act relatively independently of the wishes of influential member states. The major mechanisms exercised by these institutions are granting representation to new political groups through involving them in collective activities and legitimating their actions; contributing to social learning by educating actors through the provision of new information; and generating new national and international institutions.

"UN conferences" (Chapter 9) looks at the influence of global environmental conferences in generating patterns of social change. Global conferences, in particular the major global environmental conferences, are the largest and most heterogeneous institutions in world politics. They are vast, sprawling events with widespread representation. Under favorable conditions they are capable of fostering new social dynamics through enabling the circulation of new information, facilitating networking, and encouraging the creation of new national and international level organizations to deal with the novel substantive issues raised by the conferences.

"International institutions and social learning" (Chapter 10) explores the ways in which negotiations and regimes may empower new actors and promote the learning of new policy lessons. Lessons are not necessarily progressive or good. Lessons may be about how to better protect the environment. But lessons may be politically, and environmentally, perverse, such as when the USA drew the lesson from the successful ozone negotiations that science panels should be subject to closer political control.

"Addressing the global governance deficit" (Chapter 11) investigates the role of the United Nations Environment Programme (UNEP) in world politics. It focuses on a networked model of global governance, and UNEP's role as a hub in that network. It identifies UNEP as a source of learning amongst governments, as well as a catalyst for NGO organizing. It also offers some design suggestions for improving UNEP's contributions to sustainability, in part defending redundant international institutions as an insurance policy against funding sources seeking to exercise control over individual institutions.

"Learning to learn" (Chapter 12), written with my late father, Ernst B. Haas, was originally solicited by the Commission on Global Governance, for whom I served as a consultant. This piece analyzes the contributions of international institutions to broader processes of social learning. It identifies the properties of international institutions that are able to learn: that is, to redefine their mission in the face of a changing policy environment. It also identifies the mechanisms available to the international institutions to disseminate their views to states and to other international institutions.

9

UN CONFERENCES AND CONSTRUCTIVIST GOVERNANCE OF THE ENVIRONMENT*

Peter M. Haas

Introduction

In this article I review the history of global environmental conferences and draw political lessons about their broader role in constructing efforts at global environmental governance, and in particular regarding the future of such global conference diplomacy for the environment, in particular Rio+10 in Johannesburg in 2002 and the prospects for reaching UNCED goals for sustainable development. Global conferences are oft-used policy instruments, and thus deserving of careful evaluation and assessment. Jacques Fomerand expresses justifiable skepticism that most global conferences are momentary media events that provide sound bite opportunities without lasting effects on policies or the quality of the environment (Fomerand 1996). Gallarotti and Barnett and Finnemore offer similar skeptical judgments about the potential for effective state-based international governance (Gallaroti 1991; Barnett, Finnemore et al. 1999). Yet Fomerand also points out, as do I below, that many conferences provide indirect effects that may be beneficial for inducing states to take more progressive steps towards governance and towards sustainable development.

Governance and constructivism

Governance has recently become a popular catch phrase of international relations. Without the prospects of hegemonic leadership, and in light of the substantial growth of influence of international institutions and non-state actors, international rule making has become the domain of multiple overlapping actors and regimes, rather than the clear-cut leadership by one state or multilateral conformity with a small and homogeneous set of shared rules backed by enforcement mechanisms. Anne Marie Slaughter defines it as "the formal and informal bundles of rules, roles and relationships that define and regulate the social practices of states and non-state actors in international affairs" (Slaughter, Tulumello et al. 1998). Sustainable development requires multilateral governance because without well-defined rules and expectations most countries are incapable of unilaterally protecting themselves from transboundary and global environmental risks

Constructivist scholars of International Relations have been focusing on the institutional, discursive and intersubjective procedures by which international governance develops. John Ruggie writes that

> Social constructivism rests on an irreducibly intersubjective dimension of human action.... constructivism is about human consciousness and its role in international life...Constructivists hold the view that the building blocks of international reality are ideational as well as material; that ideational factors have normative as well as instrumental dimensions; that they express not only individual but also collective internationality; and that the meaning and significance of ideational factors are not independent of time and place.
> *(Ruggie 1998; Adler 2001; Haas 2001)*

Constructivists look at the mechanisms and consequences by which actors, particularly states, derive meaning from a complex world, and how they identify their interests and policies for issues that appear new and uncertain.

It is now widely accepted by most scholars of International Relations that governance increasingly occurs in a decentralized manner, through a loosely tied network of multiple actors, states, functional state agencies, and non-state actors who interact frequently, including, at times at global conferences (Held, McGrew et al. 1999; Keohane and Nye 2001; Slaughter 2004). Governance of the environment is no different.

Constructivists focus on such distinctive processes as socialization, education, persuasion, discourse, and norm inculcation to understand the ways in which international governance develops. Typically these are complex procedures, involving multiple interacting actors that accrue over time and contribute to transformational shifts in perceptions of national identity, international agendas, and the presumptive ways by which national interests are to be attained.

UN conferences contribute to governance and sustainable development by establishing and reinforcing some of these constructivist themes in international relations. As I argue in greater length below, international conferences seldom have direct causal influences on member states' behavior, but their outputs are part and parcel of this broader process of multilateral governance, and may contribute to stronger and more effective environmental governance by states.

Accumulated conferences over the last thirty years have contributed to an aggregate shift in international politics by extending participation and access to environmental diplomacy to national environmental agencies and to nongovernmental organizations (NGOs) and networks of scientists—a process that Fomerand describes as a "large-scale process of social mobilization" (Fomerand 1996). Over the last thirty years governments have added the inspirational norm of ecological integrity to the traditional goals of wealth and power.

The most successful conferences have promoted broader processes of social learning and the construction of new, more comprehensive conceptual frameworks for global environmental governance through issue clarification, popularization of issues, and introducing new approaches to environmental policy making to governmental officials. Through this institutionalized constructivist process of participation and education new environmental norms of environmental protection have been diffused, and participating states have been encouraged to endorse them and to apply them nationally. Gradually, many of these norms have been converted to new institutionalized practices by states. Many states were socialized to appreciate new styles of understanding of relations between economics and ecology, and were encouraged to apply new policies to achieve economic development that is more environmentally sustainable than past doctrines (Haas 1999; Haas 2000; Haas 2001). Global environmental conferences have contributed to aggregate substantive changes in environmental governance. The Founex preparations for the UN Conference on the Human Environment (UNCHE) contributed to transcending the environment/development dichotomy in the framing of international environmental policy. As scientific consensus has crystallized around comprehensive forms of ecological management doctrines, the frames and dominant discourses of the environmental conferences have shifted from concerns about resource scarcity and depletion to efforts to understand and protect ecosystem integrity. The new consensus over sustainable development that was forged at the 1994 population conference in Cairo that population growth could not be considered in isolation of social issues shaping family planning choices, such as women's role in society, a clear example of the development and application of a new policy discourse at an international conference (Sanwal 1993; Fomerand 1996).

Later, UNCED's Agenda 21 was organized and designed around a matrix of issues, so that policies would be developed to address the interconnections between human activities (industry, agriculture, styles of decision making, consumption patterns, and technology) and the environment, as well as between global ecosystems (the atmosphere, freshwater, oceans, and land) with chapters of Agenda 21 designed to capture the intersections located in each cell of the matrix (A/CONF.151/Pc/42 9 1993). The earlier UNCHE framework was organized around the more traditional tripartite administrative framework of environmental assessment (evaluation and review, research, monitoring, information exchange), environmental management (goal setting and planning, international consultation and agreements), and supporting measures (education and training, public information, organization, financing, technical co-operation) (A/Conf.48/14 3 1972; UNEP 1981). This had the effect of establishing programs with the specific functional issue or international problem that justified their creation. Environmental monitoring, for example, would be a free-standing activity, unrelated to the specific contaminants warranting monitoring. Similarly, problems were defined in isolation, and management responses were associated with each distinct problem. While UNCHE was organized around a conceptual framework of traditional administrative functions, UNCED was set up to capture the newly appreciated analytic attributes of the issues to be addressed.

Thus, internationally endorsed policies and responses were designed to address the interplay of environmental problems, including their underlying causes. At UNCED, problems were delimited in terms of the interacting array of social forces that caused them, and thus policies were designed to address the social causes—in contrast to the UNCHE approach, which devised a standard set of monitoring and administrative reforms for problems that were addressed individually and outside of their social context.

Consequently, when combined with the other array of other institutional and participatory reforms introduced at various UN environmental conferences over the last thirty years these new frameworks and agendas have led to a much broader shift in discourse, as new institutions were developed that are responsible for verifying and carrying out the elements of the agenda, as well as popularizing the language and policy ingredients for the policy communities worldwide.

Ultimately, international conferences are weak institutional features of international relations that lack many of the properties that constructivists expect to contribute to transformed state beliefs and practices. Notable among these features are iterated interactions, autonomous secretariats staffed with professionals recruited on merit, independent and capable executive heads, free and easy access to independent experts, significant institutional resources for meaningful technological and resource transfers, and adequate institutional budgets. Few international conferences are endowed with these properties by sponsoring states. The most successful conferences emerged from repeated preparatory meetings; were supported by secretariats recruited on merit, led by adroit and experiences UN Diplomats, such as Maurice Strong; maintained a porous flow of information with independent experts outside the UN system; and had sufficient resources to support the preparations. Few international conferences had the ability to provide resources transfers to encourage states to participate, although some of the more successful conferences created institutions that had that capability.

Governments generally closely follow the preparatory activities in order to assure that they are not confronted with any unpleasant political surprises at the actual conferences. Because they are one-time events, there is limited access to top-level officials, and it is difficult to maintain long-term pressure on governments through national reporting, information circulation, oversight, or lobbying. Thus it is difficult for transnational policy networks to organize and consolidate influence through global environmental conferences. Global environmental conferences generally lack significant political or financial resources for inducing change on states, and lack any lock-in mechanisms by which decisions become deeply institutionalized within the legal and political systems of attending countries. Occasionally, though, some conferences are able to generate significant outputs or mobilize individual forces that have longer-term repercussions internationally.

UN conferences, though, are quite different from the G-8 summits in this regard. Unlike the one shot nature of UN conferences, the G-8 summits are part of smaller institutionalized discussions amongst trade and finance ministers and bureaucrats, who maintain frequent interactions at G-8 summits, Organization for Economic Cooperation and Development (OECD) working groups, International Monetary Fund (IMF) working groups, Bank for International Settlements (BIS) working groups, and private conferences. Unlike these small ongoing private group meetings, UN conferences enjoy greater broader political legitimacy by virtue of their universal representation and the opportunity for middle-level powers to have a say.

The effects of the most successful conferences have been to increase national concern, and to increase government capacity to politically and technically address problems by means of agenda setting, consciousness raising, expanding participation, monitoring, knowledge generation and diffusion, target setting, norm development and diffusion, and administrative reforms. In addition they have helped to channel financial, technological and scientific resources to needy countries (Haas, Keohane et al. 1993; Fomerand 1996; Reinicke, Deng et al. 2000)

UN conferences on the environment

Global UN conferences on the environment are widely understood as an institutional innovation of the 1970s. With mounting concern about the degradation of the physical environment, governments approached the UN to convene a number of global conferences to address the host of human activities with transboundary and global environmental consequences. These environmental conferences were part of a broader effort at global problem solving for a new class of global problems associated with international interdependence. As global interdependence became increasingly politicized in the 1970s, the UN system turned to global conferences as a way to highlight the interconnections between issues that had previously been treated in isolation. The topics of the global conferences were new to the international agenda, as previous multilateral conferences had principally addressed international economic topics, human rights and arms control (Kaufmann 1988; Willetts 1989). The UN, as the only venue with global participation, was the logical forum for such meetings.

The 1972 UN Conference on the Human Environment (UNCHE), and 1992 UN Conference on Environment and Development (UNCED) directly addressed the subject of environmental protection, but special UN conferences devoted to aspects of human impact on the environment became commonplace in the 1970s. The frequency with which such global conferences were held diminished in the 1980s and 1990s. What has remained constant are the decadal meetings of conferences on population, women, and food, as well as the follow-up annual reviews on UNCED commitments, and the more comprehensive and high-profile UNCED + 5 meeting in 1997 and UNCED + 10 to be held in 2002.

These global conferences performed multiple functions. They were intended to mobilize concern about new problems, to coordinate national actions to study and monitor environment quality and human activities with environmental consequences, and to develop joint measures to prevent various sources of environmental degradation and attenuate the effects of human actions on the environment. Economic and equity concerns cut across most of the other specialized conferences.

Typically the conferences last for several weeks, with high-level diplomatic attendance for the last two or three days to overcome political deadlocks and to sign legally binding resolutions and other commitments developed at the conference. Decisions are generally reached by consensus, so negotiations are slow. Preceding the conferences though are often several rounds of sessions of ad hoc Preparatory Committees ("Prep Coms"), often spread over one or two years, at which national delegations are presented with background papers and preliminary negotiations are conducted on the documents intended to be approved at the conferences. Most of the arduous work of reconciling political differences occurs during the Preparatory Committees sessions.

Generally the global UN conferences on the environment have produced declarations and action plans for subsequent activities. The most influential conferences endorsed new policy doctrines and policy targets for the international community, authorized the creation of new international organizations, approved legal commitments and generated new financial resources. The most productive, in terms of their administrative accomplishments, have been UNCHE, the 1974 World Food Conference, UNCED, and the 1994 International Conference on Population and Development (Weiss and Jordan 1976; Caldwell 1996). Others such as the 1977 Desertification Conference, 1979 Conference on Science & Technology for Development, and the Conferences on Human Settlements, have failed to spark international concern or to catalyze robust international commitments and action.

The 1972 UNCHE, held in Sweden, was the first major global environmental conference. Sponsored by the UN, it convened 113 countries to discuss contemporary environmental issues. UNCHE adopted the Stockholm Declaration, establishing 26 principles of behavior and responsibility to serve as the basis for future legally binding multilateral accords; the Action Plan for the Human Environment that specified 109 recommendations in the areas of environmental assessment, environmental management, and supporting institutional measures (Schmidt 1973). Implementation was intended for governments and international organizations (IOs).

The 1992 UNCED, held in Rio de Janeiro, marks the high-water mark of these outputs. UNCED adopted the Framework Convention on Climate Change, the Convention on Biological Diversity, and the State of Forest Principles. In addition to those three pieces of hard law, UNCED adopted the Rio Declaration, with 287 principles guiding action, and a sweeping action plan to promote sustainability called Agenda 21, with 2,509 specific recommendations elements applying to states, international institutions, and members of civil society. The Commission on Sustainable Development was created to ensure effective follow-up of UNCED; to enhance international cooperation and rationalize intergovernmental decision-making capacity; and examine progress in Agenda 21 implementation at the local, national, regional and international levels (Grubb 1993).

The variation in the degree of influential outputs from conferences is due to a number of factors. The more productive conferences were free of profound political schisms or geopolitical tensions amongst major parties, including Cold War animosities. The environment was not nested in a politically irreconcilable frame of profound North–South cleavages. The issue at hand appealed to the immediate interests of the industrialized countries because of either popular concern within the countries or perceived linkages between the subject and material national interests, leading major donor states to commit resources to the issue. Robert Putnam and Nicholas Bayne inferred a number of similar background conditions from successful G-7 summits (Putnam and Bayne 1987).

UNCHE, for instance, was held at a fortuitous moment. Domestic environmental movements were just becoming active in the United States and Europe. Potential North–South disagreements were avoided by prior high-level discussions that rejected the conceptual dichotomy between economic growth and environmental protection, extended the international agenda to include environmental concerns of the South regarded natural resource policy, as well as the pollution concerns of the industrialized countries, and provided a notional commitment to "additionality" and financial assistance on behalf of the North (Rowland 1973; Strong 1973). Environmental protection was not seen as being inconsistent with other established goals in international negotiations, including national security and economic liberalization. UNCHE also provided the first opportunity for China to stake a position in international diplomacy. North–South relations became more acrimonious with the New International Economic Order (NIEO) discussions in the late 1970s, making it harder to forge consensus at international conferences. Even with these factors, Cold War divides still modestly influenced the conference, as the Soviet Union and the Eastern bloc countries withdrew at the last minute over the participation of West Germany; yet, because the superpowers were in a period of détente such tactical linkages were not perceived as provocative and freighted with Cold War significance.

Effectiveness of international conferences

It is difficult to evaluate the effectiveness of many of these conferences, in part due to weaknesses and gaps in our ability to monitor progress in achieving such goals. The record is generally mixed, at best, in terms of achieving the targets and aspirations expressed in the action plans and declarations of the conferences. It is difficult to directly measure effects on the environment, and the record of states in complying is mixed or uncertain. The goals are often ambiguous. State reporting about compliance is generally weak and incomplete, and few provisions for verification of state compliance are made at the conferences. Most assessments of conference successes remain impressionistic and anecdotal, although some conferences generated new doctrinal consensus or new institutions to help advance the conference goals (such as UNCHE, with the UN Environment Programme [UNEP]; the 1974 World Food Conference, with the World Food Programme and the International

Fund for Agricultural Development [IFAD]; the 1994 World Population Conference, with its strong endorsement of new population policy albeit with strong institutional support; and UNCED, with its support for the new doctrine of sustainable development but still with a weak Commission on Sustainable Development). At UNCED + 5 the General Assembly and the Commission on Sustainable Development tried to evaluate overall progress achieved since UNCED. It determined, among many observations, that production and consumption patterns had become more energy efficient in industrialized countries; that land use conflicts are more acute in developing countries between competing demands for agriculture, forest cover, and urban uses; and that water scarcity remains a major threat to development and human health in developing countries (E/CN.17/1997/2 1997).

In short, it is difficult to evaluate the effectiveness of the conferences on state policies and on observable environmental impacts. It would be unreasonable to expect such conferences to yield lasting and clear effects on states and on the environment. It is equally unreasonable to assign blame to conferences for failing to reverse environmental decline.

A full list of global environmental conference is presented in Table 9.1.

TABLE 9.1 Global environmental conferences since 1970

Year	*Name/location*	*Product/outcome*
1972	United Nations Conference on the Human Environment (Stockholm)	Declaration of Principles Action Plan UNEP
1974	World Food Conference (Rome)	Universal Declaration on the Eradication of Hunger and Malnutrition World Food Council IFAD
1974	World Population Conference (Bucharest)	World Population Plan of Action
1975	Second Women's Conference	
1977	UN Water Conference (Mar del Plata)	International Drinking Water Supply and Sanitation Decade (1981–1991)
1977	UN Conference on Desertification (Nairobi)	Plan of Action to Combat Desertification
1978	UN Conference on Human Settlements (Vancouver)	UN Centre for Human Settlements Global Strategy for Shelter to the Year 2000
1979	UN Conference on Science and Technology for Development (Vienna)	Vienna Programme of Action on Science and Technology for Development
1979	World Climate Conference (Geneva)	
1981	UN Conference on New and Renewable Sources of Energy (Nairobi)	Nairobi Programme of Action for the Development and Utilization of New and Renewable Sources of Energy
1984	World Conference on Agrarian Reform and Rural Development (Rome)	Programme of Action on Agrarian Reform and Rural Development
1984	2nd World Population Conference (Mexico City)	

1985	3rd Women's Conference	
1990	2nd World Climate Conference (Geneva)	IPCC
1992	UNCED (Rio de Janeiro)	Rio Declaration Agenda 21 Framework Convention on Climate Change Convention on Biodiversity Forest principles UNCSD
1994	International Conference on Population and Development (Cairo)	Programme of Action
1995	Fourth World Conference on Women (Beijing)	Beijing Declaration and Platform of Action
1996	Habitat II (Istanbul)	The Habitat Agenda and Istanbul Declaration on Human Settlements
1996	World Food Summit (Rome)	Rome Declaration on World Food Security and World Food Summit Plan of Action
1997	UNGA Special Session on Sustainable Development	

Source: Caldwell 1996; Fomerand 1996; Weiss, Forsythe et al. 1997.

Functions of conference diplomacy

Global environmental conferences also have a number of indirect effects with longer-term effects on national policies affecting international governance and the prospects for sustainable development. Without a strong theory of state interests it is not possible to draw clear causal inferences about the influence of international conferences on state interests and practices. Theorists across paradigmatic divides, with the exception of staunch rational choice theorists, should be able to agree that conferences that are able to mobilize more of the functions I discuss later in this article will have a stronger impact on member states than will conferences unable to mobilize as many. Some variables are of interest to neoliberal institutionalists because they influences state assessments of the economic cost of environmental pollution, the ecological benefits of its solution, and the political coalitions associated with each functional issue on the agenda. Conferences thus influence international linkage politics. For constructivists, important variables are the informational channels and actual pieces of information that shape states' appreciation of how their citizens are affected by environmental degradation and the political coalitions that support environmental protection. The casual mechanisms by which institutional factors influence state choice and predictability of their effects are highly contingent, depending upon national administrative characteristics and domestic state/society relations. Moving beyond a systemic level of analysis, variation in an individual state's sensitivity to these functions of conference diplomacy would probably vary by at least the following national level factors: freedom of the press, literacy, access to the media, and democratic institutions enabling citizens to express concern to governments (state/society relations) (Haas 1998; Haas 2000).

While it's not possible to directly stop human activities that degrade the environment through universal declarations or at conferences, global UN conferences have served a number of distinctive political and social purposes that influence governments' concern about the environment and their willingness to commit scarce political and financial resources to its protection.

Agenda setting

Global environmental conferences can place new issues on the global agenda and galvanize national concern by publicizing new issues. The conferences often have the effect of reframing issues for decision makers, locating the issue within a new political matrix and thus making possible new tactical and substantive linkages by which policies may be developed (Haas 1980; Aggarwal 1998). For instance, environmental protection was firmly placed on the international agenda at the UNCHE conference, and the preliminary Founex meeting effectively reconciled North–South differences about the priority accorded to environmental considerations in economic planning. The meeting established the principal that the two goals could be compatible, especially with concessionary finance from the North to pay for incremental pollution control costs in the developing countries. UNCHE also helped inform Northern governments of Southern countries' concern about resource deterioration, deforestation and water quality, and the underlying problem of insufficient money for sewage treatment and effective resource management. The North gradually came to include the South's concerns and still supplement the North's primary focus on industrial pollution, waste management, and transboundary environmental threats.

The 1994 International Conference on Population and Development shifted the public debate or discourse on population issues to a focus on the underlying social, political and economic forces that influence population growth. The Programme of Action marked a distinctive shift in population policy toward promoting cooperation to eradicate poverty, encouraging universal access to health care services, and empowering women (Ashford 1995; Sen 1995).

Popularizing issues and raising consciousness

Conferences provide a brief window of opportunity for educating the mass public and government officials about environmental issues. Conferences spawn publicity about the declarations and statements of principles the meetings produce. Because many journalists attend the conferences, they provide an opportunity for NGOs and the media to publicize issues in their national media as well as for educating members of the media about environmental issues. For instance, at UNCED the Natural Resource Defense Council (NRDC) sent one person whose responsibility was to court the media and frame the presentation of the daily reporting in a way that would be critical of the United States. NRDC hoped to provoke the United States into taking a more environmentally sympathetic role at the conference.

Generating new information and identifying new challenges for governments

Preparation for conferences often generates information for countries about their environmental problems, the array of policies available for addressing such issues, and the political coalitions organized around them. States are invited to submit national reports about conditions in their countries. This process can lead states to learn of new problems, clarify their recognition of their national interest, and identify the political landscape potential for compromise. These reports are often synthesized by the secretariats for subsequent dissemination (A/Conf.48/14 3 1994).

Providing general alerts and early warning of new threats

Conferences help focus attention on new problems, and also help to identify institutional gaps and needs in addressing such problems. The "Assessment of the World Food Situation," presented to the 1984 World Food Conference, helped focus attention on the "world food gap" that threatened developing countries. UNCHE helped to identify the urgency of addressing land-based marine pollution, as well as identifying the institutional need of creating a global environmental monitoring system, that subsequently became one of UNEP's core activities (Gosovic 1992).

Galvanizing administrative reform

Conferences also prompt governments to create or reform national bodies responsible for forms of environmental protection. National administrative bodies serve as the nodes of transnational environmental policy networks. At the time of UNCHE only 26 governments had administrative agencies responsible for environmental protection (15 in the developing countries, 11 in developing). The preparation for UNCHE led many governments to recognize the need for creating national environmental agencies. By 1982, the total number was up 144 (34 in developed, 110 in developing countries). UNCED led to the establishment of Sustainable Development (SD) committees and bodies in nearly 150 countries (E/CN.17/1997/2 1997).

Adopting new norms, certifying new doctrinal consensus, and setting global standards

Global conferences are sites of doctrinal contestation. UNCHE developed new principles of soft law that have been interpreted and applied by international lawyers to inform a generation of international environmental lawyers (Weiss, Barstow et al. 1992; Weiss 1996). Specific programmatic action, such as the 2,509 specific proposals in Agenda 21, set the stage for legitimate responses to international conferences. The identification of numbers of people at risk from malnutrition, and targets for official development assistance (ODA) and hunger reduction stipulated at World Food Conferences similarly established standards and aspirations for subsequent governmental practices.

Promoting mass involvement of new actors

International environmental conferences contribute to the participation of new actors in international environmental politics by inviting new groups of actors to attend international conferences. Environmental conferences have been leaders in the introduction of NGOs to international diplomacy. These meetings developed the practice, introduced at UNCHE, of holding parallel NGO conferences and governmental conferences and admitting NGO participants as observers at the governmental conferences. Roughly 178 NGOs participated at UNCHE (Feraru 1981). Over 1400 were represented at UNCED.

Despite the vast increase in numbers of NGOs attending international environmental conferences, the participation is still heavily tilted towards the North, where NGOs have greater financial support and are better able to find resources to attend conferences. At UNCED, 70 percent of the registered NGOs came from industrialized countries.

Conferences provide the potential for networking and developing transnational issue networks to coordinate international campaigns, and NGOs may subsequently provide information to governments and apply pressure on governments.

Conferences often invite participation from major non-state groups, including NGOs, the transnational scientific community, and, since UNCED, multinational corporations (Schmidheiny 1992). Such groups are invited to attend expert group meetings in advance of the conference, participate in parallel NGO events and even attend government meetings as observers. Participation is often, particularly in preliminary meetings, by expert advisory groups of specialists such as the Joint Group of Experts on Scientific Aspects of Marine Environmental Protection (GESAMP), the International Council of Scientific Unions (ICSU), and umbrella industry NGOs, such as the International Chamber of Commerce (ICC). Mass public NGOS tend to not participate in the early stages.

Global environmental conferences may be deliberately designed to foster new coalitions more generally, and to build support for environmental protection at the national level by including the political influence of transnational policy networks. Maurice Strong, the Secretary General of UNCHE and UNCED, coined the phrase "the process is the policy" to capture the idea that through conference diplomacy more actors and perspectives could be introduced to international environmental policy making.

There is still wide variation in the extent of NGO influence at conferences. The rules of participation remain set by states' decisions in ECOSOC, and the organizations are continually constrained (if not totally hamstrung) by state choices to allocate resources and set rules of behavior for the organizational dealings with NGOs. NGOs are often more influential at national and community levels, but participation and recognition at international conferences reinforces or establishes their domestic claims to authority. Yet even while states cling to formal sovereignty, the exercise of practical sovereignty erodes from NGO participation (Biermann 1998; Moltke 1998; Newell 1998; Whalley and Zissimos 1998; Hochstetler, Clark et al. 2000). Still, Realists would be quick to point out that the willingness to extend participation to NGOs is given by states, and is always subject to being reversed.

Prospects for Rio+10 and sustainable development

The aggregation of UN conferences and constructivist forces has been to create a diffuse array of pressures on states militating for forms of sustainable development. Rio+10 provides the next major opportunity for reforming and streamlining multilateral environmental governance. It is intended to refocus international attention on Sustainable Development and assess accomplishments since 1992.

Yet, as with the writing of this piece, it lacks most of the properties of conferences that led to productive outputs that contributed to improved international environmental governance. Rio+5 was widely regarded as a failure in this regard, as it did not mobilize any long-standing interest. Mass public interest in sustainable development remains weak, and the US appears to be developing a new global diplomatic posture of skeptical multilateralism, at best, as seen by the abandonment of the Kyoto Protocol and the Anti-Ballistic Missile (ABM) Treaty. Consequently, there is little political impulse for a productive conference. Multilateral financial and technological transfers for sustainable development have dwindled since the early 1990s. Moreover, there is growing disenchantment with UNEP's remote location in Kenya and its lack of resources. The Commission for Sustainable Development lacks the administrative autonomy or financial resources to be able to reach out to civil society to develop any of the conference functions discussed above that could potentially influence state policies and environmental quality, and states appear increasingly concerned about controlling NGO participation at the meetings.

The best prospects are probably institutional reforms. The international environmental governance system has not been significantly overhauled in three decades. After UNCHE, UNEP was the only international institution responsible for environmental protection. Since then, however, most international institutions have assumed some environmental responsibilities. Recent evaluations suggest that there are administrative overlaps in the system and inefficiencies, as institutions have assumed new responsibilities for the environment (Esty 1994; Biermann 2000; Downie and Levy 2000; Juma 2000). Suggestions for improvements focus on reforming UNEP and on the creation of a Global Environmental Organization (GEO).

A GEO should be established to fulfill the policy and technology-based functions that provide institutional support for multilateral environmental governance.

A GEO would consolidate environmental policy research, technology databases and clearinghouses; conduct training; and centralize the secretariats that administer current environmental regimes. Centralizing these secretariats would facilitate the creation of a broader global policy network across specific environmental issues and justify the creation of national environmental embassies to represent states and participate in future negotiations. A GEO could also serve as a legal advocate for environmental protection and regulations to counterbalance the World Trade Organization (WTO) by collecting a roster of international environmental lawyers to participate in WTO panels. The GEO should have high-profile annual ministerial meetings to address all environmental issues to assure widespread involvement in environmental policy networks and galvanize rapid responses to new alerts. Ongoing efforts would continue to be addressed through the existing secretariats and conferences of parties. The GEO could even have a panel of environmental inspectors available to verify compliance by states and firms with multilateral environmental agreements. UNEP would be retained as the monitoring and research hub of the UN system, as it was initially intended by its architects at UNCHE. The UN Commission on Sustainable Development, as well as some other institutional bodies within the UN and Bretton Woods systems could be absorbed into the GEO.

Conclusion

UN environmental conferences have helped contribute to a broader shift in international environmental governance through educating governmental elites, exposing them to new agendas and discourses, and providing them with added resources to pursue sustainable development. While Rio+10 lacks many of the conditions that have accompanied successful conferences, Rio+10 may at least encourage multi-level participation, improve contact between civil society and states, and streamline institutional responsibilities within the UN and Bretton Woods systems for sustainable development.

While the political preconditions appear modest for any dramatic achievements and cognitive transformations at Rio+10, we must remember that the conference is part of a thirty-year-long era of multilateral environmental protection. The conference can continue to legitimate the participation of NGOs and scientists in international environmental governance, improve contact between civil society and states, and streamline institutional responsibilities within the UN and Bretton Woods systems for sustainable development. Even in the absence of strong political support by member governments for significant multilateral commitments, progressive governments and other conference participants can still press for reforms to existing arrangements that will ensure more reporting on their movement toward sustainable development, create information clearinghouses about green technologies, and endow UNEP, a new GEO, or other international institutions with verification authority to monitor international movement towards sustainable development.

Note

* This chapter was first published in *Global Governance*, Vol. 8, No. 1, pp. 73–91. Copyright © 2002 by Lynne Rienner Publishers, Inc. Used with permission of the publisher.

References

A/Conf.48/14 3 (1972). "Report of the United Nations Conference on the Human Environment."

A/Conf.48/14 3 (1994). *Synthesis of National Reports on Population and Development Based on Submissions by 150 Countries for the International Conference on Population and Development*, United Nations.

A/CONF.151/Pc/42 9 (1993). *Structure and Organization of Agenda 21*, United Nations.

Adler, E. (2001). "Constructivism in International Relations". *Handbook of International Relations*. W. Carlsnaes, T. Risse-Kappen and B. A. Simmons. Beverly Hills, CA, Sage.

Aggarwal, V. K., Ed. (1998). *Institutional Designs for a Complex World*. Ithaca, Cornell University Press.

Ashford, L. S. (1995). "New Perspectives on Population." *Population Bulletin* 50(1).

Barnett, M., M. Finnemore, et al. (1999). "The Politics, Power, and Pathologies of International Organizations." *International Organization* 53(4): 699–732.

Biermann, F. (1998). "The Emerging Debate on the Need for a World Environmental Organization." *World Politics* 51(1): 45–55.

Biermann, F. (2000). "The Case for a World Environment Organization." *Environment* 42(9): 22–31.

Caldwell, L. K. (1996). *International Environmental Policy*. Durham, NC, Duke University Press.

Downie, D. L. and M. A. Levy (2000). "The United Nations Environment Programme at a Turning Point". *The Global Environment in the Twenty-First Century*. P. Chasek. Tokyo, UN Press: 355–375.

E/CN.17/1997/2 (1997). *Overall Progress Achieved Since the United Nations Conference on Environment and Development*, United Nations.

Esty, D. (1994). "The Case for a Global Environmental Organization". *Managing the World Economy: Fifty Years after Bretton Woods*. P. B. Kennen. Washington D.C., Institute for International Economics: 287–309.

Feraru, A. T. (1981). "Stockholm and Vancouver: The Role of ISPAs at UN Conferences". *Knowledge and Power in a Global Society*. W. M. Evan. Beverley Hills, CA, Sage Publications.

Fomerand, J. (1996). "UN Conferences; Media Events or Genuine Diplomacy." *Global Governance* 2(3): 361–375.

Gallaroti, G. (1991). "The Limits of International Organization." *International Organization* 45(2): 183–220.

Gosovic, B. (1992). *The Quest for World Environmental Cooperation: The Case of the UN Global Environment Monitoring System*. London, Routledge.

Grubb, M. (1993). *The Earth Summit Agreements: A Guide and Assessment*. London, Earthscan; Royal Institute of International Affairs.

Haas, E. B. (1980). "Why Collaborate? Issue Linkage and International Regimes." *World Politics* 32(3): 357–405.

Haas, P. M. (1998). "Compliance with EU Directives: Insights from International Relations and Comparative Politics." *Journal of European Public Policy* 5(1): 17–37.

Haas, P. M. (1999). "Social Constructivism and the Evolution of Multilateral Environmental Governance". *Globalization and Governance*. A. Prakash and J. A. Hart. London, Routledge: 103–133.

Haas, P. M. (2000). "Choosing to Comply: Theorizing from International Relations and Comparative Politics". *Commitment and Compliance*. D. Shelton. Oxford, Oxford University Press: 43–64.

Haas, P. M. (2000). "Institutionalized Knowledge and International Environmental Politics". *Handbook of International Relations*. G. J. Ikenberry and V. Parsi. Rome, Gins, Laterza & Figlie.

Haas, P. M. (2001). "Epistemic Communities and Policy Knowledge." *International Encyclopedia of Social and Behavioral Sciences*: 11578–11586.

Haas, P. M., Ed. (2001). *International Environmental Governance: Managing a Globalizing World*. Washington D.C., The Brookings Press.

Haas, P. M., R. O. Keohane, et al., Eds. (1993). *Institutions for the Earth: Sources of Effective International Environmental Protection*. Global Environmental Accords Series. Cambridge, The MIT Press.

Held, D., A. McGrew, et al. (1999). *Global Transformations*. Stanford, CA, Stanford University Press.

Hochstetler, K., A. M. Clark, et al. (2000). "Sovereignty in the Balance: Claims and Bargains at the UN Conferences on the Environment, Human Rights, and Women." *International Studies Quarterly* 44(4): 591–614.

Juma, C. (2000). "The UN's Role in the New Diplomacy." *Issues in Science and Technology* 17(1): 37–38.

Kaufmann, J. (1988). *Conference Diplomacy*. New York, UNITAR.

Keohane, R. O. and J. P. Nye (2001). *Power and Interdependence*. New York, Longman.

Moltke, K. V. (1998). "The Organization of the Impossible." *World Politics* 51(1): 23–28.

Newell, P. (1998). "New Environmental Architectures and the Search for Effectiveness." *World Politics* 51(1): 35–44.

Putnam, R. D. and N. Bayne (1987). *Hanging Together*. Cambridge, MA, Harvard University Press.

Reinicke, W. H., F. Deng, et al. (2000). *Critical Choices : The United Nations, Networks, and the Future of Global Governance*. Ottawa, International Development Research Centre.

Rowland, W. (1973). *The Plot to Save the Earth*. Toronto, Clarke, Irwin.

Ruggie, J. G. (1998). "What Makes the World Hang Together? Neo-Utilitarianism and the Social Constructivist Challenge." *International Organization* 52(4): 855–885.

Sanwal, M. (1993). "Sustainable Development, the Rio Declaration, and Multilateral Cooperation." *Colorado Journal of International Environmental Policy and Law* 4(1): 45–68.

Schmidheiny, S. (1992). *Changing Course: A Global Business Perspective on Development and the Environment*. Cambridge, The MIT Press.

Schmidt, E. (1973). *The Results from Stockholm*. Berlin.

Sen, G. (1995). "The World Programme for Action: A New Paradigm for Population Policy. The Cairo Conference." *Environment* 71(1): 10–15; 34–17.

Slaughter, A.-M. (2004). *A New World Order*. Princeton, Princeton University Press.

Slaughter, A.-M., A. S. Tulumello, et al. (1998). "International Law and International Relations Theory: A Dual Agenda." *American Journal of International Law* 92(3): 367–397.

Strong, M. (1973). "One Year After Stockholm." *Foreign Affairs* 51(4): 690–707.

UNEP (1981). *In Defence of the Earth, the Basic Texts on Environment: Founex, Stockholm, Cocoyoc*. United Nations Environment Programme.

Weiss, E. B. (1996). "The Changing Structure of International Law". *International Environmental Law*. E. B. Weiss, P. Szasz and D. Magraw, Transnational Publishers.

Weiss, E. B., D. Barstow, et al. (1992). *International Environmental Law*, Dobbs Ferry.

Weiss, T., D. Forsythe, et al. (1997). "Ecodevelopment and the United Nations". *The United Nations and Changing World Politics*, Westview Press.

Weiss, T. G. and R. S. Jordan (1976). *The World Food Conference and Global Problem Solving.* New York, Praeger.

Whalley, J. and B. Zissimos (1998). "What Could a World Environmental Organization Do?" *World Politics* 51(1): 29–34.

Willetts, P. (1989). "The Pattern of Conferences". *Global Issues in the United Nations' Framework.* P. Taylor and A. J. R. Groom. New York, St Martin's Press.

10

INTERNATIONAL INSTITUTIONS AND SOCIAL LEARNING IN THE MANAGEMENT OF GLOBAL ENVIRONMENTAL RISKS*

Peter M. Haas

This article investigates the role played by formal international institutions in the broader process of international efforts to respond to and manage global and transboundary environmental risks. Because few international institutions are designed to deal with the broad nature of environmental risks, it focuses on institutional learning. By analyzing the experiences of the United Nations Environment Program, World Meteorological Organization, and other international institutions involved with global warming, this article identifies institutional properties (or functions) that encourage or inhibit social learning in the management of global environmental risks by international institutions, and that influence the adoption of such lessons by their constituent members.

How can governments better deal with global and transboundary environmental risks with irreversible effects that may not be experienced until the future? Conventional wisdom now suggests that more effective management or responses to such risks requires coordinated comprehensive environmental policies that reflect an appreciation of the complexity of global environmental risks, efforts for which functionally differentiated modern states are notoriously poorly equipped. While new policies may ultimately be adopted by states at the national level, in practice, international institutions play a role in harmonizing national efforts and, at times, in educating national officials about new policies and styles of environmental management. In this article we focus on the ways in which formal international institutions have recognized global environmental risks, the ways in which they have developed new techniques for their management, and mechanisms by which such new techniques are disseminated more broadly.

International institutions generally serve as intermediaries in international relations, taking inputs from governments, which in turn are responsive to pressures from other governments and from within their own societies. I treat international institutions as the array of "persistent and connected sets of rules (formal and informal) that prescribe behavioral roles, constrain activity, and shape expectations)" (Keohane 1989). In practice, such institutions generally take the form of more conventional organizations; that is, formal international organizations (IOs). Following Arild Underdal, I differentiate between the conscious and unconscious influence of such organizations on shaping national policies towards addressing transboundary and global environmental risks:

> International organizations can serve at least two major functions in international environmental management: that of being an arena for the exchange of information, discussion and decisionmaking, and that of being an actor in the policymaking or the policy implementation process. All intergovernmental organizations ... serve as arenas ... (T)o qualify as actor, an organization must (also) provide independent inputs into the policy process, or somehow amplify the outputs of the process.
>
> *(Underdal 1993)*

This article takes an inductive approach to the study of institutions and social learning by studying a selective but illuminating set of vignettes of new directions pursued by international institutions and lessons learned within the institutions by their constituent members. The origins of innovative ideas are regarded as exogenous; in this chapter we focus on the factors associated with their adoption and diffusion by and through international institutions. I draw conclusions about the significant factors associated with learning by institutions, although we are unable to identify their relative frequency and scope. In these cases international institutions do not make policy, but rather they enhance international dialogues about possible policies, thus leading to the formulation of new policies and their adoption by governments (Linder and Peters 1995).

I differentiate between the institutional properties that enable organizations to learn; that is to "change ... behavior due to a change in perception about how to solve a problem" (Haas 1990; Haas 1991). I also identify some functional activities performed by the institutions that increase the likelihood that institutional innovations will spread, by helping actors identify and apply new techniques for the management of environmental problems as well as contributing to the capacity of such actors to improve their own management.

The article, part of a larger study examining international experience with stratospheric ozone depletion, acid rain, and climate warming, focuses primarily upon international experience with the threat of global climate warming (The Social Learning Group 2001). It begins with a brief history of institutional involvement with global environmental risks, provides a brief overview of the primary actors involved in the multilateral management of these issues, and then analyzes the application of knowledge to collective management of the climate warming problem.

Institutional histories

Although concern in the international political community about global environmental risks had previously been awakened, the 1972 United Nations Conference on the Human Environment (UNCHE) was a catalyst in engaging wider interest. Even in the wake of the UNCHE, few of the global risks mentioned here received widespread attention by governments, and most intergovernmental organizations continued to focus on particular environmental problems relevant to their general mandate rather than engaging the full magnitude of such problems. The United Nations Environment Program (UNEP) has been the only international institution with a formal mandate of encouraging action on a wide dimension of environmental risks.

National governments are the primary actors in the management of global environmental risks; however, none of the institutions involved is dominated by a small group of countries, and all, to varying degrees, are open to input from business groups and environmental nongovernmental organizations (NGOs). Business had its heaviest presence at international meetings in the climate change case; this was in part due to the fact that its material interests were most acutely affected in that case—both in terms of potential market gains as well as potential for regulation—and, perhaps, because of growing familiarity and sensitivity to the potential for international regulation, which it had failed to observe in prior international policy activities (Schmidheiny 1992). Firms and industry groups have preferred, and achieved their greatest impact by, lobbying their governments directly, although some individuals have enjoyed observer status on national delegations (Faulkner 1993).

Climate change

Climate change has been addressed by a number of international institutions. The World Meteorological Organization (WMO) began monitoring weather by satellite in the early 1960s. Its World Weather Watch (WWW) was established in 1963 with a mandate to improve the understanding of the physical basis of climate and large-scale weather modification, and to improve weather forecasting. In its operational programs, however, the climate aspect was missing and/or underdeveloped; this reflects the WMO's programmatic focus on shorter-term weather forecasting versus longer-term inadvertent climatological effects of anthropogenic activities (Weiss 1981; Weiss 1983).

The WMO

The WMO convened the World Climate Conference in 1979, following a shift in attention from weather monitoring to climate change by the scientific community stimulated by international concern over the disastrous effects drought had brought to the Sahel region of Africa throughout the previous decade. The World Climate

Program (WCP) has collected data and coordinated climate change research at the national level. Most relevant for the study of global environmental risks of the WCP's four components is the World Climate Research Program (WCRP), which is jointly administered by the WMO and the International Council of Scientific Unions (ICSU) and is the successor to Global Atmospheric Research Program (GARP). The WCRP's research focus includes studies of long-term (several decades) responses of climate to natural and anthropogenic influences. The WMO held the Second World Climate Conference (SWCC) in 1990, at which international scientific evidence of climate change was exchanged and examined. By the early 1990s climate and environmental activities accounted for 30% of the WMO's scientific and technical budget and 13.5% of its overall budget, revealing the underlying shift in orientation from weather prediction to climate change (Perry 1993). The World Health Organization (WHO) worked closely with the ICSU in administering its monitoring and research programs, as well as preparing for the first World Climate Conference and the SWCC.

The IPCC

The Intergovernmental Panel on Climate Change (IPCC) was established by the UNEP and the WMO in November 1988, at the behest of national governments, to organize the scientific background, appraise the risks from climate change, and evaluate possible mitigation strategies in preparation for negotiations for a climate change treaty (United Nations General Assembly Resolution 42/184 1987). The IPCC's approach was dominated by its method of assessing the costs and benefits of the response strategies associated with various emission scenarios. Through 1992 (the period covered in this article) the IPCC was composed of three working groups: scientific analysis, socioeconomic impacts, and policy responses. The IPCC released its first major Assessment Report in August 1990 and a supplement in 1992. The entire effort was steered by a bureau composed of IPCC chairman Bert Bolin, the panel's vice-chairman and rapporteur, the chairs of the three working groups, and the vice-chairs of the working groups (two each from working group (WG) I and II and five from WG III) (Ramakrishna and Young 1992).

Actual negotiations for a climate change treaty were conducted under the auspices of the Intergovernmental Negotiating Committee (INC) for a Framework Convention on Climate Change, established by the UN General Assembly in December 1990 with an autonomous secretariat (United Nations General Assembly Resolution 45/212 1990). The Climate Change Convention was adopted in June 1992 and entered into force in January 1994.

Epistemic communities and the management of global environmental risk

A similar international pattern is evident in the management of climate change, ozone, and acid rain. In each, some effort was undertaken to develop a set of equivalencies between potential environmental threats to human well-being by which risks could be compared and policies formulated. As these approaches were developed under the auspices of international institutions, collective action came to be framed in terms of the new framework and many countries based policy on them. In the climate change case a "tolerable rates" approach was developed to stipulate the degree of potential harm that the global environment could sustain without intolerable social costs.

Collective framing and policy identification in each case was the result of a small transnational network of experts already actively involved in policy-relevant science who gained access to the process through the timely intercession of international institutions. Scientific knowledge was not immediately accessible or apparent to decision makers; it had to be provided by a set of respected advisors. Elsewhere I call these groups "epistemic communities" (Haas 1989; Haas 1990; Haas 1992/1997). Yet the management of global environmental risks occurs within a broader interactive international context in which international institutions adopt and transmit new policy approaches to the national level. For an approach to diffuse broadly it must be institutionally sanctioned. This institutional backdrop confers authority on the views of individuals whose claims would otherwise lack substantial legitimacy.

Climate change: the Villach meetings

In the climate change case, the activities of a small group of scientists conferred urgency and focus to discussions that had lacked leadership and focus. The Study of Men's Impact of Climate (SMIC) report was a one-shot study prepared for the UNCHE, but the ICSU was capable of providing the infrastructure for long-term mobilization. A scientific community was first mobilized under the auspices of ICSU's Scientific Committee on Problems of the Environment (SCOPE). The SCOPE had a long-standing interest in global biogeochemical cycles. Scientists in the USSR, many of whom had been students of Vernadsky, had performed much of the relevant research, and SCOPE planners tried hard to include them in SCOPE projects, although few were involved in the seminal SCOPE 29 report that aired in 1985 at Villach, Austria.

In November 1980, as a joint activity of the WCP, SCOPE, the UNEP, and the WMO sponsored a meeting of 11 experts to discuss the role of CO_2 on climate and its impacts. The group reported that CO_2-induced climate change was a major environmental issue, but that, because of scientific uncertainties, it was premature to promote limits on CO_2 emissions (World Health Organization, United Nations Environment Program et al. 1980; Kowalok 1993). With the encouragement of the UNEP and principally at the UNEP's expense, the SCOPE Executive Committee authorized a report on the Greenhouse Effect, Climatic Change, and Ecosystems. Interim results were reviewed at Villach in 1983, and the full report was presented at a meeting in Villach from October 9–15, 1985. The meeting was held under the auspices of the WCP jointly implemented by the WMO, the UNEP, and the ICSU, and the report was published by the WMO as part of the WCP series of reports on behalf of the other sponsors, giving its conclusions more authority with national governments than would have been the case if it had simply been issued as a SCOPE report. For the first time, the participating experts concluded that "it is now believed that in the first half of the next century a rise of global mean temperature could occur which is greater than any in man's history" and recommended actions that included "support for the analysis of policy and economic options should be increased by governments and funding agencies. In these assessments the widest possible range of social responses aimed at preventing or adapting to climate change should be identified, analyzed and evaluated" (World Meterological Organization 1986; Bolin, Jager et al. 1996).

The 1985 Villach meeting emphasized that climate change was likely the consequence of more than just carbon dioxide, and the chair, Jim Bruce of the Canadian Atmospheric Environment Service, pushed the group to endorse the conclusion that it was time to move from research to action by developing risk assessments and response assessments for the issue. It took several years for this message to be articulated in the WMO's climate change-related activities. The heads of the UNEP, the WMO, and the ICSU formed the Advisory Group on Greenhouse Gases (AGGG) in 1986 to advise them on global warming issues. The AGGG was made up of six senior scientists appointed by the heads of the three institutions: Bert Bolin, Ken Hare, G. Golitsyn, S. Manabe, G. Goodman, and M. Kassas (Boehmer-Christiansen 1993). Three working groups were established under the auspices of the AGGG in 1988, with funding from the Rockefeller Brothers Fund, Stockholm Environment Institute (SEI) core funds, and the W. Alton Jones Foundation. WG I focused on the analysis of limitation strategies, WG II focused on the indicators of climatic change, and WG III focused on assessments of adaptation and limitation strategies. However, the final AGGG report was not released until 1990, by which time it was overshadowed by the IPCC assessments released that same year (Clark 1990; Fisher 1990; Jager 1990; Rijsberman and Swart 1990).

The "Villach Group" and uncertainty

Under the initial auspices of the AGGG, a group of younger scientists coalesced into an active science/policy network. This "Villach Group" organized an additional set of policy-related workshops on a faster timetable. The group was bound together by a common approach toward dealing with uncertainty. While they did not believe that the science was as yet definitive, they believed that more vigorous approaches to understanding and possibly delaying or avoiding climate change was needed. They were comfortable with developing heuristic approaches to understand and manage large complex and uncertain systems, such as the climate system (Keepin, Mintzer et al. 1986; Clark 1989; Fisher 1990; Jager 1990; Jager 1992; Oppenheimer 1992; Bolin, Jager et al. 1996). They were concerned with maintaining the momentum within international institutions toward policy response that was established at Villach in 1985, which had brought together the core members of the group. A subset of experts attended the major small planning meetings and were involved in the drafting committees of the final statements of the 1988 Toronto Conference and the 1990 SWCC. Contrary to Boehmer-Christiansen's suggestions that the group was principally elitist and motivated by a desire for research funding, interviews and careful readings of their documents suggest that they were concerned and motivated more fundamentally by an abiding desire to address environmental risks they deemed urgent subject to the best scientific and management techniques available to them, which included NGO and citizen participation in the decisionmaking process.

In July 1986 Professor Gordon Goodman of the Beijer Institute Stockholm, together with Michael Oppenheimer of the Environmental Defense Fund (New York, New York) and George Woodwell (Woods Hole Research Center, Woods Hole, Massachusetts) initiated a project to fulfill the policy mandate from the 1985 Villach meeting. The first workshop, held in Villach in September 1987, was attended by about 50 scientists and technical experts who examined how climatic change resulting from increases of greenhouse gas concentrations in the atmosphere could affect various regions of the earth during the next century. The participants also discussed the technical, financial, and institutional options for limiting or adapting to climate change (World Meterological Organization and United Nations Environment Program 1990).

To a follow-up workshop in Bellagio, Italy in November 1987, the Villach Group invited a small number of experts believed to be willing to attempt translating science into policy options (Oppenheimer 1992). The 24 participants used the technical material from the Villach workshop as background material and explored policy steps that might be considered for implementation in the near term and what institutional arrangements would be needed to achieve them. The Villach Group developed the concept of "tolerable rates" of environmental impact at the 1987 Bellagio workshop and proposed a target rate of 0.1° C of temperature change per decade—based on observed historic rates of temperature and sea level change and on expected consequences for ecosystems and societies. While UNEP and WMO representatives expressed interest in using such material in preparation for a climate change treaty modeled possibly on the successful Montreal Ozone Protocol experience, governments drew slightly different conclusions from the process, as discussed below. The spirit of the Villach Group's tolerable rates approach lives on in the United Nations Framework Convention on Climate Change (UNFCCC), although no formal commitments consistent with the approach are elaborated. The objective of "the stabilization of greenhouse gas concentrations in the atmosphere at a level that would prevent dangerous anthropogenic interference with the climate system" expressed in the UNFCCC also reflects the Villach Group's commitment

> to achieve … stabilization of greenhouse gas concentrations in the atmosphere at a level that would prevent dangerous anthropogenic interference with the climate system. Such a level should be achieved within a time-frame

> sufficient to allow ecosystems to adapt naturally to climate change, to ensure that food production is not threatened and to enable economic development to proceed in a sustainable matter.
>
> *(UNFCCC article 2)*

The Villach Group was also instrumental in organizing the 1988 World Conference on the Changing Atmosphere in Toronto, the first international conference on climate change to deliberately include a political perspective (World Meterological Organization 1989). The conference, financed by the government of Canada, the UNEP, and the WMO, invited a number of scientists and government leaders in their private roles. The 1988 Toronto Conference proposed that governments cut 1988 CO_2 emission levels by 20% by 2005: 10% by conservation and 10% by fuel switching. The 20% figure was a dramatic innovation; up to this point, no national governments had proposed such targets. The proposal emerged from the energy group and was included in the conference declaration by a number of the Villach Group's core members who also served on the drafting committee and who saw the 20% figure as an economically and politically tractable interim step toward the 50% cuts necessary to meet the tolerable rate of 0.1° C warming per decade that had come out of the 1987 Bellagio and Villach workshops.

The Toronto 20% cut target diffused internationally. It was subsequently discussed in Germany, for example. While the 1988 Enquete report preceding the Toronto Conference contained no CO_2 targets, the Enquete report following Toronto, released in 1990, recommended the goal of 30% CO_2 and CH_4 cuts by 2005. Later in 1990 the German government adopted the goal of a 25% CO_2 cut by 2005. The specific Toronto target did not diffuse to become policy in all states, and it was not endorsed by the INC, largely due to the diplomatic efforts of the United States and the Organization of Petroleum Exporting Countries (OPEC). However, it did stimulate national government activity regarding climate change. It was a key event in stimulating concern in Canada, and it prompted congressional bills, industry goals, and strategy formulation in the United States.

The backlash

This successful mobilization of science for global climate change policy precipitated a political backlash. Governments wanted options, not formal recommendations, regarding what they should do. The publicly aired advice tendered by the Villach Group and the Toronto Conference usurped what many government officials felt was their formal role. Governments recognized that unrestrained scientists could press governments for measures they were unwilling to adopt, or for action more quickly than was deemed politically desirable at home. Reflecting on the role of scientists and secretariats in both the ozone and climate change experiences, the U.S. State Department and the Europeans designed new institutions for climate change through which they could better control the political agenda by greater restraint over the use of science in the discussions. A number of experiences contributed to this political suspicion of unconstrained science. The UNEP's ability to designate individual experts was seen by the State Department as a loss of control over the diplomatic process. The Toronto Conference's endorsements of 20% cuts, contrary to U.S. policy wishes, demonstrated to politicians that activist scientists could shame them publicly. Moreover, many of the diplomats from industrialized countries had grown weary of Tolba's strong leadership and hectoring ways (Nitze 1989; Nitze 1992; Reifsnyder 1992)

The IPCC was formed (via an United Nations General Assembly [UNGA] resolution) as a consequence of the lesson drawn by some national governments that their interests would be better served by exerting direct control over risk and response assessment functions. The UNGA decision was adopted 6 months after the UNEP's Governing Council had instructed Tolba to consult with the WMO about establishing a coordinative scientific assessment of climate change (United Nations Environment Program 1987). The UNGA resolution removed from the UNEP and the WMO the authority to designate experts, and firmly lodged the responsibility for drafting a legal document with the governments rather than an international institution. IPCC experts were designated by governments, and the major bureau positions, particularly in WG III on policy, were high-level foreign ministry officials (Fitzgerald 1990). The Response Strategies WG III, under the chairmanship of the United States, was charged with considering "legal" issues as part of its broader agenda. The United States continually used the IPCC as an instrument to demonstrate the efficacy of U.S. domestic efforts and the absence of any urgency for further action (Gray and Rivkin 1991; Barrat-Brown, Hajost et al. 1993; Hatch 1993). The tolerable rates heuristic approach was brought from the Villach Group to IPCC WG III by Pier Vellinga, who served on both bodies. However, the innovation failed to diffuse between the institutions. The IPCC did not develop the concept, as its effects-based logic ran counter to the IPCC's focus on cost-based assessments of strategies. Further, this approach generated stringent emission control scenarios that the United States sought to exclude from consideration. The direct involvement of governments in the IPCC also prevented the UNEP and the WMO secretariats and the Villach Group from driving the international agenda on climate change.

Political responses and scientific framing

Governments were not entirely successful in their effort to divert the political impact of the scientific community. Government designation of scientists appears to have had no noticeable effect on the output of the scientific working groups (I and II) because of the large number of scientists involved, the voluminous background of peer-reviewed scientific literature, the extensive peer review process that was followed, and the large number of nongovernment agency scientists involved (Houghton, Jenkins et al. 1990; Tegart, Sheldon et al. 1990; World Meterological Organization and United Nations Environment Program 1990). Despite governmental reluctance, the 1990 IPCC report did catalyze governmental concern and precipitated the establishment of the INC. The whole enterprise has been met with extensive suspicion from less-developed countries (LDCs), who were incapable of participating in the technical discussions because they lacked the indigenous science to run their own Global Circulation Models (GCMs). Politically, they were distraught that they were unable to extract significant concessions for technology transfer and financing.

The INC itself was formed under UNGA auspices. Again, this distanced the UNEP and the WMO from exerting influence and derailed their fledgling plans to sponsor negotiations (Anderson and Aldhous 1991). Developing states, led by Malta, sought to move the climate negotiations under General Assembly auspices in order to enhance the bargaining position of the South and revive New International Economic Order (NIEO) discussions (Borge-Olivier 1992). Environmental NGOs sought to block this move, for fear that it would inject intractable NIEO discussions and make substantive negotiations impossible (Anderson and Aldhous 1991; Nitze 1992). The United States supported the move to the UNGA as a way of preventing Tolba and the activist UNEP secretariat from exerting entrepreneurial leadership and cajoling laggard states toward a substantive agreement (Young 1991; Nitze 1992). Although the UNEP continued to play a role in the climate convention process, it was never again a leading one.

Science provided the background for what was primarily a political process after 1988. Overriding political concerns about the costs of control overshadowed the influence of science in the INC (Taplin 1996). Although the 1990 IPCC report was a catalyst in the formation of the INC, a much more powerful magnet for negotiations was the United Nations Conference on Environment and Development (UNCED) timetable. The INC virtually ignored inputs from concurrent risk and response assessments, and most governments' responsiveness to the IPCC declined once negotiations began. The emphasis on political concerns at the expense of scientific input was foreshadowed at the SWCC in late 1990, where the ministerial declaration differed markedly from the more activist scientific declaration on which it was putatively based.

Reprise

Similar results emerged in the cases of climate warming, acidification, and ozone depletion. Core elements of environmental management were articulated by small transnational networks of scientists. While the existence of these groups preceded the emergence of the issue, their membership was reinforced and tightened through their collective involvement in seeking to develop ways to manage the risks, although networks had more enduring influence in the ozone and acid rain cases than in the climate change case. International institutions helped to legitimate the networks' ideas, consolidate and strengthen the networks, and provide the logistics by which governments actually converted the management concepts into concrete policy measures. The institutions facilitated the persuasion of politicians, as well as supplementing the immediate appeal of the policy concepts with additional institutional incentives. Specific lessons were transmitted by members of these networks whose professional profile and standing attracted attention by institutions.[1] However, the narrow membership of these networks hampered the transmission of lessons between issues and between institutions.[2]

International institutions and lessons drawing

How have international institutions contributed to the management of global environmental risks? How have they contributed to learning about these problems? In this section we look first at changes in the performance of programmatic functions by the institutions relating to the management of global environmental risks, both in terms of initiatives from the institution itself as well as new mandates charged by member governments. We follow with an analysis of the interventions available to international institutions that may improve the capacity of national governments and nonstate actors to cope with environmental risks.

The most striking difference between the cases is the more rapid pace of movement from early scientific warning to international action in ozone than in climate change. In the ozone case it took 4 years from the first scientific warning (publication of the Rowland and Molina hypothesis) to a political international planning meeting (1977 Washington Conference); 8 years until intergovernmental negotiations began; and 14 years until strong international measures (Montreal Protocol). The climate change case was much more protracted. The first scientific warnings appeared in the 19th century. It took 21 years from the renewal of scientific concern (Keeling's Moana Loa study is representative) to a political international planning meeting (establishment of the INC) and 23 years from first warning until weak international measures (UNFCCC).

Several institutional factors may account for some of this variation. The UNEP's institutional design emphasized its role as a catalyst and did not tie it to a single constituency or single mission, as is the case with the WMO. This enabled the UNEP to be more independent and flexible than the WMO, and thus to recognize and respond to problems in a more rapid manner.

Possibly more important for understanding the variation in institutional impact on overall environmental management across the three issues may be the characteristics of the issues themselves. Climate change was simply a much more intractable political problem, with less scientific certainty, greater anticipated economic costs of suggested responses, and more painful social adjustments asked from citizens of the industrialized countries than in the ozone case. The acid rain case in this regard falls between the ozone and climate change cases, and the involved international organizations guided international responses at a pace that fell between that of the ozone and climate change cases. Consequently, the manipulable variables associated with different institutions and their capabilities for guiding multilateral environmental management may be easier to invoke in cases that are structurally closer to ozone and acid rain than to climate change.

Institutional learning

The most responsive institutions in which learning occurred were the UNEP and the Villach Group. The WMO experienced some learning, but this was fairly delayed because of the rigid structures through which information to the institution flowed. The INC and IPCC learned very little.

The most common processes by which learning occurred involved the transmission of information to the institution from outside sources. In most cases discussed here this information came from the scientific community because it had the closest established ties to the institutions. In the absence of these ties, major new lessons are likely to be rejected or ignored. There is no reason to believe that other groups—environmental NGOs and business—could not provide valuable new information as well. Learning occurred as new knowledge or understanding was imparted to the institutions and converted into new programmatic activities. In order for an institution to be able to engage in this process it must be able to have timely access to relatively impartial information, be able to effectively process the information internally, and be capable of converting such new ideas into new activities.

Adaptation occurred in institutions without these features. However, potential lessons for dealing with environmental risks that run counter to the wishes of the dominant coalition, or are seriously at odds with major institutional routines, are unlikely to be adopted by such institutions or adapted very gradually. For example, the long gestation involving the WMO's weather to climate shift reflected the persistence of organizational inertia. Adaptation generally occurred for reasons of political expediency; the new ideas were adopted because they corresponded to broader desires of the parties involved, such as the IPCC's move to encourage greater LDC participation and the 30% SO_2 cuts in the acid rain case.

For an institution to respond promptly to new information and to develop new programs it must also be able to act independently of the direct control of member governments. An institution's ability to act independently of the direct control of member governments tends to be a function of a number of factors. Prior research and conventional wisdom in international relations suggests that at a general level, for international institutions to be able to effectively operate independently of the control of their most influential member governments there must be the absence of fundamental political schisms about world order ideologies among the member governments (Haas 1990). North–South splits often proved insuperable to consensus formation among member governments, although the Cold War did not impede low profile policy coordination on nuclear nonproliferation, limiting atmospheric nuclear radiation from testing atomic weapons, meteorological research, and Antarctica (George, Farley et al. 1988). The secretariat must have autonomy from governmental hiring choices and some discretion in programmatic choice, as well as possessing sufficient autonomy and technical capacity in order to be able to assimilate new information and actively promote its acceptance and diffusion. The institution must have relatively porous organizational boundaries, which facilitate the flow of information from outside, generally from NGOs and the scientific community. Finally, the executive head must have a management style that is appropriate to the political makeup of the dominant coalition of member countries. Other factors are important as well, such as the utility of the new ideas for satisfying other demands facing the institution.

The UNEP has the strongest institutional capabilities for learning, and repeatedly demonstrated this capacity. It was able to translate new information emanating from the scientific community into effective policy-oriented programs. Its executive heads (Strong 1973–75, Tolba 1976–92, and Dowdeswell 1993–) have been vigorous exponents of environmental protection and research in public, in private with heads of state, and also in private negotiations. They were generally able to effectively cope with disagreements among member states and avoid institutional deadlock. Tolba served as a strong advocate for stringent ozone and climate change targets and cuts during negotiations on the Montreal Protocol and the UNFCCC. Tolba also acted as the developing countries' representative at the meetings, as well as periodically interceding in his own capacity to press for more stringent measures The secretariat is chosen by merit, and actively solicits input from NGOs and the scientific community, who serve on advisory boards, receive funding to provide input into UNEP programs, and are invited in their personal capacity to appear at technical meetings. In contrast with the WMO secretariat, whose recruitment from national meteorological services reinforces WMO's existing perspective, members of the UNEP secretariat are recruited from a broader disciplinary base and often bring lessons with them from other institutions or international policy coordination efforts. This recruitment pattern underlies the UNEP's.

The WMO has been a slow learner. For instance, it took the WMO over a decade to shift its priorities from shorter-term weather issues to longer-term climate concerns. The WMO did undertake some background efforts, but was unable to launch a more vigorous influence on multilateral environmental governance because the WMO faced structural limits to its ability to significantly influence collective environmental management. The WMO has been partially constrained by the constituent meteorological services that control its activities, its status as a specialized agency that has reinforced its weather-oriented mandate and structure, and its betrothal to a single scientific constituency. The WMO heads (Davies 1949–84, Obassi 1984) have played much more reserved roles than UNEP heads, responding to pressures from the G77 on the Executive Council and overseeing a secretariat that was responsive to programmatic expressions of need by the member governments, but not engaging in the type of proactive, training and educational activities that the UNEP secretariat performed in both ozone and climate change. This role is spurred by the UNEP's organizational mission of catalyst, rather than specialized agency, within the UN system. The provision of new information and scientific initiatives to the WMO filters through the permanent representatives of national governments to the WMO. Thus, the WMO absorbs scientific input primarily through formal and informal consultations with national representatives, following the slow percolation up of information from individual scientists through their national meteorological offices, which act as gatekeepers between national experts and the WMO. The ICSU is closely involved in the administration of many of the WCRP and International Geosphere–Biosphere Program (IGBP) projects (Morel 1990). The WMO has focused on information related to atmospheric science; it is not porous to knowledge generated in other disciplines. This institutional reliance on a single disciplinary base, stemming from the demands of its core constituency of national weather service chiefs, impairs its ability to rapidly assimilate information about new environmental risks and stimulated the formation of the multidisciplinary IGBP by the ICSU in 1986. While the WMO may well be a good meteorological learner, it is much less adroit at learning about environmental risks.

The INC and the IPCC are hobbled by the limits imposed on secretariat autonomy by member governments. The INC secretariat's responsibilities are highly circumscribed—due to the insistence of China, India, the United States, and the OPEC states—in order to avoid the potential for an independent secretariat that could advance international goal and strategy efforts more aggressively than they desired (Djoghlaf 1994). The small secretariat relies only on governments for data and provides data only to governments. The secretariat cannot exceed these constraints out of fear of antagonizing the developing countries (Dolzer 1994). Although governments are responsible for submitting reports to the secretariat on their emissions and planned reduction policies, their review is to be "facilitative, non-confrontational, open and transparent" (A/AC. 237/46 paragraph 18). Deep North–South political schisms have further hampered the potential for learning.

Control over the IPCC is left entirely up to the member governments serving on the Bureau. Flawed organizational design further inhibited the openness of the IPCC to the transmission of scientific information: IPCC risk assessments and response assessments did not take account of one another's findings as the groups worked in parallel rather than in sequence, so that response policies failed to take much account of evolving risk assessments. This institutional design was deliberately constructed by the U.S. chair of WG3, which sought to influence the results of WG3. Although WG1 and 2 were less directly influenced by state concerns than WG3, they were still disappointing in their lack of responsiveness to innovations elsewhere, or to their ability to disseminate risk and response assessment innovations more broadly.

Further, the IPCC deliberately excluded alternative approaches that based analysis of response options on environmental targets. For instance, the Villach Group's tolerable rates innovation was excluded by IPCC WG3 as a basis for the formation of emission scenarios as its effects-based logic ran counter to the IPCC's cost-based assessment of strategies. Overriding political concerns about the costs of control also precluded the use of science in the INC. The INC virtually ignored inputs from concurrent risk and response assessments, paying no attention to the 1990 scientific declaration of the SWCC or the IPCC's 1992 supplemental appraisal. Once negotiations on the climate change treaty were under way, governmental responsiveness to the IPCC declined.

The United Nations Economic Commission for Europe (UNECE) staff has circumscribed autonomy due to the limited financial resources of the institution. The staff is limited to supporting individual meetings and travel; it lacks time or money to initiate additional activities. It has worked with some European NGOs in publicizing reports of forest dieback and works closely with the International Institute of Applied Systems Research (IIASA) in applying the Regional Acidification Information and Simulation (RAINS) model to developing regulatory controls. While the European Monitoring and Evaluation Program (EMEP) Trust Fund established by the 1984 Protocol provides important autonomous funding for atmospheric monitoring and modeling activities, the work on monitoring of effects (on freshwaters, forests, materials, ecosystems, and crops), critical loads, and mapping has depended on voluntary national funding.

Institutions as teachers

Institutions vary in their capacity to foster social learning by groups participating in the institution. Ultimately, innovative ideas are adopted by other actors for their own reasons, through their own methods. Thus, institutions' ability to diffuse lessons is largely a consequence of their ability to influence other actors' willingness to change their behavior and the capacity of these actors to absorb lessons.

Institutions that foster learning exhibit different features than do institutions that learn themselves. To foster learning, institutions must be capable of working directly with national figures in the field as well as providing financial resources to enable others to apply the lessons, or to reward them for doing so. The UNEP, the Global Environmental Facility (GEF), and the UNECE (in conjunction with IIASA) have been some of the most effective institutions in this regard, because they have the financial, technical, or intellectual resources that enable them to influence other actors.

Institutional properties that favor policy teaching

A lack of institutional legitimacy will hinder the acceptance of new information from an institution. Few results diffused to LDCs, because many LDCs were suspicious of the IPCC and unable to independently appraise its findings

The UNEP's strength is its ability to work with counterparts, both national groups and other international institutions, in countries where the innovations may be adopted, and to work with NGOs and firms who may also wish to adopt innovations. The UNEP also has a proactive engagement with NGOs, providing funding to permit their participation in UNEP assessments and program development. Lack of resources can inhibit the spread of ideas between institutions. The UNEP had little success in catalyzing the WMO to undertake more comprehensive activities in climate change, despite the joint sponsorship of the World Climate Impact Studies Program (WCIP) and Background Air Pollution Monitoring Network (BAPMoN) climate monitoring activities. The WMO approached the climate issues with the same organizational routines with which it addressed weather. It was resistant to the interdisciplinary features of the issues, which were inconsistent with its organizational mission. The UNEP lacked the financial clout to be able to dictate to the WMO, and the WMO was relatively immune or impervious to information provided by the UNEP because the WMO depended almost exclusively on information from the ICSU and from its constituent Meteorological Services. The GEF's primary resource is its ability to financially reward governments and firms who are willing to embrace its policy innovations.

Functional activities that encourage diffusion of policy lessons

In addition to encouraging the dissemination of innovations to other actors, institutions also perform a number of functions that build national capacity to learn. International organizations can set the agenda for members, distribute information, build national monitoring and research capacity, assist industry and societal groups identify new practices that contribute to effective environmental management, train and assist governments to enforce international commitments, structure bargaining fora, and empower new national and transnational political coalitions. Many institutional activities contribute to several of these effects, and the effects are also interactive in their influence on governments and other actors to modify or change their practices.

Sponsorship of international meetings is a primary technique in performing these functions. In addition to international institutions, environmental NGOs were active in this regard. International meetings can forge a policy-relevant scientific consensus, setting the agenda for states or other societal actors; this was the case with Wurzburg regarding the use of ozone depletion potential (ODP) and Villach regarding the magnitude and likelihood of climate change. They can also encourage research coordination and national institution building. The WMO established the WCP to coordinate national research regarding climate variability and change, and the UNEP sponsored international ozone assessments. An important effect of the WCP was the creation of national climate programs in states that had previously lacked this infrastructure.

International meetings spread new scientific knowledge among meeting participants and subsequently to their home states, and they publicize information that can increase public and national government concern regarding an environmental risk. Ozone information generated by the UNEP was used by Greenpeace in Germany to pressure its own government. The Toronto Conference had these effects on public opinion and national government activity in the climate change case.

International meetings also provide national governments with a source of scientific information. Rather than rely on the scientific expertise of a single state, states that lack indigenous scientific expertise often look to international sources. Developing states often accept scientific expertise that stems from international institutions more readily than that from an individual metropole. Even states with significant capacity utilize the results of international efforts. For example, Germany's Enquete panel used the IPCC sea-level rise estimates and impact assessments, and used the Villach workshop's and Villach Group's temperature change estimates in making its recommendations (Jager and Cavendar 1992). In the United States, the 1991 National Academy of Sciences (NAS) Effects Panel used the scientific results of the IPCC to bolster the authoritative legitimacy of its report (Dickson and Clark 1993). The Long-Range Transboundary Air Pollution Convention for Europe (LRTAP) forest surveys also heightened regional concern about forest dieback.

Financial assistance to developing states builds national capacity to implement environmental management strategies. Financial support for ozone is provided primarily through the Montreal Ozone Fund (MOF), while climate change is provided primarily through the GEF. Other international institutions—particularly the United Nations Development Program (UNDP) and the World Bank—are involved with supporting developing states' overall environmental management capacity, through training, purchase of equipment, and funding projects that will reduce emissions. Although it had limited resources itself, the UNEP was very effective at enlisting the support of these larger, better funded institutions in an effort to spread innovative practices, particularly for mobilizing funding to seek commercial alternatives to chlorofluorocarbons (CFCs).

The UNEP's Industry Environment Program Activity Center (IE/PAC) conducts seminars and workshops on energy efficiency to assist industry to respond to climate change (United Nations Environment Program and International Petroleum Industry Environmental and Conservation Association 1991). Participation in UNEP and WMO global research and monitoring programs has the ancillary effect of building low levels of indigenous scientific capacity. The UNEP's Ozone Action Programme sponsors country programs to assist developing states establish baseline surveys and prepare CFC replacement and control strategies. It also publishes a series of technical publications designed to assist developing states comply with the Protocol. The Industry and Environment Program Activity Center in Paris alerts firms to opportunities for improving energy efficiency and helps developing-country officials to establish baseline surveys of CFC use and to prepare strategies for controlling and replacing CFC use. The UNEP thus helps relatively inefficient companies to improve their performance, as well as allowing more efficient manufacturers of energy-efficient technology to expand their markets.

International organizations can stimulate the spread of innovative ideas during state bargaining by playing leadership roles. For example, in the ozone case, the UNEP convened the Wurzburg meeting to develop a scientific consensus regarding ODPs. It then channeled this concept into the bargaining forum that it was sponsoring, facilitating agreement on control measures. Conversely, an innovative bargaining proposal, such as pledge and review in the climate change case, may fail to take root in the absence of an independent, third-party initiative, especially given the ease with which both governmental and nongovernmental actors can kill or maim new initiatives before their possibilities can be developed.

Monitoring and research projects in member countries can also build receptivity for social learning. The governments will be more prone to accept new ideas that accompany activities they value, such as training activities and technology transfer, as well as the prestige that accompanies such activities. National participants in such activities tend to become advocates of the ideas with which such activities are associated, and thus contribute to transnational social learning. The use of local researchers in coordinated international activities is critical. Thus, institutions with significant local counterparts are more likely to encourage learning by doing in member countries. For instance, the WMO, the IPCC, and the IIASA all contributed to weaker countries' (eastern Europe, some LDCs) ability to independently assess computer models developed elsewhere. Institutions can also accord publicity to scientists' views and thus help to set the overall international agenda. The UNEP did this in ozone by convening the 1977 experts meeting and attracting attention to the Rowland–Molina hypothesis. The hypothesis was also legitimized by its recognition by an authoritative group such as the UNEP.

Institutional activities may provide impetus for indigenous learning through socialization in member countries, or among actor groups in those countries. As a consequence of capacity-building activities by institutions, groups in society have an enhanced chance to learn and to disseminate their own lessons throughout their country. Capacity-building exercises, such as joint monitoring and training exercises, can contribute to learning by doing, as other actor groups gain knowledge from new experiences resulting from obligations entailed from the institution. Enhancing domestic capacities may enable countries to learn that otherwise were unable. For instance, the UNEP training programs and the IPCC Third World outreach efforts provide information and training in information processing. Risk assessment and risk management functional capacities are also enhanced by IOs, although not significantly. By improving the cooperative environment, the potential for disseminating lessons abroad is enhanced. By building concern, governments become more sensitive to lessons propounded by members of their society.

Notes

Financial support came from the National Science Foundation (SES 9010101 and SBR9123033) and the German Marshall Fund of the United States. A version of this paper was presented at the University of Bologna Summer Symposium on The Innovation of Environmental Policy, 21–25 July 1997. A lengthier and more comprehensive treatment of the arguments and history presented in this article will appear in Peter M. Haas and David McCabe "Amplifiers and Dampeners: International Institutions and Social Learning in the Management of Global Environmental Risks" in *The Social Learning Group, Social Learning and the Management of Global Environmental Risks*. Cambridge, MA: MIT Press, 2000.

1 In one collaborative work by the UNEP and the IIASA for the World Climate Impact Programme assessing the vulnerability of food production to climate change was adopted in the IPCC WG2's 1990 Impacts Assessment chapter on agriculture because its author

served as a consultant to WG2 (Parry, Carter et al. 1985; Parry 1990). The widespread use of the framework convention/protocol sequence is another exception, and this is due to internal communications in the UNEP and also the UNEP's demonstration effect for the INC and the UNECE. This sequence was first proposed for ozone protection at a UNEP meeting of legal experts in Montevideo in 1981 as a result of its success in UNEP's regional seas program and was subsequently adopted in the climate change case as well. However, the diffusion of the convention/protocol sequence and the ODP–GWP concept are cases of adaptation rather than learning, as we defined them above. The critical loads, tolerable rates, and chlorine peak concepts are instances of learning, in that they all entail a reconceptualization of the risk and new goals. Learning has diffused much less frequently than adaptation, due in part to the barrier of institutional inertia, as well as the idiosyncrasies of the issues and political opposition.

2 For example, despite Pier Vellinga's efforts, the Villach Group's tolerable rates innovation, which based analysis of response options on environmental targets, was excluded by IPCC WG3 as a basis for the formation of emission scenarios, as its effects-based logic ran counter to the IPCC's cost-based assessment of strategies. Few examples exist of developments in one institution in one issue that were successfully adopted elsewhere. The ODP concept was adopted by the IPCC in its formulation of Global Warming Potentials (GWPs), but it was eventually found technically inadequate for the climate change case.

References

A/AC. 237/46 paragraph 18.

Anderson, C. and P. Aldhous (1991). "Third World muscles in on climate treaty." *Nature* 349(6312): 727.

Barrat-Brown, E., S. A. Hajost, et al. (1993). "A forum for action on global aiming: The UN framework convention on climate change." *Colorado Journal of International Environmental Policy and Law* 4(1): 103–115.

Boehmer-Christiansen, S. (1993). "Science policy, the IPCC and the climate convention." *Energy and Environment* 4: 362–406.

Bolin, B., J. Jager, et al. (1996). "The Greenhouse effect, climate change, and ecosystems". *The Greenhouse Effect, Climate Change, and Ecosystems: SCOPE 29.* B. Bolin, B. R. Doos, J. Jager and R. A. Warrick.

Borge-Olivier, A. (1992). Interview conducted by Anilla Cherian. New York.

Clark, W. (1989). "Managing planet earth." *Scientific American* 261(3): 45–54.

Clark, W. C., Ed. (1990). *Usable Knowledge for Managing Global Climate Change.* Stockholm, Stockholm Environmental Institute.

Dickson, N. and W. C. Clark (1993). "Global climate change: A historical perspective of risk management in the United States". *Contribution 1.15, version 2 to the project on Social Learning in the Management of Global Environmental Risks,* Mimeo: 21–22.

Djoghlaf, A. (1994). "The beginnings of an international climate law". *Negotiating Climate Change: The Inside Story of the Rio Convention.* I. M. Mintzer and J. A. Leonard. Cambridge, Cambridge University Press.

Dolzer, R. (1994). Interview conducted by Peter M. Haas. Cambridge, MA.

Evans, P. B., H. K. Jacobson, et al., Eds. (1993). *Double Edged Diplomacy.* Berkeley, University of California Press.

Faulkner, H. (1993). "Some comments on the INC process". *Negotiating Climate Change: The Inside Story of the Rio Convention.* I. M. Mintzer and J. A. Leonard. Cambridge, MA, Cambridge University Press: 229–238.

Fisher, D. (1990). *Options for Reducing Greenhouse Gas Emissions.* Stockholm, The Stockholm Environment Institute.

Fitzgerald, J. (1990). "The Intergovernmental Panel on Climate Change: Taking the first steps toward a global response." *Southern Illinois University Law Journal* 14: 231–256.

George, A. L., P. J. Farley, et al. (1988). *U.S. Soviet Security Cooperation*. Boulder, CO, Westview Press.

Gray, B. C. and D. B. J. Rivkin (1991). "A 'no regrets' environmental policy." *Foreign Policy* 83: 47–65.

Haas, E. B. (1990). *When Knowledge is Power: Three Models of Change in International Organizations*. Berkeley, University of California Press.

Haas, E. B. (1991). "Collective learning: Some theoretical speculations". *Learning in U.S. and Soviet Foreign Policy*. G. W. Breslauer and P. Tetlock. Boulder, CO, Westview Press.

Haas, P. (1989). "Do regimes matter? Epistemic communities and Mediterranean pollution control." *International Organization* 43(3): 377–403.

Haas, P. (1990). "Obtaining international environmental protection through epistemic consensus." *Millennium: Journal of International Studies* 19(3): 347–364.

Haas, P. M. (1992/1997). "Knowledge, power, and international policy coordination." *International Organization* 46(1).

Hatch, M. T. (1993). "Domestic politics and international negotiations." *Journal of Environment and Development* 2: 1–40.

Houghton, J. T., G. J. Jenkins, et al. (1990). *Climate Change: The IPCC Scientific Assessment*. Cambridge, Cambridge University Press.

Jager, J. (1990). *Responding to Climate Change: Tools for Policy Development*. Stockholm, Stockholm Environment Institute.

Jager, J. (1992). "From conference to conference." *Climatic Change* 20.

Jager, J. and J. Cavendar (1992). *The Issue of Climate Change in the F.R. Germany. Contribution to the project on Social Learning in the Management of Global Environmental Risks*, Mimeo.

Karns, M. P. and K. A. Mingst (1992). *The United States and Multilateral Institutions*. London, Routledge.

Keepin, W., I. M. Mintzer, et al. (1986). "Emission of CO_2 into the atmosphere". *The Greenhouse Effect, Climate Change, and Ecosystems*. B. Bolin, B. R. Doos, J. Jager and R. A. Warrick. Chichester, England, John Wiley & Sons.

Keohane, R. O. (1989). *International Institutions and State Power: Essays in International Relations Theory*. Boulder, CO, Westview Press.

Kowalok, M. E. (1993). "Research lessons from acid rain, ozone depletion, and global warming." *Environmental* (July/August).

Linder, S. H. and B. G. Peters (1995). "The two traditions of institutional designing". *Institutional Design*. D. Weiner. Boston, Kluwer.

Morel, P. (1990). "Overview of the World Climate Research Programme". *Proceedings of the Second World Climate Conference, Geneva*.

Nitze, W. A. (1989). "The intergovernmental panel on climate change." *Environment* 31(1): 44–45.

Nitze, W. A. (1992). Interview by Peter Haas.

Oppenheimer, M. (1992). Interview conducted by David McCabe. New York Environmental Defense Fund.

Parry, M. (1990). *Climate Change and World Agriculture*. London, Earthscan.

Parry, M., T. Carter, et al. (1985). "Climatic change: How vulnerable is agriculture?" *Environment* 27(1): 4–5.

Perry, J. (1993). Telephone interview with J. Perry.

Ramakrishna, K. and O. Young (1992). "International organizations in a warming world". *Confronting Climate Change*. I. M. Mintzer. Cambridge, Cambridge University Press.

Reifsnyder (1992). Personal communication.

Rijsberman, F. R. and R. J. Swart (1990). *Targets and Indicators of Climatic Change.* Stockholm, Stockholm Environment Institute.

Schmidheiny, S. (1992). *Changing Course: A Global Business Perspective on Development and the Environment.* Cambridge, The MIT Press.

Taplin, R. (1996). "Climate science and politics: The road to Rio and beyond". *Climate Change: Developing Southern Hemisphere Perspectives.* T. W. Giambelluca and A. Henderson-Sellers. New York, John Wiley & Sons: 377–395.

Tegart, W. J. M., G. W. Sheldon, et al. (1990). *Climate Change: The IPCC Impacts Assessment.* Canberra, Australian Government Publishing Service.

The Social Learning Group (2001). *Social Learning and the Management of Global Environmental Risks.* Cambridge, MIT Press.

Underal, A. (1993). "The roles of IGOs in international environmental management: Arena or actor?" *The Role of Regional Organizations in the Context of Climate Change.* M. H. Glantz. Berlin, Springer Verlag.

UNFCCC article 2 *The United Nations Framework Convention on Climate Change.*

United Nations Environment Program (1987). "Ad hoc scientific meeting to compare model generated assessments of ozone layer change for various strategies for CFC control". UNEP/WG. 167/INF. 1. Nairobi, United Nations Environmental Program.

United Nations Environment Program and International Petroleum Industry Environmental and Conservation Association (1991). *Climate Change and Energy Efficiency in Industry.* London, IPEICA.

United Nations General Assembly Resolution 42/184 (1987). *International cooperation in the field of the environment. Adopted by vote: 149 for, 1 against (Israel).* New York, United Nations.

United Nations General Assembly Resolution 45/212 (1990). *Protection of global climate for present and future generations of mankind. Adopted by consensus without vote.* New York, United Nations.

Weiss, E. B. (1981). "The international implications of seasonal climate forecasting." *Stanford Journal of International Law* 17(2): 315–345.

Weiss, E. B. (1983). "Management of weather and climate disputes." *UCLA Journal of International Law and Policy* 3: 275–309.

World Health Organization, United Nations Environment Program, et al. (1980). *Assessment of the role of carbon dioxide on climate variations and their impact.* Austria, Meeting Report.

World Meterological Organization (1986). *Report of the international conference on the assessment of the role of carbon dioxide and of other greenhouse gases in climate variations and associated impacts.* Villach, Austria WMO No. 661.

World Meterological Organization (1989). *The changing atmosphere: Implications for global security.* Geneva, World Meteorological Organization. Toronto, Canada, June 27-30, 1988 WMO No. 710.

World Meterological Organization and United Nations Environment Program (1990). *Climate Change: The IPCC Response Strategies.* Washington, D.C., Island Press.

Young, O. R. (1991). "Political Leadership and Regime Formation." *IO* 45(3): 281–308.

11

ADDRESSING THE GLOBAL GOVERNANCE DEFICIT[1]

Peter M. Haas

The Brundtland Commission report wrote, "The globe is one, but the earth is not" (World Commission on Environment and Development 1987). The challenge for effective governance is how to encourage governments to pursue comprehensive policies to achieve Sustainable Development within an international political context that has inhibited the pursuit of comprehensive and long-term goals. An international society of states founded on the principle of national sovereignty alone discouraged serious state attention to international environmental externalities, as well as suppressing the voices of those in ecologically threatened areas, often the poor within societies, and led states to ignore the global commons outside national jurisdictions. While developing countries have not been pollution havens for the rich, they do lack the ability to forcefully present their positions at international negotiations on sustainable development and environmental protection.

Brave new world

The contemporary international political system faces two new geopolitical realities that challenge the old geographical principles of national sovereignty (Schaeffer 1997; Annan 1998; Keck and Sikkink 1998; Held, McGrew et al. 1999; Keohane and Nye 2000; Reinecke and Deng 2000; Hoffmann 2002). Consequently there is the potential for replacing the traditional dichotomous concepts of global governance organized hierarchically or anarchically (Bull 1977; Oye 1986) with a network model of complex decentralized global governance performed by multiple actors, whose interactive effects in practice would yield more effective global coordination and performance of major governance functions (Haas and Kanie 2004). This is a political project or vision of incremental multilateralism, as more parties become part of a growing project of globalization over which each has an interest and a say; thus establishing the institutional mechanisms for promoting the beneficial features or globalization while minimizing the more egregious negative effects.

First is the complexity of a globalizing world, whose management requires more holistic or comprehensive policies to address environmental externalities (a diplomatic term for ecological collapse) and to support sustainable development. Most international and national institutions were designed historically to address discrete problems, whereas the current globalized agenda consists of intertwined issues (or what organizational theorists term non-decomposable or partially non-decomposable problems) whose effective management requires procedures for responsible agencies (either nationally or internationally, or states as a whole) to think about how their actions will affect the responsibilities of other autonomous agencies and how their policy domain may be affected by decisions taken in or by other bodies (Haas 1975; Simon 1981; Perrow 1984; Axelrod 1997; Jervis 1997).

Second is the proliferation of new political actors and the diffusion of political authority over major governance functions, particularly in the environmental sphere. These new actors include NGOs, MNCs, organized transnational scientific networks known as epistemic communities, global policy networks, and selective international institutions that are capable of exercising discretionary behavior independently of the wishes of their dominant member states, such as UNEP, the World Bank since 1987, the ECJ, and possibly the EU Commission.

International responses to global complexity

Since 1972 there have been many efforts to design international institutions to better harmonize international decision making to promote environmental protection, and, later, to promote sustainable development (Kennan 1970; National Academy of Sciences 1972; Skolnikoff and Kay 1972; Haas and Haas 1995). The United Nations Environment Programme (UNEP) was established in 1973 to be "catalytic" and performed remarkably well with fairly scarce resources. It helped to develop a significant body of international environmental law, encouraged other international institutions to take account of the environmental consequences of their programmatic activities, and trained hundreds of developing country officials in techniques of ecological resource management.

But things have changed dramatically in the last 30 years with the spread of environmental consciousness and the proliferation of Multilateral Environmental Agreements (MEAs). UNEP is now underfunded, overloaded and remote (Holmberg 1992; Downie and Levy 2000; Gehring and Buck 2002). It is relatively obsolete, eclipsed in resources and prestige by other international institutions that have taken on new environmental responsibilities, such as the World Bank, the EU for Europe, and even, to the dismay of many environmentalists, the WTO. Indeed many NGOs such as the World Resources Institute, Greenpeace and TRAFFIC have assumed some of the functions of global governance including environmental monitoring and policy verification.

Recent years have seen several major reform proposals for redesigning the United Nations and Bretton Woods systems, which recognize these new political realities and address the perceived gaps in the performance of some key governance functions in the realms of environmental protection and sustainable development. In other words, these proposals identify a governance-deficit. The primary functions of environmental governance are presented in Table 11.1.

TABLE 11.1 Matrix of functions

Function	*Formal/direct*	*Informal/indirect*
Issue linkage	• By inter-governmental negotiations • By new information provided by epistemic communities • Through financial mechanisms (GEF) • BY IOs (GEO/WEO)	• By scientists • By business/industry
Agenda setting	• By IOs and member states • By scientists	• By NGOs • By media • By scientists
Developing useable knowledge	• By scientists	• By scientists • By NGOs • By business/industry
Monitoring	• By IOs • By committees nominated by MEA secretariat • By MEA signatory governments	• By NGOs (particularly in developing countries) • By scientists
Rule making	• Negotiations by national governments • By NGOs (principled standards)	• By business/industry (de facto standards) • By NGOs (principled standards)
Norm development	• Epistemic communities	• By NGOs (equity and environmental preservation) • By business/industry (efficiency)
Policy verification	• Governments	• NGOs • IOs
Enforcement	• (Hard) Law • WTO and MEA rules	• NGO campaigns
Capacity building (technology transfer)	• Official technical assistance (national and local government) • Business/industry • Science community (education/training)	• By business/industry
Promote vertical linkage	• IOs • National and local governments	• NGO • Scientific community
Financing	• Government (ODA) • Regional development banks • Multilateral bodies	• By business/industry

Why reform?

Three broad arguments are provided for the need for such reform (Dodds 2000; Dowdeswell 2001).

The first focuses on redundancy and overlapping responsibilities and tasks amongst international institutions. These are presumed to be a bad thing out of a general rationalist impulse for simplification and centralization, because of the inefficient use of scarce resources by competing international institutions, and the logistical difficulties faced by small bureaucracies who have to attend and prepare for too many meetings at the institutions and the MEA Conferences of Parties. For the rationalist redundancy, inefficiency and logistical difficulties constitute profound impediments for effective international governance. I find this argument unpersuasive because I think that some degree of redundancy is actually desirable in the international system, as it provides insurance against the decline of any individual international institution and fits better with an ecological institutional design vision of requisite diversity. Moreover, redundancy provides for more contact and linkage between institutions. If the governance deficit is due to performance gaps then responses should be addressed through capacity building. If the governance deficit is due to redundancies between international institutions then responses should await a clear inventory and assessment of the performance of vital governance functions.

The second argument is a straightforward efficiency argument. More activities could be conducted if there was less competition for resources between organizations. I find this unpersuasive because it would have the effect of consolidating political influence in the international system in a smaller number of major IOs, increasing the possibility of political capture and the actual decline in the efficient use of financial resources due to capricious national budgetary cycles, as has been a recurrent problem for UNEP. If the governance deficit is simply a problem of UNEP's resources, then it should be elevated to UN Specialized Agency status and given more money.

The third argument is the need for a strong environmental presence in the international system, especially as an environmental advocate at the WTO, or as a counterweight to the WTO in trade and environment disputes. I find this argument persuasive but it is not clear that a counterweight to the WTO's presumptive bias towards trade liberalization over environmental protection in its trade and environment arbitration decisions requires the creation of a massive countervailing institution, particularly since the WTO's recent dispute panel record has upheld some environmental protection in its decisions, such as the protection of sea-turtles (Williams 2001; DeSombre and Barkin 2002). Rather, as I argue below, such a counterweight can be more pragmatically pursued through institutional reforms that amplify environmental voices within WTO trade and environment dispute resolution panels, because the political will behind the creation of a mammoth new international organization is clearly lacking in the United States, which would have to be largest funding source for such an institution.

Proposals for reform

The most ambitious reform is the creation of a Global Environmental Organization, or World Environmental Organization. This idea has been proposed most forcefully by the German Advisory Council on Global Change, the German academics Frank Biermann and Udo Simonis, and Dan Esty and the Yale Center for Environmental Law and Policy (Esty 1994, Biermann 2000, Biermann 2001; German Advisory Council on Global Change 2001; Biermann 2002; Charnovitz 2002; Esty and Ivanova 2002; Najam 2003).[2] Chancellor Schroeder and President Chirac have publicly supported this initiative, although it has absolutely no support in the US government. The United States remains selectively committed to most elements of multilateral environmental diplomacy, despite its reversal on the Kyoto Protocol, but it is not interested in potentially expensive institutional reform or the creation of new international institutions until 2005 at the earliest.

In 2002 a Governance review commission organized by UNEP called for the creation of a Global Ministerial Environment Forum, essentially a periodic set of Summits for Environmental Ministers. This would be a good way to encourage the adoption of high-sounding commitments when the domestic climate is favorable in the major countries, but would lack any ongoing administrative abilities or institutional memory for how to conduct effective multilateral environmental diplomacy.

A more modest suggestion came from the 1997 Task Force on Environment and Human Settlements, which, not surprisingly, has been adopted by UNEP's Governing Council. It suggests strengthening UNEP by elevating it to a specialized agency (and thus being entitled to a fixed and regular budget) and by improving its ability to coordinate activities with other specialized agencies. However, no clear guidelines were given about how such coordination was to be achieved in the absence of strong political will by member governments or the heads of the agencies. France is currently circulating a slightly more comprehensive proposal for a strengthened UNEP that would conduct more scientific activities without shedding any of its present responsibilities. It is doubtful that there is much US government support for this proposal, or much concern by the US electorate. Positions outside the United States have not yet been clearly formulated.

These proposals constitute an overly narrow conception of potential responses to the governance deficit. UNEP reform still puts too much reliance on a small and remote organization. Yet there is no political future for eliminating UNEP, because developing countries insist on having the headquarters of a UN agency in a developing country. The question then is what to do with UNEP, since it is currently overstretched.

On the other hand creating a new centralized GEO appears utopian, given the lack of political will in the United States and abroad, lack of popular concern in the mass public, and the general distraction for the international community provided by the fight against terrorism.

Proposals for increased centralization of responsibilities or the creation of a new monolithic body run counter to the key insight of the most sophisticated current organizational theorists about the best institutional design for managing complex problems like global environmental issues. The best designed institutions for dealing with complex and uncertain policy environments are loose, decentralized, dense networks of institutions and actors that are able to quickly relay information, and provide sufficient redundancies in the performance of functions so that the elimination or inactivity by one institutions does not jeopardize the entire network (Aggarwal 1998; Ansell and Weber 1999; Ostrom 2001).

In short, strong centralized institutions are fundamentally unecological. They run counter to the ecological principal of requisite diversity or flexibility; inhibit random mutation, or policy innovation; and are easily captured by single powerful parties.

A better way of reforming the global environmental governance system, I argue, takes this new decentralized governance design principle seriously. In international circles it is now referred to as multilevel governance. The United Nations Global Compact, for example, is a recent effort to institutionalize multilevel governance within a network of networks that includes a variety of nonstate actors without sacrificing the principle of national sovereignty (Ruggie 2001).

Such a network model would involve the streamlining and improvement of the performance of existing governance efforts, rather than creating new governance bodies. A lot of governance is clearly already going on; the trick is to improve it and to enhance the synergies between the performance of these different functions. A clearer map is necessary of the actual division of labor between governments, NGOs, the private sector, scientific networks and international institutions in the performance of various functions of governance; their comparative advantages; and how well they actually perform these activities.

Effective governance rests on the performance of multiple governance functions. Some functions are formally performed; that is, the international community directly tasks some agent to explicitly perform them. Others may be performed indirectly: action is not the consequence of explicit instructions by those contracting some set of activities to be performed by the relevant actors. Some activities may have multiple indirect effects. For instance, by publicizing issues, norms and standard setting may be achieved. By verifying and providing resources, one may achieve compliance. By mobilizing civil society governance, efforts may promote agenda setting and framing, and thus define new national preferences that narrow the range of feasible negotiated outcomes. Educating elites and governments may have similar effects.

Addressing the governance deficit effectively should take serious account of these new ideas. Governance should rely on a differentiated division of labor among elements of international civil society, with clearer attention paid to coordinating the efforts, assuring rapid and accurate information flow between the various actors involved in governance, and to addressing the real existing gaps in current effective governance. Seriously applying this network vision of governance would entail some reorganization, some consolidation, and the creation of a limited set of new organizations.

I think that some core set of responsibilities should be left with UNEP associated with its initial scientific research and monitoring responsibilities, with other governance functions redistributing amongst other international actors. UNEP could help draft a global ecosystem assessment seeking to develop a priority list of global environmental threats of interest to the international community, and coordinate ongoing standing international scientific panels to conduct environmental research and monitoring. Such a concentrated and reinvigorated UNEP would contribute to improved agenda setting for international environmental governance.

Reorganization

Agenda setting

A more systematic early warning system is needed for alerting the international community to impending environmental threats, such as suggested above for UNEP. Current agenda setting is largely performed by NGOs, who often provide exaggerated claims or false warnings. The challenge is to develop early warning signals that are accurate—that is, that do not miss threats or cite potential urgent threats that subsequently prove unfounded. Better environmental monitoring might improve agenda setting, but the creation of standing international scientific panels responsible for evaluating the state of the environment would be a valuable reform, similar to the already existing IPCC, the Millennium Ecosystem Assessment, and the Ozone Trends Panel.

Alternatively more sophisticated measurement techniques using social indicators of populations at risk could provide an early warning system of ecosystem threats. Migration patterns of groups living in target ecosystems is an example, as well as tracking prices of scarce resources or measuring keystone species for signs of threats to marine species are examples. NGOs and scientists monitoring coral reefs as an early warning sign of climate change is another example of alternative measurements techniques for agenda setting.

Verification

Current arrangements for verifying state compliance with international environmental obligations are very weak. NGOs could help keep track of governmental adherence to their international obligations, and a revitalized Earth Council could serve this purpose. The Earth Council was created after the 1992 Rio Earth Summit, modeled on Amnesty International and its verification role in human rights regimes, but has subsequently languished.

Technology transfer and financial resource transfers

There is clearly a gaping and continued need for financial transfers in support of sustainable development. The GEF exists for this purpose, as does the Montreal Ozone Fund, but the financial resources for these institutions are inadequate to the task. Recent trends in ODA and foreign investment also suggest that insufficient amounts of money are going to countries in need of significant capacity building for sustainable development. Incentives from industrialized country governments could encourage MNCs to transfer green technologies to developing countries, and informational clearinghouses about green technologies in the public domain could serve a public good here, created either by international institutions such as the International Atomic Energy Agency (IAEA), NGOs, or the private sector.

Enforcement

It is a common lament that many MEAs are not enforced, or that governments don't submit data on enforcement so that there is insufficient information to enable informed judgments about the extent of enforcement with MEAs. NGOs can help monitor enforcement, as well as the creation of impartial third-party inspectors who would be able to inspect facilities for compliance, such as the IAEA does, though with limited success.

Consolidation

Efficiency gains and the creation of new usable knowledge could be achieved through the consolidation of the way that many of the current scientific (i.e. research and monitoring) functions are performed at present.

Environmental monitoring

Environmental monitoring should be consolidated by environmental medium, to be conducted by consortia of international institutions and scientific networks, and possibly NGOs, and even MNCs if they would be willing to provide emission data. For instance, in the area of ocean monitoring there is a plethora of monitoring activities conducted by the IOC, GESAMP, UNEP's Regional Seas Programme, and US NOAA. And this is only a partial list. Different bodies are responsible for monitoring different oceans. All of these activities should be formally consolidated, although the logic of consolidation has not yet been established: should it be by simple environmental medium, by geographic region, by common cause of the environmental problem, or by some other justificatory logic?

Rule making and MEA administration

Many countries complain of regime saturation that exceeds their ability to effectively participate in the management and development of particular environmental regimes. This is largely because the treaty secretariats are spread around the world and the schedules of major meetings are not coordinated. Indeed, there are over 200 international organizations involved with administering MEAs, although the number of major influential ones is probably under a dozen. Similarly the actual secretariats

TABLE 11.2 Locations of major MEA secretariats

Location	*Number*	*Comments*
Geneva	11	Many of the interim secretariats for MEAs that are not yet in force are in Geneva
London	9	Largely shipping related by IMO, OSPARCOM
Vienna	4	Related to nuclear safety
Rome	4	Administered by FAO
Montreal	3	ICAO, Montreal Ozone Fund, CBD
Bonn	3	CCD, CMS, UNFCCC

Source: Stokke and Thommessen 2002.

The rest of the 19 secretariats are spread over 17 locations.

are not as widely spread, as many critics would imply. Table 11.2 shows the headquarters of 54 of the major current MEAs (not all of which are yet in force).

Consolidating the MEA secretariats in one location makes sense in this regard. Geneva, London, Bonn and many other cities would fulfill these criteria. Consolidation would make travel easier for government officials, and would facilitate joint activities between the environmental regimes and their secretariats. Most importantly, perhaps, it would also have the effect of creating the equivalent of standing environmental embassies at this centralized location which would improve national foreign environmental policy making as well as elevating the profile of environmental policy makers within their own foreign ministries and governments. The basic institutional requirements for this clustering proposal are a location with good telecommunications, sufficient office space, and ample conference facilities. It is unclear to what extent a new organization body would be required, or whether it could simply consist of a MEA department store, all under one roof. When the GATT became the WTO there was a massive change in institutional influence and design, but in practical terms it merely entailed changing the sign over the front door and printing new business cards for the secretariat.

Creation

Several governance functions are inadequately performed, and probably require the creation of new institutions to improve their performance. It is not clear that one institution needs to serve all these functions, or whether they could be assigned to different bodies.

Norm setting

A High Commission for the Environment should be created so that there would be a high-profile figure able to help develop normative principles for environmental protection and sustainable development, akin to the UNHCR (Refugees) or UNHCHR (Human Rights).

Protecting the environment from the WTO

The environment needs an advocate before WTO trade and environment arbitration panels. There are two different institutional options for performing this function. A more modest one involves the creation of a roster of potential trade and environment lawyers who would be invited by the WTO to participate on arbitration panels. A more ambitious option would be to assign that function to the clustered MEA body.

Proximity and distance

The institutional design model here is one of multilevel decentralized governance. Some activities still require old brick and mortal type institutions, such as consolidated MEA secretariats which can coordinate intergovernmental activities and help foster intergovernmental trust. Other functions or activities may be performed through a looser and more decentralized network—a technological network of networked formal nodes and actors. A switchboard institution must be created to facilitate information flow between the different networks and levels of actors engaging in environmental governance. A centralized information coordination and diffusion body is still necessary, but it could be small, so long as it is technologically sophisticated and able to make use of current communications technologies to rapidly transfer information between the various bodies engaged in performing these key functions of global environmental governance. It would be largely virtual, and it is not clear that it need be affiliated with the environmental policy body. The major monitoring and verification functions would be coordinated through the switchboard, and the findings would be circulated from the switchboard. Yet the actual collection of information would be done by diffused units around the world.

Kyoto Protocol

These arguments about multilevel governance may be applied to the Kyoto Protocol as well. The reality is that the Kyoto Protocol is dead. President Bush made it clear that the US will never sign the Protocol, and read the obituary to the Protocol in his 2003 State of the Union Address. Without US participation few other countries will seriously pursue their commitments, even if the Protocol enters into force.

But alternative ways to mitigate global climate change still exist, from the perspective of the decentralized institutional design features and multilevel governance presented here. Major corporations including BP and Royal Dutch Shell have begun to apply Kyoto targets to their own corporate operations, so there is clearly some political will in the private sector for private environmental governance (Levy and Newell 2000; Hoffman 2001; Skjaerseth and Skodvin 2001; Levy and Newell 2002). Carbon taxes would encourage private sector innovation in cleaner technologies. Government support for investment in alternative fuels and cleaner technological processes would also help. President Bush announced that the US would spend $1 billion on green technology R&D. These are both proposals that do not require international cooperation, and can be based on public–private partnerships by individual countries. After a period of private governance of climate change governments would find it easier to mobilize constituencies for a public climate change regime that would be founded on more mature technologies.

Conclusion

We live in a new world. By taking advantage of a decentralized network of governance functions, global governance may be improved, and the prospects for achieving sustainable development advanced. By clinging to models based on an obsolete, exclusively state centric model of governance, claims of governance deficits will be exaggerated, and corrective designs erroneously applied that neglect new political realities.

Notes

1 Based on a presentation to ECOLOGIC on 5 November, 2003 in Berlin, Germany. The paper was also presented at the Mershon Center, Ohio State University, Columbus, Ohio on 22 April 2004, and at Wellesley College 1 May 2004. My thanks to Andreas Kraemer and the participants at the dinner presentation organized by the Heinrich Boll Foundation at which these ideas were further discussed and to M. J. Peterson, Craig Murphy, and Jennifer Clapp for comments on the final written draft. For an elaboration of these arguments see Haas and Kanie 2004. Copyright © 2004 The MIT Press. This chapter was first published in *Global Environmental Politics*, Vol. 4, No. 4 (2004) pp. 1–19. Reprinted with permission of the MIT Press.
2 Websites include www.globgov.org, www.yale.edu/envirocenter; and www.ias.unu.edu.

References

Aggarwal, V. K., Ed. (1998). *Institutional Designs for a Complex World*. Ithaca, NY, Cornell University Press.
Annan, K. (1998). "The Quiet Revolution." *Global Governance* 4: 123–138.
Ansell, C. K. and S. Weber (1999). "Organizing International Politics: Sovereignty and Open Systems." *International Political Science Review* 20(1): 73–93.
Axelrod, R. (1997). *The Complexity of Cooperation*. Princeton, NJ, Princeton University Press.
Biermann, F. (2000). "The Case for a World Environment Organization." *Environment* 42(9): 22–31.
Biermann, F. (2001). "The Emerging Debate on the Need for a World Environment Organization." *Global Environmental Politics* 1(1): 45–55.
Biermann, F. (2002). "Strengthening Green Global Governance in a Disparate World Society." *International Environmental Agreements* 2: 297–315.
Bull, H. (1977). *The Anarchical Society*. New York, Columbia University Press.
Charnovitz, S. (2002). "A World Environment Organization." *Columbia Journal of Environmental Law* 27(2): 323–362.
DeSombre, E. R. and J. S. Barkin (2002). "Turtles and Trade." *Global Environmental Politics* 2(1): 12–18.

Dodds, F. (2000). "Reforming the International Institutions". *Earth Summit 2002*. F. Dodds. London, Earthscan.

Dowdeswell, E. (2001). "Design for the Real World: Ideas for Achieving Sustainable Development." Text of a speech delivered at Harvard University, 12 December 2001.

Downie, D. L. and M. A. Levy (2000). "The United Nations Environment Programme at a Turning Point". *The Global Environment in the Twenty-First Century*. P. Chasek. Tokyo, UN Press: 355–375.

Esty, D. (1994). "The Case for a Global Environmental Organization". *Managing the World Economy: Fifty Years after Bretton Woods*. P. B. Kennen. Washington D.C., Institute for International Economics: 287–309.

Esty, D. and M. H. Ivanova, Eds. (2002). *Global Environmental Governance*. New Haven, CT, Yale School of Forestry and Environmental Studies.

Gehring, T. and M. Buck (2002). "International and Transatlantic Environmental Governance". *International Environmental Policymaking*. M. Buck, A. Carius and K. Kollman. Munich, Oekom Verlag: 21–43.

German Advisory Council on Global Change (2001). *World in Transition 2: New Structures for Global Environmental Policy*. London, Earthscan.

Haas, E. B. (1975). "Is There a Whole in the Hole?" *International Organization* 29(3): 827–876.

Haas, P. M. and E. B. Haas (1995). "Learning to Learn: Improving International Governance." *Global Governance* 1: 255-285.

Haas, P. M. and N. Kanie, Eds. (2004). *Dynamics of Multilateral Environmental Governance*. Tokyo, United Nations University Press.

Held, D., A. McGrew, et al. (1999). *Global Transformations*. Stanford, CA, Stanford University Press.

Hoffman, A. J. (2001). *From Heresy to Dogma*. Stanford, CA, Stanford University Press.

Hoffmann, S. (2002). "The Clash of Globalizations." *Foreign Affairs* 81(4): 104–114.

Holmberg, J., Ed. (1992). *Making Development Sustainable*. Washington D.C., Island Press.

Jervis, R. (1997). *System Effects*. Princeton, NJ, Princeton University Press.

Keck, M. E. and K. Sikkink (1998). *Activists Beyond Borders*. Ithaca, NY, Cornell University Press.

Kennan, G. (1970). "To Prevent a World Wasteland." *Foreign Affairs* 48(3): 401–413.

Keohane, R. O. and J. S. Nye (2000). "Globalization: What's New? What's Not? (And So What?)." *Foreign Policy*(118): 104–119.

Levy, D. L. and P. Newell (2000). "Oceans Apart? Business Responses to the Environment in Europe and North America." *Environment* 42(9): 8–20.

Levy, D. L. and P. J. Newell (2002). "Business Strategy and International Environmental Governance." *Global Environmental Politics* 2(4): 84–101.

Najam, A. (2003). "The Case Against a New International Environmental Organization." *Global Governance* 9: 367–384.

National Academy of Sciences (1972). *Institutional Arrangements for International Environmental Cooperation*. Washington D.C., National Academy of Sciences.

Ostrom, E. (2001). "Decentralization and Development: The New Panacea". *Challenges to Democracy: Ideas, Involvement and Institution*. K. Dowding, J. Hughes and H. Margetts. New York, Palgrave Publishers.

Oye, K., Ed. (1986). *Cooperation Under Anarchy*. Princeton, NJ, Princeton University Press.

Perrow, C. (1984). *Normal Accidents: Living With High Risk Technologies*. New York, Basic Books.

Reinecke, W. H. and F. M. Deng (2000). *Critical Choices*. Toronto, International Development Research Centre.

Ruggie, J. G. (2001). "The Global Compact as Learning Network." *Global Governance* 7: 371–378.

Schaeffer, R. K. (1997). *Understanding Globalization*. Lanham, MD, Rowman & Littlefield.

Simon, H. (1981). *The Sciences of the Artificial*. Cambridge, MA, MIT Press.

Skjaerseth, J. B. and T. Skodvin (2001). "Climate Change and the Oil Industry." *Global Environmental Politics* 1(4): 43–64.

Skolnikoff, E. B. and D. Kay, Eds. (1972). *World EcoCrisis*. Madison, University of Wisconsin Press.

Stokke, O. S. and Ø. B. Thommessen (2002). *Yearbook of International Co-operation on Environment and Development 2002/2003*. London, Earthscan.

Williams, M. (2001). "Trade and Environmental Commitments." *Global Environmental Politics* 1(4): 1–9.

World Commission on Environment and Development (1987). *Our Common Future*. New York, Oxford University Press.

12

LEARNING TO LEARN

Improving international governance[1]

Peter M. Haas and Ernst B. Haas

The character of contemporary international relations is far more complex than that confronted by the architects of the UN in 1945. Problems are interlinked to an extent previously unimagined by those designers, but the organizations they helped create were charged with solving problems as they were defined at the time of the founding. The UN was designed to cope with problems of the recent past that the architects felt could be treated as if they were discrete: collective security was designed to avoid the presumed military causes of World War II; balance-of-payments stability was pursued to prevent the competitive devaluations of the Great Depression. The current agenda now includes issues that were not imagined at San Francisco, including macroeconomic management, sustainable development, ecological disaster avoidance, and nuclear proliferation. The new problems reflected by these issues exist, to a large extent, by virtue of the successful reduction of barriers to trade and improved individual "quality of life" that the UN system helped to develop. Moreover, virtually all of the original concerns apparent at San Francisco—collective security, stable world commodity markets, public health, expanding trade, and stable currencies—remain with us.

The intertwined and interdependent dimension of contemporary international existence, characterized as the global "problematique," was recognized during preparations for the 1972 United Nations Conference on the Human Environment (UNCHE). In a preparatory report for the conference, the United States National Academy of Sciences observed:

> Environmental problems are beginning to lay a heavy and unfamiliar burden upon the organizations through which societies determine policies and make decisions. This happens at a time when the magnitude and complexity of social problems already critically tax the capacity of our institutions to deal with them effectively. Yet we are less

> likely to be faced with a sudden collapse of natural systems than with an accelerating decline into chaos with recourse to arbitrary power, as institutions falter under the pressures of problems that increasingly tend to become unmanageable.[1]

We wish to help in the design of international organizations (IOs) to ensure that they are able to respond quickly and effectively to these new challenges. The original organizational missions and aims of IOs need to be amended once it becomes apparent that those missions and aims have lost relevance because new problems and actors have arisen to claim a place on the global agenda, and even to shape it.

In the absence of a dominant state willing to lead, a strong shared universal vision, or a world government, collective responses to the global problematique depend on international institutional mechanisms. Only flexible institutions with expanding organizational visions can respond effectively to these problems and help guide their member states toward more productive governance likely to benefit the international community as a whole.

While ad hoc and disjointed responses to these challenges are likely to occur through most processes of international relations, robust and resilient responses are possible in multilateral settings characterized by well-developed processes of organizational learning. In such cases, many problems should still be resolvable through existing organizations, without having to construct new superorganizations.

We catalog the features of organizations that have successfully learned to improve their programmatic activities for environmental management in order to provide a positive example of organizational design for encouraging effective governance after the Cold War. We seek to determine to what extent lessons derived from these organizations' experience in dealing with environmental problems may be generalized to other international issue areas. Our approach differs from most prevailing theories of international institutions and governance because of its avowedly constructivist stance.[2] We focus on the shared beliefs that inform the practices of institutions, thus augmenting attention to the formal rules by which an exogenously determined set of values is authoritatively determined and applied. We regard a problematique as an intersubjective phenomenon and look at the social process by which knowledge informs visions and contributes to organizational practices. We also take seriously the organizational factors that enable a given institution to translate a shared vision effectively into actual governance. We develop a model to indicate the process by which epistemic communities contribute to organizational learning, and we develop a coding scheme to specify organizational factors that are likely to facilitate the conversion of shared visions into broad patterns of action.

The problematique: The nature of the challenge to governance

The setting in which international governance occurs has become more complex since 1945. In simpler times, the main role players were virtually only governmental delegates, especially when the main issues on the agenda of IOs were kept in almost watertight compartments: collective security, balance-of-payments stability, development aid and lending, commodity prices, decolonization, and human rights. New nonstate actors are becoming numerous and vocal, and experts no longer invariably mirror the preferences of their governments. Because of the world's growing reliance on the knowledge of specialists, networks of experts as learners and transmitters of knowledge have acquired enormous significance. States themselves are porous and often unable to satisfy their citizens by relying on their national capacities alone. State actors will remain the most important ones on the world scene, but their centrality and range of autonomous choice will decline in the face of transgovernmental, transnational, and non-governmental actors.

The scope of many current problems is now felt across space, time, and functional domain. A given set of interdependent problems, places, acts, and policies—such as those associated with sustainable development— is what we call a problematique. Because of their close linkages, several problematiques become the focus of analysis rather than each problem being taken discretely.

Many kinds of problems were traditionally viewed as purely local or, at the most, national in scope. Now they are seen as infecting entire regions or the entire globe, and the atmosphere and stratosphere as well. Effects of actions have both long-term and more immediate consequences.[3] The functional dimension for characterizing problems overlaps with both space and time. For instance, when we seek to specify the "problems" of peace, economic growth, sustainable development, or intergenerational equity, we must show how such diverse "places" as water, atmosphere, and terrestrial ecosystems interact with such human "acts" as agricultural and industrial production, armaments and the arms trade, and demographic trends. In turn, we explore how these look in various social time frames. Policy spaces, such as public health measures, macroeconomic choices, and military operations, now demand to be studied in terms of these complications. Issues no longer stay in tightly sealed compartments.

Not all issue packages in the global problematique are equally intertwined. Nondecomposable problems are those for which effective solutions must take account of all linkages. Partially decomposable problems have solutions that ignore some of the links and concentrate on others.[4] Full decomposability facilitates action, but the action is less and less frequently effective under conditions of complex substantive issue linkage. Decisionmakers in a learning organization must be able to tell the difference between various degrees of decomposability.

When things are seen as more complexly linked, the attribution of desired effects to single causes and linear forces of influence becomes untenable. Old organizational technical routines and patterns of resolving political differences become obsolete. For instance, when the continued convertibility of currencies becomes causally linked to the absence of budgetary deficits, low inflation, and economic growth at the national level, and the ensemble is linked to a certain volume of international lending, the definition of "macroeconomic management" becomes more complex. When the avoidance of aggression is seen as related to stopping civil wars, and domestic peacemaking to the guarantee of human rights, the causal network informing the so-called problem of peace becomes denser. Earlier notions of causality do not disappear; they are subsumed under the more complex picture.

Increased complexity may well increase uncertainty about effective remedies, as the certainties of the past are discarded. One core consequence of a more complex picture of causality is the need to link issues on negotiating agendas—issues that in simpler and earlier times were kept in their separate compartments. Complex causality implies greater uncertainty. The attempt to cope with greater uncertainty calls for a more complex linking of policy issues into comprehensive issue packages, which, in turn, complicates international negotiations because it calls for large package deals.

Why focus on learning?

We are concerned with describing the experience of IOs that grappled with the environmental problematique, illustrating activities that failed to do so, and offering suggestions on how to improve IO capacity to deal with this and other problematiques. Because most IOs in the past failed to cope adequately, learning to learn—that is, learning to do better—is the core of our concern.

True, states—or, more accurately, those who act in their name—may be steadfast about what they want to achieve by means of multilateral cooperation. They may merely wish to continue a pattern of action that motivated them when they created the IO. If actors' values and the interests to which they give rise do not change much, no learning may be called for. We live in the presence of the global problematique. Whoever is concerned with the interplay of resource use, population growth, processes of democratization, the desire for higher living standards and a better quality of life for all, and the management of civil and interstate conflict is engulfed by it. Any multilateral effort to cope with even a subset of these problems cannot escape its logic. Consequently, if we wish to understand how multilateral activity can address the problems that constitute the global problematique, we must show how learning can take place in IOs and how that learning can be diffused to the membership. IOs can help guide states through the challenging maze posed by the current international agenda by developing effective road maps for governance. IOs can become agents for the redefinition of global problems.

Learning defined and described

Learning is a political process whereby "consensual knowledge" is applied by policymakers to change their policy projects. We use the term learning as verbal shorthand to describe a process by which actors, playing roles, engage in institutionalized behavior that shows their recognition that problems currently on the

IOs' agenda are much more complexly linked than was recognized at an earlier time. Organizational learning is the process by which the learning becomes ingrained in IOs. Institutional learning is the broader international process by which state entities and other actors learn and assimilate some of these lessons. With regard to the environmental issues, learning entails an appreciation of complexity and an effort to integrate problems into a more comprehensive whole. In other issues it is possible that learning might entail an appreciation of simplicity as well as a process of issue subtraction on the policy agenda.

The capacity to learn is based on the willingness to make use of available (or obtainable) knowledge—the structured authoritative information about causes and effects. That knowledge is not simply uninterpreted data or information; it is information that has been subjected to a methodical analysis and arrangement, albeit epistemologically biased toward one of several possible notions of causality. As the content of the several problematiques grows denser and the presumed causal connections among pieces of knowledge more numerous, mastery of the totality of knowledge for purposes of global policy becomes an immense task.

That is why the degree of consensus about that knowledge is of vital importance. Consensual knowledge is structured information about causes and effects among physical and social phenomena that enjoys general acceptance as true and accurate among the members of the relevant professional community. To become consensual, information must be analyzed, arranged, and structured in accordance with epistemological principles that command wide acceptance in society. In our day and age, this has meant that the various strands of positivism enjoy a preferential position. Still, no consensus remains unchallenged for long. No body of knowledge is built on a permanent consensus regarding its substance or the procedures used to create it. We might expect an international order resulting from the process of institutional learning in any given policy area to be stable for perhaps a decade.

There are, of course, alternative modes of learning to the methodical analysis of the relationship between causes and effects using positivist principles.

A change in problem-solving behavior can also be induced by means of trial-and-error experimentation, without involving any profound study of causality. It can also follow a simple change in perceived interest on the part of actors, not necessarily the result of deep cogitation about cause and effect. Probably these more superficial modes of learning are encountered far more commonly than the reliance of consensual knowledge in multilateral problem solving. Neither of these processes is likely to provide adequate collective responses to the challenge of the problematique, because they yield gradual and partial policy approaches to a more complex shared vision.

Consensual knowledge is very helpful in shaping the program of an IO, but it is not absolutely essential. Competing bodies of knowledge, none of which is yet hegemonic, are also likely to precipitate self-reflection about past program failures, which is infinitely superior to trial-and-error learning and the mindless addition of new objectives and programs to the failed earlier ones.

Knowledge and multilateral problem solving

We think of the organizational learning process as animated by small groups performing roles in IOs on behalf of national bureaucracies, interest groups, business firms, and nongovernmental organizations (NGOs) that serve as advocacy groups. Typical roles include providing representation, articulating interests, furnishing informational feedback to the clients back home, and (occasionally) making decisions that matter. The role players include (1) instructed delegates representing their governments, (2) uninstructed expert consultants who speak for their professions, (3) lobbyists who articulate their clients' values and interests, and (4) members of IO secretariats interacting with all of these but receiving instructions (in principle) only from their executive heads. In simpler times, when the objectives of states did not change much, these role players only mirrored the interests, perceptions, and forces external to the IO. In our day we are concerned also with showing how role-playing inside the IO can shape events.

The most important role players purveying consensual knowledge are groups of like-minded professionals, usually self-recruited around some paradigm linking their lore to some aspect of a problematique. We call them "epistemic communities."[5] Not only do the members generally accept a common causal paradigm, but they also strive to insinuate that aspect of consensual knowledge into the public bureaucracies and legislative channels that produce public policy. One epistemic community, the Club of Rome, originally articulated the idea of the global problematique, which we regard as describing the character of contemporary international relations.[6]

Epistemic communities may function exclusively within single countries or as transnational entities, such as communities of atmospheric scientists or Keynesian economists. The advice of a dominant and authoritative epistemic community may temporarily monopolize the initiation of the learning process in an IO. Members of the group in national and international bureaucracies take the first steps in suggesting a new way to envision a problematique and offer possible ways to manage it.

In the absence of such a knowledge monopoly, competing epistemic communities seek to take over bureaucracies in order to promote their preferred knowledge scheme. Eventually, one group may capture the key national bureaucracy with regard to a particular issue, or perhaps even the entire national governmental apparatus responsible for the linked issues that make up the relevant problematique. Eventually, other governments are also "captured." They then form a coalition with the first captured to make the relevant multilateral organizations act as the agents of the new lore. Alternatively, they may capture an international secretariat and use it as a springboard to reach national governments by way of technical and financial assistance.

The process is ragged and almost never complete. Several problematiques usually coexist and may even compete in the nested system of IOs. Four problematiques have flourished in the UN since 1950: the devotion to collective security, economic development and the redistribution of wealth from North to

South, decolonization/democratization/human rights protection, and protection of the environment. The commitment to sustainable development suggests that a new metaproblematique may be forming from some of these four, as certain epistemic communities try to repackage elements of knowledge drawn from these earlier bodies of lore. It remains to be seen how tightly linked the causal chains among elements in that problematique turn out to be, or whether there remains a possibility of decomposing the elements into smaller clusters of linked issues.

Organizational learning requires that the efforts of epistemic communities be accepted and advocated by a coalition of hegemonic member states rather than being endorsed merely by majorities of weak states. After the programs have been validated by such a coalition, a learning IO then becomes an active transmitter of new ways of defining and solving problems by persuading most member governments of the appropriateness of the consensual knowledge involved. The ultimate stage in the evolution of learning comes when the IO is given something akin to executive power to induce member governments to accept the implications of that knowledge. Perhaps the fact that this stage is rarely reached is due to the fact that knowledge hardly ever remains consensual once it passes out of the control of the initiating epistemic community.

We acknowledge that our conceptualization of learning contains a rationalist bias, more consistent with the Western intellectual tradition than with cultural matrixes from other parts of the world. The appreciation and the use of structured knowledge are dependent on the learners' exposure to a modern education. If not science in its full cornucopia, then an appreciation of scientific research methods is probably a prerequisite for the utilization of consensual knowledge. Learning, in short, is associated with the prevalence of a culture in which specialists and experts with very modern educations (regardless of geographic location) are honored and respected by officials similarly educated. If decisionmakers in IOs or their constituencies do not share this bias, then they are unlikely to find persuasive our design for institutions that are able to learn. We are persuaded that our bias is justified because the epistemological presuppositions that accompany our notion of consensual knowledge seem to be emulated widely by non-Western cultures. They are the subject of demands and programs urged on UN specialized-agency by spokespersons from Asia and Africa.

How have IOs learned in the past?

Learning is rare. To learn is to put consensual knowledge to work defining and solving problems seen as interconnected. In most organizations, decisionmakers have failed to apply consensual knowledge to a more effective policy enterprise. Instead, they have tended to respond to environmental challenges through a process we call "adaptation," a response that fails to recognize the significant links within the problematique. To adapt is to change routines of problem solving without bothering to reexamine one's beliefs about cause and effect. No effective scan of the technical and scientific communities for new ideas is undertaken to muster political support for organizational reforms; rigid operating procedures are not adjusted to recognize a changed task domain.

Adaptation in IOs

Adaptation is by no means to be despised as suboptimal, or even as irrational behavior. When faced with disappointment about organizational effectiveness, actors typically respond by first altering the means they use to realize their common interests. If that change does not do the trick, the ends the program are to serve may come under scrutiny and be altered or mixed with new ends. What distinguishes adaptation from learning is the absence of any searching self-reflection about the proper way to compose, decompose, or recompose problem sets. It is unlikely to generate effective new organizational routines to cope with partially decomposable or nondecomposable problematiques. Adaptation typically follows one of two patterns: incremental growth or turbulent nongrowth. Incremental growth proceeds slowly, as an IO gradually acquires new tasks—while under conditions of turbulent nongrowth, the IO lurches from one highly politicized problem to another. The differences in organizational characteristics of the two patterns are described in the following section.

Characteristics of an IO capable of learning

Learning by and in the IO is not possible unless there is an unimpeded flow of ideas and information "upward" from universities, think tanks, national bureaucracies, and advocacy groups. Such groups serve as an early warning system of potential challenges to the organization as well as a conduit of new responses. Ideas and information circulate freely in a learning IO, as opposed to one that adapts. Contact between secretariat members and purveyors of ideas is continuous; executive heads do not attempt to limit or structure such contacts to suit their agendas. Successful learners are able to scan the technical or scientific community effectively, muster political support for organizational and program reforms, apply new approaches to rigid operating procedures, and affect significantly the domain for which those procedures were designed.

Controversy between advocates of rival world order ideologies is the essence of change. Learning IOs thrive on such controversy, provided no more than two ideologies confront one another, and provided the two are not so different as to prevent compromise altogether. Acceptance of sustainable development is compatible with liberalism and with dependency reduction, but liberalism probably cannot compromise with antidependency policies. Learning IOs are led by a dominant coalition of the member states that are most important for the enactment of the victorious ideology. The privileged coalition allows NGOs allied with them to represent their interests too but seeks to exclude NGOs that oppose the coalition's leadership. Decisions are made on the basis of consensus in order to avoid the divisiveness of majority voting that prevails in turbulent-nongrowth organizations and the qualified majorities and vetoes that dominate incremental-growth IOs.

Executive heads of learning IOs take strong initiatives in alerting the world to new problematiques. While being careful to remain close to the dominant coalition of important governments, they seek to persuade their leaders to back the initiatives proposed. Executive heads excel in mediating between important governments and in fashioning compromises between them. The secretariats may be somewhat penetrated by member governments but resist being staffed with political exiles. On balance, staff members are recruited on the basis of merit alone and remain largely autonomous despite some dependence on their states of origin. Expert consultants in learning IOs are members of epistemic communities who have captured all or part of the secretariats. NGO representatives are also increasingly used as consultants in fashioning new programs, as participants in program implementation, and as whistle-blowers and monitors.

Because of the unreliability of revenues raised on the basis of annual assessments, learning IOs prefer to rely on capital subscriptions or voluntary contributions from members of the dominant coalition, whether large or small states. They would gladly accept a form of taxation or user fee as an additional source of revenue. Budgeting avoids logrolling and disjointed incrementalism while stressing various forms of program budgeting. While these IOs, like incremental-growth and turbulent-nongrowth organizations, are able to administer their own programs with their own staffs, they also delegate administration to member states. More than the others, however, learners like to arrange for a sharing of tasks among members of the secretariat and member states on whose soil the program is being implemented. The "teaching" role of the IO is best carried out in this manner.

Learners also differ from adapters in the manner in which they construct consensual knowledge, combine knowledge with political goals, make decisions, strike bargains, and construct new problem sets. Learning IOs depend on consensual knowledge for upgrading their programs, while the adapting IOs do not. That is why epistemic communities and professional NGOs are so vital to their work. The relationship between member states' political goals and consensual knowledge is also different in learning IOs from the dominant pattern among adapters. We can conceive of such goals as discrete and stable at one extreme, and as rapidly changing into ever larger interconnected clusters at the other. The second situation obtains in learning IOs dedicated to mastering various problematiques. Decisionmaking must respect this state of affairs. Hence, decisions take into account knowledge-defined long-range factors rather than political expediency or immediate benefit. Decisions feature bargains based on linking diverse issues on the agenda; this is done by actors taking seriously the substantive causal connections among separate issues instead of linking them opportunistically merely to construct minimum winning coalitions. Because of the vital role of consensual knowledge, the various coalition partners can negotiate with one another in such a manner as to take for granted that most participants are motivated by similar objectives and act on the basis of generally shared and understood information.

One vital consideration in examining consensus among decisionmakers about the applicable problematique is the question of its decomposability. Most bureaucrats, especially in the adapting IOs, prefer to consider all sets as decomposable because this allows them to split a complicated problem set into small subsets that established units feel able to solve on the basis of established routines. Extreme decomposability is good for uninnovative bureaucrats and bad for solving problems. But complete nondecomposability means that every issue and problem is linked to every other issue and problem. Hence, only a very complex problematique does justice to the complexity of links. Taking it apart is to condemn reform to fail. But if we fear that no organization or epistemic community today has the intellectual and political power to solve such high-level clusters or nests of problems, then to insist on nondecomposability is the equivalent of wishing to fail. Learning IOs, therefore, avoid the Scylla of decomposability and the Charybdis of non-decomposability in favor of partial decomposability (or near nondecomposability), a halfway house of provisionalism likely to offend purists.

What about the authority and the legitimacy likely to be enjoyed by learners? Learning IOs increase their authority to mount programs that are more ambitious, more intrusive of state sovereignty, and more likely to make states dependent on them. It cannot be taken for granted, however, that the legitimacy enjoyed by such IOs in the eyes of their members will increase in proportion. Increases in authority may also trigger fear and dislike among clients and supporters.

Organizational learning and environmental management

Investigating organizational responses to the environmental crisis provides an ideal empirical test for our learning propositions because environmental media transmit problems broadly, creating new organizational challenges for organizations responsible for particular domains. Operational international organizations were challenged to upgrade their programs to reflect the growing concern about and appreciation for the problematique as heralded by the widespread environmental disasters of the 1960s and growing public demand in the West for a concerted response, especially after the 1987 publication of the Brundtland and Commission report highlighted this new concern with controlling the collateral environmental damage of economic growth. The Brundtland Commission report and the ensuing preparations for the 1992 United Nations Conference on Environment and Development (UNCED) constituted an organizational crisis for many UN agencies, who were publicly expected to demonstrate their green credentials. All organizations were forced by their various constituencies to reform their procedures in order to reduce environmental harm. Organizations also feared a loss of institutional turf to other agencies.

Below we consider the environmental response since 1987 of thirteen IOs: the nine principal operational agencies that are also the core group of the secretary-general's Inter-Agency Committee on Sustainable Development as well as the United Nations Fund for Population Activities (UNFPA), the International Maritime Organization (IMO), the World Food Programme (WFP), and International Fund for Agricultural Development (IFAD). Together, these organizations' budgets equaled $7.65 billion for the 1990–1991 biennium and were more than $6.4 billion for 1994.[7]

Organizations that learned

Yet only a small number of organizations can be said to have learned. Only the UN Environment Programme (UNEP) and the World Bank can be said to have fully learned to integrate environmental considerations with their traditional responsibilities. The original mandate of UNEP charged it with catalyzing the UN system into integrating environmental considerations into its activities while developing new approaches for sustainable development, monitoring environmental quality, training national officials in more environmentally benign development techniques, and developing international environmental law. In addition to fulfilling these charges, it now develops and publicizes ecosystem-based management for regional seas and river basins.[8] Since 1989, the World Bank requires environmental impact assessments on all its major projects, funds environmental protection projects, and helps to administer the Global Environment Facility for financing the share of development projects likely to improve global environmental quality.[9]

The World Meteorological Organization (WMO) and the World Health Organization (WHO) each demonstrates some learning but have not moved as far as UNEP or the World Bank. In 1990, following the Second World Climate Conference, the WMO redirected a significant proportion of its activities to research and monitoring of global climate change—a dramatic shift from its prior focus on weather monitoring. By 1993, climate and environmental activities accounted for 30 percent of the organization's scientific and technical budget and 13.5 percent of its overall budget. The WHO reveals its reorientation through the introduction of a wide variety of projects aimed at preventing a number of environmentally caused threats to public health. Innovative programs have been developed by UNESCO to harmonize species preservation with indigenous peoples' survival through the establishment of biosphere reserves as well as support widespread research on marine environmental quality. Most other organizations merely adapted to the environmental crisis by adding a few disjointed activities to their traditional package of activities. Table 12.1 summarizes the major organizational changes that have occurred.

The widespread introduction of environmental impact assessment procedures is actually a weak indication of learning, because these activities are seldom well integrated into the organization's overall activities.

Conducting such assessments often remains the responsibility of small and marginalized environmental units that lack leverage over the rest of the organization. It is only IOs that have installed environmental experts in operational divisions that can learn.

The World Bank provides an example of how these institutional characteristics help to promote learning. U.S. NGOs sounded the alarm in the early 1980s that large-scale Bank-funded projects in Brazil were contributing to massive destruction

TABLE 12.1 Efforts of organizations to internalize environmental considerations into their programs

Organizations that formally changed their organizational mandate to control environmentally destructive effects of its activities	Organizations that adopted new programmatic activities that capture many of the consensually identified causal links between traditionally mandated responsibilities and new problems identified within the problematique	Organizations that introduced Environmental Impact Assessment procedures
IBRD (1989 Operational Directive on Environmental Assessment)	UNEP (programs draw links between environmental quality and a wide variety of human activities)	UNEP, UNFPA (1989), IBRD (1989), FAQ (1989) UNDP, WFP, UNESCO, WHO, WMO, IFAD (1990)
UNEP (1973 original mandate)	WHO (new programs draw links between environmentally mediated threats to public health arising from a wide range of social activities)	
WMO	IBRD (considers environmental packages within structural adjustment lending)	

Sources: United Nations, *Everyman's United Nations: A Handbook on the United Nations. Its Structure and Activities* (New York: various years); Food and Agricultural Organization, *FAO Policies and Actions, Stockholm 1972–Rio 1992* (Rome: FAO, 1992); United Nations Development Programme, *The Challenge of the Environment: 1991 UNDP Annual Report* (New York: UNDP Division of Public Affairs, May 1992); Timothy Rothermel, "UNDP Plays Its Part," World Health (April 1986); UNESCO, *The Intergovernmental Oceanographic Commission: A Strategy for the Ocean*; UNESCO, *An Initiative of the Intergovernmental Oceanographic Commission: Global Ocean Observing System*; UNESCO, *Marine Science and Ocean Services for Development* (Paris: UNESCO, 4 January 1985); UNESCO, *Environment and Development* (informational packet distributed at Rio); Michael Mercier and Morrell Draper, "Chemical Safety: The International Outlook," *World Health* (August–September 1984); "Watchdog," World Health (March 1985); *Twenty Years after Stockholm: 1972–1992* (Berlin: Erich Schmidt Verlag, 1982); Ludwik A. Teclaff and Eileen Teclaff, "International Control of Cross-Media Pollution—An Ecosystem Approach," *Natural Resources Journal* 27 (winter 1987): 21–53; Alexandre Kiss and Dinah Shelton, *International Environmental Law* (New York: Transnational, 1991); Lee A. Kimball, *Forging International Agreement* (Washington, D.C.: World Resources Institute, 1992).

Notes: FAO-Food and Agriculture Organization; IBRD—International Bank for Reconstruction and Development; IFAD—International Fund for Agricultural Development; UNDP UN Development Programme; UNEP—UN Environment Programme; UNESCO—UN Educational, Scientific and Cultural Organization; UNFPA—UN Fund for Population Activities; WFP—World Food Programme; WHO—World Health Organization; WMO—World Meteorological Organization.

of the Amazon rain forest. With the U.S. government, they pressed the International Bank for Reconstruction and Development (IBRD) governing board to pay greater attention to environmental consequences of its funded development projects. Bank president Barber Conable hired new environmental personnel—at first establishing an environmental department, which became marginalized and lacked any leverage over the operational divisions, and later placing environmental staff in each of the Bank's operational divisions in which they drafted new environmental impact assessment procedures—and introduced training programs in environmental management for borrowers. By 1994, nearly two hundred environmental specialists had been added to the staff, many of them members of an ecological epistemic community with beliefs about the relations between environment and development that are significantly more holistic than those of the traditional economists and engineers on the staff. Since 1989, nearly 75 percent of the Bank's projects have been reappraised to compensate for potential environmental damage, and a structural adjustment loan to Madagascar (formerly the Malagasy Republic) was refined to support environmental administration and conservation. Advising and consulting ties between the Bank staff and NGOs have become much closer, and ecosystems experts are increasingly consulted in project design.

UNESCO offers a curious example of how units within an organization can learn while the organization as a whole is adaptive. The environmental units are less confined by the deadlocked higher-level politics of the organization and also interact with scientist experts belonging to ecological epistemic communities who are independent of close governmental briefings, unlike many of the relationships in the rest of the organization. UNESCO learned by creating the Man and the Biosphere Programme. The program is a holistic exercise in studying conditions and developing policies that improve human interaction with the natural and social environment.

The new programmatic activities undertaken by the learning organizations capture many of the causal links between traditionally mandated responsibilities and new problems consensually identified within the problematique. The programs' grasp of spatial dimensions varies according to the geographic nature of specific issues and related organizational responsibilities. At the same time, few programs address the temporal dimensions of these problems.

Learning organizations as teachers

The learning organizations tried to disseminate the lessons they had drawn about environmental management with other organizations and with states who rely on the organizations for operational activities.

The diffusion of learning between organizations has occurred primarily through three channels: interagency coordination, programs jointly administered with other agencies, and environmental monitoring. None have performed well in inculcating other IOs with a more comprehensive approach to environmental management.

Since its creation, UNEP has been responsible for coordinating environmental activities with other UN agencies in order to encourage them to integrate environmental concerns into their package of activities. Neither the Environment Co-ordination Board (1972–1977) nor the Designated Officials on Environmental

Matters (since) has been effective at persuading other agencies to take environmental matters seriously because of UNEP' s lack of organizational leverage within the UN system and the lack of financial resources to be used as incentives for other agencies to change their behavior. The World Bank has served a similar role in the Committee of International Development Institutions on the Environment (CIDIE) since 1980, with similarly unimpressive responses by other multilateral development banks. The Bank has the potential to leverage other financial institutions into environmental learning through its influence over jointly financed projects.

Incremental changes in other agencies' activities have occurred as a result of interagency programs initiated by UNEP, WHO, and WMO. These three bodies have coordinated joint activities with a number of other agencies in the UN system, instilling a seed for more comprehensive approaches elsewhere. UNEP has more than doubled the monetary value of its expenditures on programs through partnerships with other agencies, although many of these have occurred in tandem with WHO and WMO. A distinguishing feature of UNEP's joint ventures is its enthusiasm about including the scientific community along with environmental, grassroots, and corporate NGOs in its activities. As a consequence, NGOs gain access to organizations of whom they are suspicious, yet they do not feel that they are compromised by association due to UNEP's insulating role.

Organizations like UNEP, WHO, and WMO also monitor the quality of the environment, thereby alerting other agencies to problems falling within their purview. These outreach efforts by learning organizations appear to have, at best, the effect of stimulating or reinforcing adaptive efforts in other agencies.

Learning IOs teach states

These learning organizations exert influence on the states that rely on their operational activities. Organizations that have learned have helped national bureaucracies learn in. several ways—leading in turn (at times) to changes in national policy from which individuals and firms have changed their behaviors as well, in ways that are more environmentally friendly.[10]

Organizations exercise a demonstration effect by which they identify and justify policies that national decisionmakers may adopt. For instance, the recommendation by the WHO regarding exposure limits for various chemicals has served as the basis for legislation and policy in many developing countries. While this organizational activity is, properly speaking, more often a case of imitation by governments rather than the reflection and the application of newly discovered consensual knowledge at the national level, it is an important channel by which organizational learning is converted to new measures on the ground that may more effectively manage environmental risks.

IO programs for national officials train them to adopt and apply the new techniques that have been developed or identified by the organizations. Persuaded of the viability of such measures, they may serve as a political constituency within national administrations for the adoption and enforcement of the organizationally identified measures. Public education efforts serve a long-term function of changing individual consumption habits, while contributing in the shorter term to the creation of new domestic constituencies for environmental protection, at least in democratic societies where they may have influence over governmental activity.

Finally, many national bureaucrats and scientists learn by doing as they participate in projects coordinated or funded by the organizations. Thousands of developing country officials and scientists have attended UNEP-sponsored training seminars in environmental monitoring and comprehensive approaches to environmental management. International financial institutions and institutions with resources that are highly desirable to national governments can offer linkages to encourage countries to adopt and comply with the new lessons imparted by learning organizations.

Conclusions about organizations that learned

Almost all IOs have responded to crises only after the troubles have become painfully obvious, rather than anticipating them. Ideally, of course, a learning IO would possess the ability to head off crises, presupposing the capability to recognize crisis-producing conditions before the emergency erupts and to bring flexible new exercises to bear.

The difference between IOs that functioned in the two adaptive modes and IOs that learned is in the scope of their response; learning IOs redefined their missions in light of the new interdependencies to their original mission that the crisis helped to illuminate, whereas adaptive organizations only introduced slight modifications to their standard routines. A number of identifiable features characterize the international institutions that learned to develop more comprehensive environmental management efforts in response to crises. These features are largely absent from the institutions that merely adapted. The variation between organizations can be seen in the appendix. Learning organizations were able to promptly assess changes in their policy environment through their own monitoring systems, through an open flow of environmental information from the scientific community or local NGOs, or from other IOs on whom they rely for information about environmental quality (such as the way the World Bank relies on UNEP and UNEP's joint monitoring programs, with other IOs). They were able to apply nonpartisan consensual knowledge to the problem by soliciting information from ecological epistemic communities. They were staffed by relatively autonomous and capable secretariats who were able to obtain authorization from their governing bodies (which were not riven by irresolvable political disputes) and disseminate their advice through networks and the publicity commanded by their authority and legitimacy.

It seems likely that IOs learn more effectively if they are preponderantly influenced by member states that possess a democratic culture, because such states tend to be less dogmatic and more flexible. The corollary of this assumption is the expectation that IOs learn better if they are dominated by member coalitions that favor the free flow of ideas and information, or are at least indifferent to imposing an ideological orthodoxy.

Information used by IOs frequently comes from knowledge claimants such as scientist experts and NGOs. Scientist experts should be widely consulted and should be largely independent of guidance or be part of an epistemic community. The organization should rely on consensual knowledge for designing new activities—while continuing to monitor national performance through extensive consultations with governments and NGOs—and on reporting by governments and NGOs. The organization should at least be able to engage in adverse publicity to stigmatize nonperformers and to monitor the quality of its policy domain in order to evaluate performance. If NGOs serve as complainants, new information will also be made available.

We now offer a set of tentative conclusions about how lessons learned in and by IOs can be transmitted most effectively to their member states. If an appreciation of modem education and science seems a likely prerequisite for learning in an IO, the same holds true for the culture of member states that is expected to benefit from programs that incorporate prior learning. Lessons are more likely to be learned if much of the population no longer lives in a premodern and preindustrial manner, if it is largely "socially mobilized." The existence of a democratic culture, usually associated with advanced social mobilization, is helpful but not essential. The prevalence of competing political parties and interest groups able to communicate with the bureaucracy and with elected politicians certainly favors the circulation of new ideas and modes of thinking. However, exposure to lessons transmitted by IOs is still possible in countries that also have socially mobilized populations but lack democratic cultures, if the bureaucracy is highly centralized and determined to diffuse the lessons. Conversely, a centralized-authoritarian bureaucracy can block the diffusion of lessons, whereas this is much more difficult for a democratic bureaucracy. Consensual knowledge is likely to be given an appreciative hearing by politicians, whether democratic or not, desperately looking for solutions to crisis problems. Centralized, nondemocratic states may be capable of a more rapid application of lessons learned from IOs than their democratic counterparts, which must rely on building at least a minimal domestic consensus behind their new proposals, although the lessons may be more enduring and remain more firmly "stuck" in democratic societies. An interest by political leaders in using such knowledge remains an essential prerequisite for effective transmission of lessons in any political setting. Without such a motivation, consensual knowledge will not find its way into national programs.

How widely can the environmental management problematique be generalized?

Our discussion of organizations that learned to manage the environmental problematique and successfully passed their lessons on to the member states has been confined to activities that deal with the problems of late capitalism/industrialism.

But this is a highly limited domain. Dealing with environmental degradation in the context of the economics of highly developed (and mostly democratic) countries is different from prescribing institutions capable of learning when we deal with such things as sustainable development, democratization, national self-determination, and collective security. These environmental lessons are not easily transferable to other issues because the environmental issues share a set of features that facilitate organizational learning but that are not widely encountered.

For learning to occur, there must be value consensus and a stable knowledge base

Issue areas vary with respect to the extent of consensual knowledge available for conceptualizing and managing them. They also differ with respect to the extent of the political and value agreement about the issues. Before assigning problems or problematiques to IOs that have been able to learn in the past, we must be certain that it is in the nature of the problem "to be learned." If we do not observe this stricture, we will undoubtedly overload IOs with tasks; the inability to carry them out will diminish the authority and undermine the legitimacy of international institutions. But even in the absence of the ideal learning pattern, it is still necessary that new and old tasks continue to be carried out; IOs able to adapt fall short of our ideal, but they ought still to be able to handle this kind of problem.

For learning to occur through the institutionalization of ideas held by epistemic communities, two key background conditions must be satisfied. The values of dominant decisionmakers and epistemic communities must agree, otherwise governments will not tolerate the policy innovations recommended by secretariats and executive heads. Technical consensual knowledge must exist, the claims of which can be intersubjectively evaluated by secretariats, governments, and their advisory experts. In the absence of this condition, knowledge claims for policy will lack any legitimacy and cannot claim any authority with putative institutional reformers. Overall, the domain in which our learning model is likely to command explanatory power and possess some real utility for institutional design is confined to the upper two left cells in Table 12.2 within the domain of the organization's established mandates. The diagram indicates issues and organizations in which learning may be possible.

These organizations may fit in different cells when confronting various problematiques, including the environmental problematique we discussed earlier. These organizations may fit in different cells when confronting other problematiques.

Making Sustainable Development into a problematique that is conducive to a learning experience

Few dispute the knowledge that establishes causal links between styles of economic development, pollution, ill health, and ecosystemic health. The ties between ecological problem sets and a broader economic problematique—Sustainable Development (SD)—encompass costs of industrial production and the competitiveness

TABLE 12.2 Domain of the learning model

	Extent of technical consensual knowledge		
	High	*Medium*	*Low*
Disciplinary domains	Natural science, engineering, ecology	Economics	Humanities, law, social sciences
Values of dominant decisionmakers, NGOs, and epistemic communities agree	Organizational change via leaning is most likely	Organizational change via leaning is possible	Organizational change via learning is impossible, but adaptation is possible (i.e., elimination of slavery)
Values of dominant decisionmakers, NGOs, and epistemic communities disagree	With high saliency and uncertainty, organizational learning is possible, but institutional responses may be reversed. With low saliency and low uncertainty, learning is impossible	Adaptation is possible	Adaptation is possible (i.e., human rights, collective security)

Note: Disciplines at the right end of the scale have no epistemic communities.

of a nation's industries in international trade. This problematique, however, is much more contested in value and in knowledge terms than environmental management taken alone. Equally important, to the extent that this consensus is weak, those who purvey that knowledge to politicians—scientists, engineers, economists—are less of a privileged group in their access to policymaking than are articulate and well-informed interest groups.

Despite a number of pre-Rio conferences at which ambitious statements were issued about the need to develop and apply systematic efforts to internalize environmental considerations into organizations' traditional package of activities, very few of these efforts have yet been introduced. For many organizations, Rio and Sustainable Development were merely an opportunity to repackage their traditional activities in the wrapping of environmental support, and they have only reluctantly pursued some of the measures proposed at the nationally hosted conferences in which they participated.

Sustainable Development, while now characterized by value dissensus and less-than-consensual knowledge, nevertheless might become the kind of concept that, when made part of its mandate, could enable an organization to become a true learner. Introducing global institutions to turn SD into more than a value-laden slogan, to turn it into an ongoing activity, implies upgrading environmental management by adding new issue areas and new connecting tissue among them. Such upgrading calls for the augmentation of environmental management with development economics, planned technology transfers, resource allocation, and resource planning that takes the future needs of all of humankind into account. In short, changing SD into an organizational mandate—and creating the setting of a major learning experience for the organization so blessed, and of its member states—is an act of creating a nearly nondecomposable problem set from what was previously thought to be a series of decomposable ones. UNCED suggested that the world approves of such a huge act of conceptual and programmatic aggregation.

SD then implies that hitherto separate substantive issues and disciplines be combined. It also implies that the concepts and methods needed to link these disciplines analytically be invented. Only thus can the causal patterns thought to be operative be highlighted. Yet, unlike environmental management, SD remains a highly contested concept until a value consensus about its nature emerges. How can institutions be changed to advance such an agenda?

1 Forums, such as the International Council of Scientific Unions and its committees, could be made to proliferate. To do so would be to generalize all over the world the privileged position occupied by scientists and engineers in industrialized countries. It would provide opportunities for contact and discussion from which a more global substantive consensus might emerge.
2 Local interests ought to be increasingly empowered to contribute to policy debates on SD issues. Empowerment is likely to confer increasing legitimacy and authority on IOs that seek to practice and teach SD, because the lessons imparted to national governments will have the blessing of local interests. Advice and support from IOs will no longer fall into a national vacuum. Western and non-Western attitudes, thus focused on a common and urgent problem set, may be made to overlap.
3 The representation of nongovernmental interests at the IO level ought to be enhanced, thus giving a number of such interests (labor, industry, consumers, and trade associations as well as ecological advocacy groups) a direct voice in the elaboration of international measures with direct impact on their ecological and economic interests. The model of the International Labour Organization (ILO) might be kept in mind here, or of the institutionalized role of such interests in the International Telecommunications Union (ITU).

Not all problems are as tightly linked as the environmental problematique suggests

Once the likelihood of a crisis has been recognized, care must be taken to conceptualize the resulting new problematiques so as to make eventual national and multilateral action possible. Understanding the crisis demands that all available causal schemas, no matter how unpopular, be examined for their relevance. Knowledge used must include science, but not all applicable knowledge must be "scientific" in terms of the logic, methods, and causal problems found relevant. Coping with the crisis calls for the design of policies that are not so complex as to make success depend on the effectiveness of every single component of the plan.

For example, it is currently fashionable to combine military security issues with ecological ones by speaking of an "international security" problematique, and to substitute "cooperative security" for the more familiar collective security.

Such overaggregation suggests that ecological problems escape solution unless we banish war, that peacekeeping or peacemaking are not possible unless we achieve Sustainable Development. To think of all the possible problematiques as constituting a single system of causes and remedial actions is to learn the wrong lessons.

Earlier we argued that not all issues are equally nondecomposable. Learning ought to develop the institutional ability to treat complex problems as if they were partly nondecomposable, and develop appropriate responses, rather than commit the errors of treating everything as an integral system or as totally decomposable. Effective learning should involve the institutional ability to judge the extent to which a given problematique is wholly or partly decomposable. Is each link in a functionally conceived chain of causation truly necessary as the focus of policies that seek to deal with the problematique (nondecomposability)? Or are solutions conceivable that ignore some of the links and concentrate on others (partially decomposable)? The answer is crucial in the design of public policy in complex situations involving great uncertainties. If we believe that some links matter more than others, then we simplify our problem because the policies can be concentrated. We also save organizational resources and reduce apparent uncertainty.

But if we believe that we have to understand the entire system before we can act to influence any part of it, and if we also think that every part depends causally on every other, then we cannot disaggregate or unlink any component—spatial, temporal, or functional. The latter vision is particularly gruesome for political architects of international governance: there is still no political constituency behind such a grand vision (absent gaia), and virtually no one can suggest how to formulate policy effectively when every action influences everything else, much less administer such policies.

Luckily, we feel, there is growing consensus that not all problems are equally interconnected—they are partially decomposable—and the specialized agencies appear to be learning to assemble more comprehensive measures recognizing the problematique one piece at a time. Analysts from many developing countries express legitimate concerns that environmental problematiques be cast sufficiently broadly to include the prevailing economic development styles associated with patterns of environmental degradation.[11]

There are many other actual and potential problematiques less inclusive than the ones made popular decades ago by the Club of Rome. Some seem poor candidates for any kind of learning, because they lack the necessary value consensus and consensual substantive knowledge.

When we have good reason to suspect that a value consensus remains an elusive goal, the institutional devices suggested above are unlikely to be useful in developing more comprehensive institutional programs. Our prescription for the institutional pursuit of nearly nondecomposable problem sets ought to be rejected when we have reason to fear that substantive issue aggregation—no matter how easily justified by Club of Rome-type models—is not accepted by many experts and does not enjoy widespread support. Decomposability ought to be pursued instead. The issue areas of collective security, human rights protection, democratization, and the fostering of ethnic self-determination do not warrant being made into a single comprehensive problematique, or expressed in a single highly aggregated UN program. We take this position because we are convinced that these fields are more highly contested on moral grounds than even SD and that the effort to make integrated UN programs of them would condemn the UN to overload, disappointment by member governments, disrepute, and failure.

Summary and conclusion

We have attempted to show that by no means all kinds of knowledge—and the human collectivities that provide and diffuse it—are likely to lead to learning to manage interdependence more effectively. We have stressed that the knowledge apt to lead to learning is not universally shared or even available everywhere. And we have argued that there are many kinds of knowledge and policies that are not likely to become more consensual or more interconnected and nearly nondecomposable. We have urged that organizations charged with missions that reflect sharply contested knowledge and values be left as they are currently constituted.

What lessons for institutional reform may be drawn from this study to improve global governance? The limiting conditions for organizations that can learn appear to be the absence of irreconcilable political differences among the dominant member countries and stable, impartial information flows to an effective secretariat.

Organizations that learned were ruled by a dominant coalition whose members were in agreement on the main principles of world order. Effective discussions are hamstrung, and political compromises by which more comprehensive programmatic missions may be crafted are impossible without such agreement. At a minimum, learning requires an institutional design that provides for the provision of nonpartisan scientific information about the state of the physical environment, the regularized feedback of information regarding activities by governments and firms, and the building of developing countries' capacity to conduct monitoring and research and to apply it indigenously to their policy process. Parties should be able to keep track of each other's activities and hold governments accountable for enforcing their international commitments.

Secretariats ought to be recruited on the basis of merit alone. Connections between secretariats and NGOs and the scientific community, from which new ideas and warnings can be received, should be close, especially when monitoring is delegated to NGOs. Both greater independent monitoring of national performance to augment reports by governments and more public education to inform groups about problems also benefit from close secretariat/NGO ties. Closer administrative ties to local authorities may accelerate the diffusion of institutional learning.

Learning and institutional reform must come from within. Both learning, and efforts to educate others by organizations other than UNEP, IBRD, WHO, and WMO have largely proved to be failures. Other bodies should be reformed to allow for greater input from more diverse groups, which might eventually result also in the acquisition of mandates to monitor national performance. Organizations characterized by irreconcilable disagreements over desirable world orders or ineptitude may not even be made capable of learning to manage interdependence, rather than merely adapting to it.

But it is clearly not possible to foretell with any confidence which fields are likely to become more consensual. We cannot guess which conceptual and cognitive breakthroughs that now seem unlikely may still come about in the not-too-distant future. Hence, what we really need are IOs that are flexible enough to learn new interconnections and profit from new interdependencies among functions, values, time periods, and places, even though we cannot foresee the problematiques to which they might be responding. The truest learning organization, we believe, is the one blessed with people and institutional routines that will recognize and identify such brand-new problematiques before the problems have become too serious to yield to a multilateral response.

Notes

Ernst B. Haas is Robson Research Professor of Government at the University of California, Berkeley. He writes on international institutions and international change, reason, and learning. His most recent book is *When Knowledge Is Power* (1990). He is currently working on nationalism. Peter M. Haas is associate professor of political science at the University of Massachusetts at Amherst. He is the author of *Saving the Mediterranean: The Politics of International Environmental Cooperation* (1990), editor of *Knowledge, Power and International Policy Coordination* (1995), and editor (with Robert 0. Keohane and Marc A. Levy) of *Institutions for the Earth: Sources of Effective International Environmental Protection* (1993).

This article was originally prepared for the Commission on Global Governance, Geneva. We acknowledge the valuable research assistance of Nicola Poser and Sean Kay in helping collect and code information about environmental management. We appreciate the advice about the makeup of international organizations offered by Roger Coate, Barbara Crane, Ray Hopkins, Ernest Landy, Karen Mingst, Ron Mitchell, Craig Murphy, John Perry, Larry Scheinman, Susan Sell, and Mark Zacher. The interpretations remain our own.

* This chapter was first published in *Global Governance*, Vol. 1, No. 3 (1995) pp. 255–284.

1 *Institutional Arrangements for International Environmental Cooperation*(Washington, D.C.: National Academy of Sciences, 1972), p. l.
2 *Our take on soc*ial constructions of "reality" are akin to those of Donald T. Campbell, *Methodology and Epistemology for Social Science* (Chicago: University of Chicago Press, 1988); Richard Rorty, *Objectivism, Relativism, and Truth* (Cambridge: Cambridge University Press, 1991); Hilary Putnam, *Reason, Truth and History* (Cambridge: Cambridge University Press, 1981); and Stephen Toulmin, *Human Understanding* (Princeton: Princeton University Press, 1972). For empirical confirmation of this approach, see Ronald N. Giere, *Explaining Science* (Chicago: University of Chicago Press, 1988).
3 John Gerard Ruggie characterizes this dimension of modernity as "social time." See Ruggie, "Social Time and International Policy," in Margaret Karnes, ed., *Persistent Patterns and Emergent Structures in a Waning Century* (New York: Praeger, 1986); also see Ruggie, "International Structure and International Transformation," in James N. Rosenau and Ernst-Otto Czempiel, eds., *Global Changes and Theoretical Challenges: Approaches to World Politics for the 1990s* (Lexington, Mass.: Lexington Books, 1989).

4 This entire section is an adaptation of Herbert Simon's work. See, for one of many sources, his "Rationality in Psychology and Economics," in Robin M. Hogarth and Melvin W. Reder, eds., *Rational Choice* (Chicago: University of Chicago Press, 1986).

5 Peter M. Haas, "Do Regimes Matter?" *International Organization* 43, no. 3 (summer 1989); Ernst B. Haas, *When Knowledge is Power* (Berkeley: University of California Press, 1990); Peter M. Haas, ed., *Knowledge, Power, and International Policy Coordination* (Columbia: University of South Carolina Press, 1995).

6 Donella H. Meadows, Dennis L. Meadows, Jørgen Randers, and William W. Behrens III, *The Limits to Growth* (New York: Signet, 1972); Mihajlo Mesasrovic and Eduard Pestel, *Mankind at the Turning Point* (New York: Dutton, 1974); Jan Tinbergen, *RIO: Reshaping the International Order* (New York: Dutton, 1976); Aurelio Peccei, *The Human Quality* (Oxford: Pergamon, 1977); and James W. Botkin, Mahdi Elmandjra, and Mircea Malitza, *No Limits to Learning: Bridging the Human Gap, A Report to the Club of Rome* (New York: Pergamon, 1979). More recent reports on the subject include Alexander King and Bertrand Schneider, *The First Global Revolution* (New York: Pantheon, 1991), and Jim MacNeill, Pieter Winsemius, and Taizo Yakushiji, *Beyond Interdependence: The Meshing of the World's Economy and the World's Ecology* (New York: Oxford University Press, 1991). The history of the global-modeling research program, which was closely associated with the global problematique, is reviewed in Donella Meadows, John Richardson, and Gerhart Bruckmann, *Groping in the Dark* (Chichester, U.K.: John Wiley & Sons, 1982). For critical rejoinders to the Club of Rome's formulation of the problematique, see Sam Cole, *Global Models and the International Economic Order* (New York: Pergamon, 1977); Christopher Freeman and Marie Jahoda, eds., *World Futures* (New York: Universe, 1978); Brian P. Bloomfield, *Modelling the World* (Oxford: Basil Blackwell, 1986); Richard K. Ashley, "World Modeling and Its Politics," *International Organization* 37, no. 3 (summer 1983): 495–536; Nicholas N. Onuf, "Reports to the Club of Rome," *World Politics* 36, no. 1 (October 1982): 121–146.

7 Figures for the 1990–1991 biennium from "Program Resources of the United Nations System," UN Document E/1991/42/Add. l, 10 April 1991, pp. 7–9, cited by Lee A. Kimball, *Forging International Agreement* (Washington, D.C.: World Resources Institute, 1992), p. 36. The 1994 budget figures were calculated from the Fridtjof Nansen Institute, *Green Globe Yearbook 1994* (Oxford: Oxford University Press, 1994). While budgets are not the same as resources, particularly for the multilateral development bodies, they do provide a rough indicator of the degree of organizational energy committed in this area and the extent to which it has grown more vigorous over time.

8 Mostafa K. Tolba, et al., *The World Environment 1972–1992: Two Decades of Challenge* (London: Chapman & Hall, 1992); UNEP, "UNEP Profile," Nairobi, UNEP, 1987; Peter M. Haas, "Institutions-United Nations Environment Programme," *Environment* 36, no. 7 (September 1994): 43–45.

9 Bruce Rich, *Mortgaging the Earth* (Boston: Beacon, 1994); Susan George and Fabrizio Sabelli, *Faith & Credit* (Boulder: Westview, 1994); Herman E. Daly, "Farewell Lecture to World Bank," 14 January 1994, mimeo. Strikingly, traditional critics of the World Bank have not been overly skeptical of the Bank 's efforts to integrate environmental considerations into its activities.

10 See Peter M. Haas, Robert O. Keohane, and Marc A. Levy, eds., *Institutions for the Earth: Sources of Effective International Environmental Protection* (Cambridge: MIT Press, 1993); Peter M. Haas, *Saving the Mediterranean: The Politics of International Environmental Cooperation* (New York: Columbia University Press, 1990).

11 For instance, Osvaldo Sunkel, "Beyond the World Conservation Strategy: Integrating Development and Environment in Latin America and the Caribbean," in Peter Jacobs and David A. Muqro, eds., *Conservation with Equity*, proceedings of the Conference on Conservation and Development, Ottawa, Canada, 31 May–5 June 1986; Latin American and Caribbean Commission on Development and Environment, *Our Own Agenda* (Washington, D.C., and New York: Inter-American Development Bank and United Nations Development Programme, 1991).

APPENDIX Environmental management problematique: organizational learning or adaptation, 1987–1992

	UNEP	*IBRD*	*WMO*
IO characteristic			
World order ideology	moderate dependency reversal v. liberalism, compromise possible	liberalism modified by selective incentives to LDC governments	weather prediction v. global change
Mode of representation	equality biased toward Third World, NGOs represent	sharply stratified by state power, NGOs do not represent	equality of states
Executive head's leadership	crisis manager	crisis manager	reactive to LDCS
Secretariat	recruitment by merit, unpenetrated	recruitment by merit, unpenetrated	recruitment by merit, unpenetrated
Status of experts	represent epistemic communities	represent epistemic communities	represent epistemic communities, independent
Source of revenue	annual assessments, voluntary contributions, other IOs	capital subscriptions, borrowing	annual assessments, UNDP voluntary contributions
Administration of local tasks	mostly indirect	shared	indirect, shared
Monitoring state compliance	consultations, regular reporting, publicity sanction	consultations, ad hoc reporting	irrelevant
Role of NGOs	lobbying, advisers to staff, consultants, participation in meetings	advisers to staff, consultants	lobbyists, ICSU personnel as program administrators
Knowledge	more consensual	more consensual, integration	toward consensus
Mode of IO decisionmaking			
Organizational experience	extensive programs covering many environmental/human interactions	commitment to reducing environmental externalities of loans, stress their environmental benefits	sharp increase in resources devoted to climate and environmental programs
Issue linkage	mostly fragmented, some substantive	fragmented, becoming more substantive	fragmented, some substantive
Problem definition	nearly nondecomposable sets authority	nearly nondecomposable sets	nondecomposable
Authority, legitimacy	growing, legitimacy unclear	authority growing, legitimacy disputed in parts of Third World	authority declining

(Continued)

APPENDIX (Continued)

	UNDP	*U NFPA*	*IAEA*
IO characteristic			
World order ideology	liberalism v. dependency reduction	resource conservation, welfare enhancement	AIC nuclear hegemony v. dependency reduction
Mode of representation	donor–recipient balance, no NGO representation	same as UNDP	stratified by state power, no NGO representation
Executive head's leadership		reactive/passive	reactive to AIC members
Secretariat	recruitment by national quotas, some penetration	recruitment by national quotas, no penetration	recruitment by national quotas, exile staffing, penetration
Status of experts	independent	epistemic community	instructed by states, some epistemic community
Source of revenue	voluntary contributions	voluntary contributions, UNDP	annual assessments, voluntary contributions
Administration of local tasks	shared	shared	direct
Monitoring state compliance	consultations	evaluations	regular reporting, inspections, sanctions
Role of NGOs	lobbyists, project administration	consultants, project administration	none
Knowledge	not more consensual	less consensual	consensus improving
Mode of IO decisionmaking			
Organizational experience	environmental externalities to be considered in project design	guidelines for conservation strategies, family planning program	radioactive emissions, radiation safety, reactor improvement
Issue linkage	none	none	fragmented
Problem definition	decomposable	decomposable	decomposable
Authority, legitimacy	both low	authority high, legitimacy mixed	both improving

	WHO	*ILO*	*UNESCO*
IO characteristic			
World order ideology	"primary health care," "health for all by 2000"	social-democratic reformism	liberalism v. dependency reduction, redistribution
Mode of representation	stratified by power of states, NGOs can on penetrate delegations	stratified by power of states, NGOs represent their national interests	egalitarian, NGOs can serve delegations
Executive head's leadership	crisis manager	reactive to G-77	M 'Bow reactive to G-77, Mayor reactive to West
Secretariat	recruitment by merit, full autonomy	recruitment by merit, some penetration	recruitment by national quotas and exiles, heavily penetrated
Status of experts	represent epistemic communities	NGO-instructed	independent, instructed, some epistemic communities
Source of revenue	annual assessments, UNDP, voluntary contributions	same	same
Administration of local tasks	mostly indirect	direct	shared
Monitoring state	regular reporting, some complaining,	regular reporting, complaining,	consultations, some reporting
compliance	investigations, publicity sanction	hearings, investigations, publicity sanction	
Role of NGOS	consultants, lobbyists, complainants	lobbyists, legislators, complainants, mediators	lobbyists, consultants, some administration
Knowledge	not more consensual	not more consensual	more consensual
Mode of IO decisionmaking			
Organizational experience	public health redefined to include environmental hazards	environmental threats to health/safety	programs on marine pollution, species preservation, mapping biomes
Issue linkage	little linkage	tactical	tactical
Problem definition	nondecomposable	decomposable	decomposable, some nearly nondecomposable
Authority, legitimacy	both high	authority low, legitimacy high	both low

(Continued)

APPENDIX (Continued)

	FAQ	*IFAD*	*WFP*	*IMO*
IO characteristic				
World-order ideology	Liberalism v. dependency reduction	dependency reduction		liberalism
Mode of representation	stratified by power of states	tripartite egalitarian, NGOs do not represent		stratified by maritime state power, NGOs participate in meeting
Executive head's leadership	reactive to G-77		crisis manager	passive mediator
Secretariat	recruitment by national quotas, exile staffing, partly penetrated			recruitment by national quotas, somewhat penetrated
Status of experts	mostly independent		NGO-instructed	NGO-instructed
Source of revenue	annual assessments, UNDP, voluntary contributions, subsidies from firms	voluntary contributions	voluntary contributions	annual assessments, UNDP
Administration of local tasks	shared		shared	direct
Monitoring state compliance	consultations	none		regular reporting, publicity sanction
Role of NGOS	lobbyists, consultants		lobbyists, consultants, complainants	lobbyists, consultants
Knowledge	not more consensual			toward more consensus
Mode of IO decisionmaking				
Organizational experience	environmental projects added to earlier tasks, not more consensual	applies environmental principles to projects	seeks to avoid harmful environmental effects from normal projects	ship-caused marine pollution
Issue linkage	tactical			tactical
Problem definition	decomposable	decomposable	decomposable	nearly nondecomposable
Authority, legitimacy	both low			both high

PART IV
Effectiveness

The effectiveness of international governance efforts depends on states enforcement of their commitments. These two pieces discuss the influences of the international community on national decisions.

"Choosing to comply" (Chapter 13) frames the question in terms of states' concern about, and capacity for, addressing global environmental challenges. Countries vary in terms of their administrative capacity, and in terms of their national concern about environmental problems. Thus the sensitivity of states to international pressures depends upon their domestic attributes. The chapter breaks down this insight to identify which kinds of institutional pressures are likely to induce what kinds of states to assume their international responsibilities, or to engage more vigorously in international negotiations.

"Evaluating the effectiveness of marine governance" (Chapter 14) elaborates the points made in Chapter 13. This piece was originally solicited by PEMSEA, and was presented at a policy workshop in Lisbon, Portugal in December 2008, convened by the government in Portugal. It identifies a number of international interventions which can enhance capacity and concern in different types of countries via providing information, empowering domestic constituencies, and providing material resources.

13

CHOOSING TO COMPLY

Theorizing from international relations and comparative politics[1]

Peter M. Haas

1 Introduction

How to understand compliance with soft law? In this chapter I discuss possible elements for a research program on understanding patterns of compliance with international soft law, focusing on manipulable variables which may influence individual states' decisions to comply. It begins with conceptual discussions of compliance and soft law, followed by a review of perspectives from comparative politics and international relations regarding what makes individual states choose to comply with soft law instruments. The domestic and state-based account offers a baseline estimate of individual states' propensity to comply with a particular instrument of soft law, in isolation of any other influences, but such an analysis is inadequate for most contemporary efforts because domestic influences do not operate in a vacuum; systemic and transnational factors now exercise a strong influence on the origin of many domestic factors and the political context in which they operate.

By 'compliance' I follow Jacobson and Weiss' earlier definition: "*Compliance* refers to whether countries in fact adhere to the provision of the accord and to the implementing measures that they have instituted" [Jacobson and Weiss 1995, p. 123]. As they note this definition goes further than the related notion of implementation, which involves the conversion on paper of international commitments to domestic law. Compliance entails actual behavioral change in the direction of international aspirations.

'Compliance' is conceptually distinct from effectiveness, which involves the question of whether or not the goals of the commitments are achieved, and may occur independently of compliance. In policy terms it is quite possible that effectiveness would not require universal compliance if a small number of states were the principal sources of the problem for which international laws were designed to confront.

Authors in this volume appear to mean two things by 'soft law' [chapter by Chinkin; see also Chinkin 1989, Wirth 1994, Wirth 1995]. One is a process of establishing authoritative judgments and establishing collective aspirations. The second is a set of public commitments which lack clear mechanisms for enforcement. Yet given the difficulty which international lawyers have encountered with understanding compliance with 'hard law', I focus on factors which may influence compliance with international commitments in general, out of the belief that such a focus is more fruitful than trying to deal with the legal distinctions between instruments. Indeed, Charles Lipson argues that the distinction between international and domestic law is much more significant than between forms of international law because we do not know what proportion of hard and soft law obligations are converted into legally binding domestic obligations in which aggrieved parties have the right to sue [Lipson 1991, Bull 1995]. Hard and soft law share a set of features insofar as they emerge out of the same impulse of states needing to signal their commitments to other states in a setting of international anarchy, and thus this chapter looks at factors which influence choice under such circumstances.

Very little is known about the degree to which states comply with international commitments, and empirical studies suggest that national compliance is uneven at best [on human rights see Donnelly 1989, McCormick and Mitchell 1997, Brannum 1998; on trade see Jackson et al. 1984, Hudec 1993, Jackson 1989; on arms control see Nolan 1994, Spector 1990; on environment see Sand ed. 1992, Weiss and Jacobson eds. 1998, Victor, Raustiala and Skolnikoff eds. 1998]. Even in the EU, where compliance with environmental directives should be strongest because of overriding public concern combined with strong international institutions, implementation varies by country [Haigh et al. 1986, Ward, Lowe and Buller 1997, p. 202]. Moreover, George Downs and others argue that estimations of compliance with international injunctions are systematically overstated because instances where states comply in the absence of strong incentives for non-compliance are not as important a sign of state willingness to comply as when states comply in the face of material incentives for violations [Downs, Rocke and Barsoom 1996]. The lawyers' dictum that "most treaties are complied with most of the time" is surely premature, and probably exaggerated. Studies of compliance find variation in compliance along a number of dimensions. Not all countries comply with the same legal instruments, the same country may vary in its compliance with different legal instruments across functional areas and even within the same functional domain, and such patterns may change over time.

In the absence of well-established patterns of compliance, or well-defined expectations for patterns of compliance on the part of decision makers (that is, outside the community of legal scholarship), this chapter relies on theoretically informed foundations to identify factors which may influence compliance. Overall levels of international compliance would be measured by whether most states vary their behavior consistent with treaty commitments. A possible problem with operationalizing such concepts is that some treaties are, often intentionally, ambiguous about what targets or procedures are actually expected so that it is difficult to specify with any precision whether states are in compliance or not.

National compliance could be measured in terms of state resources committed to the specified goal after ratification: whether a state changes its policy, laws, organizational routines and practices in accordance with international commitments. While the best evidence for compliance with most types of international commitments is based on changes in investment patterns, and bureaucratic budgets, such evidence is often absent or extremely difficult (either in terms of time or money) to obtain so that research is based on the weaker evidence of enforcement.

One must be careful to distinguish cases of deliberate compliance from serendipitous compliance, where states actions fit international obligations but without deliberate choice. This distinction may be accomplished through process tracing, identifying whether credible causal mechanisms affect the choice by which compliance occurs.

The actual measurement of compliance may be difficult because domestic political systems vary in the actual location of enforcement, and because states may refuse to submit data (or doctor reports) which they anticipate to be embarrassing. Highly federal systems locate authority for compliance at local levels, which are difficult to research and also far from the direct influence of the central state [Michelman and Soldatos eds. 1990, Hanf 1993, Downing and Hanf 1983, Hanf and Toonen 1985, Harrison 1995, Gross and Scott 1980]. For instance, enforcement of air pollution policy in Austria, Germany, and Switzerland occurs at the level of the state (lander or cantonal), and in Denmark water pollution standards are enforced by municipal and local water courts, over which the national government has little direct influence.

Compliance is a matter of state choice. While some choices may be easy, because compliance is manifestly in the national interest and there is little organized opposition, such as may be the case in ISO standards, in most cases the choice is potentially much more difficult. Compliance entails committing scarce resources (either in terms of staff time, political energy and attention, or money) by a state. It is a decision with distributional consequences. Even if a state may believe that signing a treaty is in its best interest, the political calculations associated with the subsequent decision to actually comply with international agreements are distinct and quite different. Like an individual pushed to testify at an old-time revival, the implications for behavior the next day are quite weak despite the fervor of the original impulse. A US diplomat, discussing WTO negotiations on National Public Radio said that "reaching agreement on the pact was relatively easy. The real negotiations began the day after, when China wouldn't enforce its intellectual property obligations." So long as states do not anticipate stringent political retribution for failing to comply with an obligation, they may well commit to things which they know they cannot meet, or which are crafted so ambiguously that their obligations cannot be uniformly interpreted. In fact there are a host of self-interested reasons for why states may sign soft law instruments, independently of any expectation of subsequent compliance. States may recognize that they are unable to comply, and commit out of a hope that others will help them comply at home (as with many Eastern European governments in the area of the environment), to signal their commitment in related areas of national importance, or to strengthen a leader's political potential for implementing at home later, or because signing is part of a broader diplomatic culture associated with the West with which leaders wish to be associated.

2 Domestic sources of compliance

Compliance is, more to the point, a matter of choice by the state to discipline civil society, often at the request of activist sources in civil society. Yet while states may wish to comply, not all are capable. Technical and political factors intervene in the choice to comply. Cases remain where capable states willfully flaunt international rules, such as Norwegian and Japanese whaling practices, and Greek practices in licensing ships.

States may lack the technical capacity to fulfill obligations because the state lacks the technical competence to develop and enforce technical regulations consistent with international commitments. Many developing countries and formerly centrally planned economies have greater difficulties in complying with international obligations than do industrialized countries due to less developed administrative systems and fewer monitoring and financial resources which can be devoted to enforcement. Technical capacity may be less of a factor when a state is asked to forego an activity rather than to halt an ongoing activity.

Some states may lack the political will to comply, either because of the magnitude of expected resistance at home or because the state lacks the political wherewithal to induce behavioral change on its citizenry. One possible indicator of the political capacity of the state to comply is a measure of the legitimacy of the state's claim to rule, operationalized through a calculation of the proportion of government revenue which is drawn from income taxes and other direct sources of taxation (and thus a measure of voluntary acceptance of the state) as against customs duties or other indirect sources of taxation which do not require voluntary acquiescence to the state. To fully flesh out this social dimension of state capacity one would also want to take account of the amount of resources which the state had to deploy in order to collect domestic taxes; it is always possible that a large majority of the tax revenues go into police and military which are necessary to enforce the order on which voluntary payment of taxes is based. Compliance with international obligations in such a country would also be poor, because a large proportion of state political energy would be channeled into domestic surveillance and suppression rather than enforcement of international commitments.

Compliance patterns would thus vary as a consequence of the political costs of inducing compliance and the state capacity to inflict those costs. Table 13.1 offers a conceptual taxonomy of the likelihood for compliance across these dimensions.

TABLE 13.1 Likelihood of state compliance

	Costly compliance	*Uncostly compliance*
State is capable and willing	Possible	Most likely
State is capable but unwilling	Unlikely	Unlikely
State is incapable and willing	State may try to comply and expect to fail in order to attract resources from international institutions to improve capacity.	State may try to comply
State is incapable and unwilling	Highly unlikely	Unlikely

Assuming that states are capable of enforcement, the willingness to comply may still vary by the anticipated political costs of eliciting compliance. Students of comparative politics and of American politics have drawn from game theory, the economic institutionalism and social choice theory to focus on such matters of social choice [Shepsle 1989, Moe 1984, Koelble 1995, Noll ed. 1985]. The state is generally treated as a unitary actor dealing with a pluralistic society. Variation in state choices, from the domestic perspective, lies largely with the ability of diffuse domestic interests to forge dominant coalitions with which to pressure the government. . The degree of political will necessary to comply varies by the anticipated degree of domestic resistance, due in part to the identity, number and influence of the number of actors who will have to change their behavior. One key potential source of variation in compliance is the nature of the issue being regulated and whose activities a state is seeking to influence because the array of interests vary accordingly, as does the array of policy networks and the ability of those bearing the costs of enforcement to resist such measures. It is not *a priori* clear which classes of issues will have a greater likelihood of compliance than others, but the dynamics of state choice should be quite different by issue, as the constellation of policy networks, actor, and potential influence varies by issue, even for the same state. For instance, the political costs of states enforcing compliance on the private sector and of individuals may be much higher than that of enforcing compliance on the activities by parastatals or the state itself. Most arms control and human rights actions are taken by the state itself, so compliance is in a sense a matter of self-regulation. Most environmental and trade activities are conducted by the private sector. On the other hand, states would also be expected to downplay compliance obligations in issues in which anticipated political backlash from powerful domestic sectors and actors would be high [Goldstein 1993, Milner and Yoffie 1989, Oye and Maxwell 1995]. In the environment domestic factors would seemingly militate against strong compliance, at least until recently, because of the heavy concentration of costs on industry and its high degree of political representation, with a diffuse concentration of benefits on more poorly organized and represented tourists and individuals. Compliance with arms control regimes would largely be a matter of jockeying for influence between arms producers and citizens groups.

Compliance with human rights treaties would be a matter of human rights activists and NGOs seeking to exercise some degree of leverage on a state.

Table 13.2 applies this framework to the various cases in this volume. The rows present the source of enforcement, the columns indicate the target entities whose activity must be regulated to secure compliance. Many cases have multiple dimensions; they are presented here according to the most salient dimension from the perspective of enforcement. Issues in which many actors are responsible for the targeted activities may elicit selective compliance according to the political costs of compliance for the state: that is, states would more easily extract compliance from state actors than private actors; it is less clear how different targets would respond to compliance demands from international organizations.

Table 13.2 suggests that there will be variation between the cases in terms of the extent of anticipated domestic resistance, and thus establishes an *ex ante* initial baseline against which the international forces discussed later in this chapter are directed.

More sweeping views of comparative politics look outside the institutional linkages within the state to patterns of state society relations. They suggest that the willingness of a state to comply has to do with the nature of its relations to civil society more broadly [Migdal 1988, Lijphart 1994, Schmitter and O'Donnell 1986] and to the administrative organization of the state itself [Noll ed. 1985, Risse-Kappen ed. 1995]. From this perspective variation in compliance would vary by state, rather than by issue. States which can command respect from domestic society would be more likely to allocate resources to enforcement than would states with feeble claims to legitimacy, where every enforcement action is politically costly. While for some states the act to enforce is a symbol of the legitimacy of the state, few contemporary states have sufficient resources to squander on signs of power when such actions engender political resentment. From this perspective legitimate states would be more likely to comply than would less legitimate states. Propositions from this perspective would also suggest that obligations worked out through antagonistic state–society

TABLE 13.2 Types of cases by source of activity and source of enforcement

	State and state-owned firms	*Firms (private)*	*Individuals*
IOs	Money laundering? Nuclear materials	Money laundering? Nuclear materials	Money laundering
State	Antarctica Nuclear materials OSCE treatment of minorities Land mines Missile technology	Antarctica ISO standards GA driftnet ban ILO labor standards Nuclear materials Land mines Missile technology	Antarctica—private science, tourism GA driftnet ban
Firms		ISO standards	
Individuals/NGOs		McBride/Sullivan Principles	

relations would be more difficult to enforce than those worked out within the state. In instances where state administrative capacity varies by issue it is further possible that the same state would demonstrate different compliance patterns for different issue areas (for instance, the US may comply with air pollution commitments but not with marine pollution commitments) since the responsible agency is not the same, and the constellation of interests whose behavior must be disciplined varies by issue.

3 International sources of states' choice to comply with soft law obligations

Compliance choices are seldom based solely on domestic considerations. States' choices are strategic: that is, they are contingent on expectations of other's independent behavior. International Relations scholars have looked at the systemic factors which influence states' expectations and choices. They argue that arrays of domestic influences are themselves intimately intertwined with international realities. Domestic groups may anticipate international effects of compliance, and international factors amplify some domestic forces while suppressing others. Even the configuration of domestic interests can be the consequence of international factors [Milner 1988, Destler and Odell 1987]. In the area of trade regimes compliance with liberal free-trade rules has been encouraged by industrial sectors which rely on free trade and foreign markets for their sales. Such accounts may help to explain recent compliance in pollution control treaties by potential leaders with highly competitive pollution control technologies who anticipate market opportunities being created by the treaties. Such an account would also suggest differential compliance within the environmental sphere between pollution control regimes, with potential economic rewards, and conservation regimes, which offer much more limited opportunities for the private sector.

International Relations theorists differ in their identification of the international causes or reasons for why states choose to comply or not [Caporaso 1992, Hansenclever et al. 1996, Krause and Williams 1996; similar displays of theoretical disagreements about the future of European security appear in Mearsheimer 1994/1995, Mearsheimer 1995, Keohane and Martin 1995, Muller 1993, Ruggie 1995, and Wendt 1995]. The most prominent contemporary efforts involve Realist and Neorealist efforts which stress the systemic distribution of material capabilities which provide the basis for interstate leverage, Neoliberal institutionalists focus on the formal organizational rules which guide strategic behavior, and social constructivists focus on the shared understandings which shape state behavior.

IR approaches differ in their essential theories of the state. Realists and institutionalists regard states as unitary rational actors, whose behavior and choices may be understood in terms of the array of incentives and choices available to the states.

Social constructivists have a different notion of the state; treating it in more sociological and Weberian terms as a weak bureaucratic animal. Thus they relax many of the rationalist assumptions which inform Neorealist and Neoliberal Institutionalist analysis. Social constructivists assume that states are not monolithic;

in fact their characteristics vary in terms of the extent to which the state is accountable to domestic society. At one extreme is the rare case of the totalitarian regime, impervious to influence. Much more common are various forms of pluralism. Moreover, states are functionally differentiated; made up of multiple competing bureaucratic elements each with its own functional jurisdiction or domain.

For constructivists, states are not substantively or procedurally rational. Rather they make decisions subject to bounded rationality, the easiest choices are taken at any one point it time, and choices persist until new state action is galvanized by political crises. Choices persist, and are not frequently returned to until political crises catalyze a response.

None of these perspectives from IR and comparative politics are deterministic: all are contingent. This contingency is based on the interplay between domestic and foreign politics: between comparative politics and IR [Keohane and Milner 1996, Evans et al. 1993, Gourevitch 1986, Gourevitch 1996, Katzenstein 1977, Risse-Kappen 1995]. The impact of systemic and domestic pressures on state choices to comply cannot effectively be understood in isolation: International pressures play off existing domestic conditions by creating, amplifying or inhibiting domestic pressures, while variations in the intensity of national preferences and state capabilities to influence others are often based, in part, on available domestic coalitions and available issue linkages.

In the following sections I elaborate a number of these factors which may affect states' compliance choices. It is possible that a country would choose not to comply, even if it had been complying earlier, if the international inducements were withdrawn. In the concluding section I consider the possibility that exposure to such inducements over time may lead to new habits on the part of states, and that they would comply even with the removal of institutional incentives to comply.

Propositions from these various approaches can be evaluated in practice by several techniques: Process tracing for a given country would focus over time on the decision to comply in a given country with a given treaty or set of treaties, seeking to determine if the institutional factor correlates with the decision to comply and if there is a plausible causal mechanisms between the institution and the decision. Aggregate analysis of treaties with these characteristics correlated with compliance levels would offer a statistical appraisal of the propositions. More ambitiously, multivariate analyses taking account of changes in institutional factors over time and of state choices over time would offer a far more convincing appraisal of the propositions, capable of taking account of changes in state behavior over time with the presence or absence of each factor, and where more factors correspond with higher levels of compliance. Finally, counterfactuals applied to individual state choices could also contribute to an appraisal of these propositions.

3.1 Realism, power-based factors, and choices to comply

Classical Realists argue that states seek to protect territorial integrity above all other goals. Compliance depends on threats of sanctions for non-compliance levied by a powerful state [Austin 1954]. Consequently, compliance patterns would vary directly with regard to the extent to which such integrity was potentially at risk and power is available to enforce compliance or deter non-compliance. Compliance, in the absence of the exercise of power or leadership by a dominant country, would be strongest in the area of human rights, slightly less strong in the environment, followed by trade and with arms control the issue with the lowest level of compliance. Within the area of trade, compliance levels would vary relative to the nature of the regulated products, where compliance would be lower for dual-use technologies and technologies with potential military applications.

Measures of power and its concentration, in terms of the measurable capabilities which would confer potential influence on a state, may well vary from issue to issue. A number of methodological and conceptual debates persist regarding the appropriate measures of influence in each issue, and the extent to which there exist modes of influence distinctive to each, as well as what distinguishes universal sources of influence applicable to all areas of international relations [Baldwin 1985, Nye 1990].

Neorealists focus on the extent to which policy autonomy is threatened, arguing that states are jealous of such autonomy and must be compelled to sacrifice it. Thus all compliance would be a matter of compulsion, but compliance might be higher in substantive domains which entail less perceived sacrifice of autonomy. While most Neorealists presume that states are loathe to cooperate out of anticipation of free-riding, the presumption is that all states would only comply if they were compelled. Realists suggest then that compliance will only occur if a dominant country—a hegemon—exercises some degree of pressure on a country to comply; either through rewards for compliance or threatened sanctions for violations. Compliance levels, possibly taking into account the possible variation between issues, would vary then according to the presence of hegemony and its use. In general, compliance would be spotty, because enforcement would require constant vigilance and the exercise of power by a dominant party over weaker participants in the regime. If we assume that hegemons are rational in their use of their capabilities, we would expect threats to be used when states appeared likely to comply (and thus the threatened would not be humiliated if not delivered) while rewards would be used only when non-compliance appeared likely, thus not having to expend them unnecessarily on parties who would probably comply even without external inducements [Oye 1993 ch. 3].

More pragmatically, it is also possible that the role of power in compliance may be directly related to the way in which the treaty was concluded. If it was adopted by truly voluntary agreement then the impediments to compliance might be less. If it was contentious, and resolved to some extent through the exercise of power, then compliance would still be difficult and power would have to be deployed to extract compliance.

3.2 Functional institutionalism and institutional inducements to comply

State choices can also be influenced by international institutions, where institutions are defined as "persistent and connected sets of rules (formal and informal) that prescribe behavioral roles, constrain activity and shape expectations" [Keohane 1989 p. 3]. Neoliberal institutionalists and similarly inclined international lawyers seek to design institutions to perform functions which may induce states to comply [Chayes and Chayes 1993, Chayes and Chayes 1995, Bilder 1981; see Schelling 1960 for many of the initial insights].

There is a strong *ceteris paribus* basis to the research from which these propositions are drawn. Most Neoliberal institutionalist analysts presume that states already desire to cooperate (or comply) and merely require reinforcement to indulge their initial inclinations. Institutions serve a therapeutic role to encourage compliance and to deter non-compliance by eliminating barriers to self-interested compliance, but may not exercise a direct influence on state preferences which were formed previously.

Research on international institutions and their potential influence on national choice has identified three principal analytic functions performed by international institutions: enhancing the contractual environment within which state choices are made (including voting rules, suffrage provisions, number of parties, frequency of meetings, etc.), building state concern, and building national capacity [Haas, Keohane and Levy 1993]. The contractual environment relates principally to the initial decision to sign or not to sign a treaty rather than subsequent decisions regarding compliance, and thus is not discussed in any greater detail below. It is difficult to weigh the influence of each institutional property (or factor) on state choice; presumably more is better. That is, each factor may affect some states' choices to comply, but institutional settings in which many factors are present may well command higher compliance levels (i.e. more countries will comply, and possibly comply more vigorously) than settings in which fewer factors are present or the institutions lack many of the more influential properties.

Clearly, some degree of institutional design improves the likelihood that each factor could contribute to state decisions to comply. For instance, public activities are best verified or enforced by private parties, and private actions by public authorities [Russell 1992 p. 219]. Monitoring and verification is easier with small numbers. Cross-checking by third parties (who need not be impartial) of monitoring and verification data may contribute to confidence in the accuracy of such data, and thus increase its potential political impact on compliance decisions. Effective monitoring and policy surveillance, often through the involvement of non-state parties, may also compensate for (willful) gaps in national reporting, such as Russia's failure to report on dumping of decommissioned nuclear submarines' core reactors in coastal waters or about whaling catches.

In the real world these variables are not so easily manipulated for policy purposes. Many international institutions were not designed to be able to perform such functions [Haas and Haas 1995] and few scholars have tried to think systematically about which issues command sufficient concern by major states that they are likely to be endowed with strong institutions and thus capable of inducing compliance [Gallarotti 1991, Murphy 1994 and Ruggie 1997 are some of the few exceptions]. Researchers, including those on this project, are thus left with inductive techniques to identify issues in which institutionally powerful international organizations are capable of inducing compliance [see also Levy, Young and Zurn 1995, Zacher with Sutton 1996].

3.2.1 Monitoring

Information may affect political will by publicizing state actions to potentially critical domestic (and foreign) audiences. It may also affect capacity by giving governments more and better information with which to act.

Monitoring efforts worldwide remain largely the domain of governments, although most treaties require the provision of periodic reports. National reporting to secretariats about their environmental protection activities (which may include monitoring environmental quality or reporting on compliance efforts) is often poor, and many secretariats lack the resources or authority to check data submitted by governments. A 1991 US General Accounting Office survey found that only about 60% of the parties to the 1972 London Dumping Convention complied with reporting obligations; only 30% of the members of the MARPOL convention on operation oil pollution submitted reports; and many reports under the Montreal Ozone Protocol and the Helsinki Sulfur dioxide protocol are incomplete and impossible to verify [Sand 1992 pp. 13–14, USGAO 1992].

Many environmental treaties contain monitoring provisions for environmental quality. Explicit concern with monitoring is a reasonably recent phenomenon; the median year for treaties requiring environmental monitoring is 1982. Table 13.3 summarizes the monitoring provisions of environmental treaties.

Monitoring data is most likely to be accurate, and command political attention when it is collected and disseminated by impartial third parties. When states or principal actors are responsible for monitoring the effects of their own actions they face too many incentives for lying or misrepresentation.

Many NGOs are now capable of monitoring environmental quality, as well as national compliance, and are becoming involved as a source of shadow verification of government obligations. Their activities help compensate for the dearth of effective environmental quality data, as well as providing an independent quality check on data collected through other sources. Greenpeace International seeks to keep

TABLE 13.3 Monitoring provisions in multilateral environmental treaties

What is monitored (n of treaties)	*Who does it*	*Voluntary or mandatory*	*Frequency*
Environmental quality (48)	Governments (69%) Governments must provide to IOs (4%) IOs (8%) Unspecified (19%)	81% mandatory 19% voluntary	Annually 17% Biannually 19% Triannually 2% Unspecified 62%

Source: calculated from Kiss 1983 and Rummel-Bulska and Osafo 1991.

track of national compliance with many treaties, and the Natural Resources Defense Council collects data on national compliance with the 1992 climate change treaty (UNFCCC) [Natural Resources Defense Council and CAPE 2000 1994]. The IUCN and Greenpeace also try to track national compliance with many of the species conservation treaties.

Much of the environment can be monitored remotely from satellites, and does not require the active collection and submission of data by governments. Remote sensing and satellite monitoring would also enhance verification of trends in natural resource use, marine pollution from organic sources and from oil, as well as in monitoring levels and production of greenhouse gases, although ground truthing is still necessary to confirm remote sensing data. Satellite- and airplane-based monitoring is less effective at monitoring inorganic marine contamination and urban air quality, for instance, which requires localized sampling and monitoring.

3.2.2 Verification

Direct verification of state compliance may directly affect state choices to comply [Fischer 1991, Chayes and Chayes 1995, UN Department for Disarmament Affairs 1991, Ausubel and Victor 1992]. By providing prompt information about other's actions, early warning of violations is available, as well as reducing the fear of free riding and confirming countries' reputations. The UK government for instance, has launched a campaign to certify non-compliance with EU laws by other member states. Also, by making available information of one's own activities, verification may indirectly deter non-compliance by increasing the likelihood of detection. To seriously influence compliance, verification data must be accurate, timely, and reliable. Verification may not be equally feasible in all cases. Verifying action is easier when the actions to be verified are large: for instance, verifying arms control measures is easier with large weapons than with small ones. Remote sensing works well for the environment and for arms control, less well for trade and human rights.

Table 13.4 summarizes the verification provisions of international environmental treaties, dealing with appraisals of state compliance. International concern about compliance, as well as monitoring, is reasonably recent. The median year for treaties requiring monitoring of national compliance is 1981.

TABLE 13.4 Verification provisions in multilateral environmental treaties

What is monitored (n of treaties)	*Who does it*	*Voluntary or mandatory*	*Frequency*
National compliance (58)	Governments 72% Governments must provide to IO 7% IO 3% Unspecified 18%	95% mandatory 5% voluntary	Annually 29% Biannually 18% Triannually 2% Unspecified 51%

NGOs are increasingly active in verifying state compliance with environmental accords. In international regimes Greenpeace now regularly monitors trade in hazardous wastes and in flora and fauna, and publicizes shipments which are in violation of international treaties. The publicity generated by the NGO monitoring is often sufficient to inform recipient governments of activities of which they may have been unaware, as well as pressuring them to enforce their international commitments and to refuse entry of such products. Many NGOs have become virtual watchdogs over private activities in the field as well, replacing or supplementing the monitoring activities of national enforcement agencies. Because governments are often unwilling to cede the semblance of authority to NGOs, private monitoring of governments' actions and of the environment may best be accomplished through independent scientific panels, which have access to a variety of sources of information. Surprise visits by independent inspectors are used in some regimes as a means of verification, and have long been a part of the nuclear non-proliferation regime, and the Antarctic Treaty System.The concept is accepted by eastern European and OECD countries, but not by LDCs.

3.2.3 Horizontal linkages

Linkages among institutions involved in an issue area may contribute to compliance. Dense networks of institutional factors, including such factors as numbers of international institutions involved in negotiations, and frequency of interactions could contribute to stronger levels of compliance. Such factors would improve the likelihood of compliance by encouraging states to build up their reputation to anticipate reciprocity in other areas of potential importance. Dense networks would also amplify the number of institutional factors affecting state compliance; with many institutions the possible rewards for compliance may be stronger, and states may have better chances to shape others choices through clever choices of which institution to encourage others to comply. For instance, the densely overlapping networks within which G7 finance ministers operate—the BIS, OECD working groups, IMF working groups, and G7 summits—provide multiple sources of influence over state choice. In the environmental area, the overlapping memberships in the Baltic and North Sea environmental regimes by Denmark, Sweden and Germany led each country to comply with both regimes (rather than considering each separately) as well as pushing other parties in each to comply [Haas 1993]. The UK proved willing to accept compromises in the EU Large Scale Power Plant Directive out of a perceived need to escape the reputation of being the "dirty man of Europe"; a reputation which had impeded UK diplomatic efforts in other functional domains [Levy 1993].

Similarly, actions taken in one area may affect (positively or negatively) the indirect potential for compliance in another area: for instance, human rights obligations which improve the domestic political influence of NGOs may also contribute to compliance with treaties in other areas as the political influence of other, non-human-rights NGOs is enhanced as well and they gain in their potential to exercise domestic influence on their governments.

3.2.4 Nesting

State choices to comply may also be affected by the issue-related context in which such choices are taken. Vinod Aggarwal refers to the array of hierarchical influences on a state's compliance with international obligations as 'nesting' [Aggarwal 1985 p. 186].

Nesting can take two forms. Nesting may be conceptual, regarding the causal connections which state decision makers believe tie together various issues. Consequently, if leaders hold a tightly coupled view of international politics, then such high-level beliefs will exercise a strong influence over state choice in lower-level conceptual areas. For instance, because many LDCs shared a tightly coupled view of international political and economic relations during the NIEO debates, they endorsed similar positions (and presumably similar compliance patterns) in the variety of issue areas which were being addressed internationally. Thus, population, environment, and trade, at the least, were all nested below this broader conceptual organization of issues. If the higher-order conception is only loosely coupled, then the higher-level cognitive map would be less likely to affect choice.

Nesting may also be legal in nature. By this I mean that choices to comply in one issue may be legally prescribed by some domain which has legal precedence, or is politically more influential, much as the Supreme Court in the US exercises ultimate determination over lower-level judgments. In the area of trade, the new WTO may be able to exercise such influence. The ECJ may be acquiring the authoritative power of an independent court to overrule national legal decisions [Burley and Mattli 1993, Sweet and Caporaso 1996].

3.2.5 Capacity building

The anticipation of the provision of various capacity-building resources accruing as a consequence of compliance may induce compliance. The opportunity of acquiring technology, training, financing, and more general resource transfers may encourage compliance. Conversely, the fear that such resources will be withheld from non-compliance, such as occurs through conditionality, would also encourage states to comply. Capacity building alone is unlikely to sway compliance decisions by any but the smallest and weakest countries, whose decisions would be unlikely to have any discernible impact on the internationally shared problem which is being collectively addressed by international soft law.

3.2.6 National concern

Concern about the issue by elites and the mass public may positively affect a state's decision to comply. In the short term institutions publicize events and engage in public education to catalyze concern on issues for which mass concern already exists, whereas in the longer term institutions may build such concern from scratch through public education and information campaigns, spreading literacy, the creation and strengthening of NGOs, and promoting the findings and individual status of epistemic community members.

3.2.7 Institutional profile

The institutional profile may also influence national compliance choices. High-level institutions, if domestic conditions are satisfied, may provide an opportunity for politically opportunistic entrepreneurial civil servants to encourage compliance if they anticipate domestic rewards for public commitments. For instance, Robert Putnam notes how the high-profile public exchange of commitments at G7 Summits may strengthen the hand of each leader in subsequent policy battles at home, while also increasing the stakes of failing to implement international commitments for the individual leader, and thus enhancing their commitment in the domestic political battle [Putnam and Bayne 1987]. Similarly, in the North Sea and Baltic seas the move from low-level bureaucratic coordination to periodic Ministerial conferences greatly enhanced the potential for making public commitments [Haas 1993].

3.3 National sensitivity to institutional incentives

Such institutional factors are unlikely to be equally influential over all states' choices. The potential influence of international institutions is thus contingent, in this regard, on whether their information functions mesh with concerns of potentially powerful domestic groups. Institutional analysis is insufficient on its own to account for patterns of national compliance. Even holding institutional factors constant, one encounters wide variation in national enforcement with the same regulations States vary in terms of their vulnerability to institutional inducements [Risse Kappen ed. 1995, Evangelista 1997]. Not all states are vulnerable to external influence, although with increasing globalization this number is probably declining. State vulnerability can be measured in terms of relatively high levels of national trade/GNP, DFI as a percentage of GFCF, and foreign indebtedness. High levels of any of these increases the degree of foreign interest and the vulnerability of target states to influence from outside. Some states are relatively impervious to such direct and indirect leverage. Without pluralistic societies and without the need to seek credit or finance from international institutions, governments of many Newly Industrializing Countries (NICs) are relatively insensitive to the array of international and domestic political influences on governments to protect the environment.

Assuming some degree of vulnerability to external inducements, the degree to which specific institutional incentives are likely to exercise a direct impact on different states' choice is elaborated in Table 13.5.

TABLE 13.5 Variation in national sensitivity to institutional incentives for compliance with environmental soft law

	State is capable of making and enforcing policy		*State is incapable of making and enforcing policy*	
State is responsive domestically	Effective institutional functions include: verification, monitoring, national concern, institutional profile Capacity building is less influential here, because state capacity is already high		Effective institutional functions include: verification, monitoring, capacity building, national concern, institutional profile	
	Argentina	**Italy**	Botswana,	**Jamaica**
	Australia	**Japan***	**Colombia**	**Mauritius**
	Austria	**Netherlands***	**Costa Rica**	**Venezuela**
	Belgium	**New Zealand**	Greece	
	Canada*	**Norway***		
	Denmark*	**Portugal***		
	Finland*	**Spain**		
	France	**Sweden**		
	Germany*	**Switzerland***		
	Ireland*	**UK***		
	Israel	**USA***		
State is not responsive domestically	Effective institutional functions include: verification, monitoring		Effective institutional functions include: verification, monitoring, capacity building	
	Brazil*	**Pakistan**	**Algeria**	Iraq
	Bulgaria	**Poland**	Afghanistan	Jordan
	Chile*	**Russia***	Albania	Laos
	China	Rwanda	Angola	Lebanon
	Czechoslovakia	**Saudi Arabia**	**Bangladesh**	Lesotho
	El Salvador	**Senegal**	Burundi	Libya
	Ghana	**Singapore**	Central African	Malawi
	Guatemala	**S. Africa**	Republic	**Malaysia**
	Hungary*	**Togo**	Chad	**Mali**
	India*	**Trinidad &**	Congo	**Nicaragua**
	Kenya	**Tobago**	**Cote d'Ivoire**	Niger
	Kuwait	Turkey*	Cuba	**Nigeria**
	S. Korea*	Vietnam	Dominican	**Panama**
	Mexico*	Yugoslavia	Republic	**Paraguay**
	Nepal	**Zambia**	**Ecuador**	**Peru**
			Egypt	**Philippines**
			Gabon	**Sri Lanka**
			Guinea	**Sudan**
			Guyana	**Tanzania**
			Haiti	**Thailand**
			Honduras	**Uganda**
			Indonesia	**Uruguay**
			Iran	Yemen

Bold type indicates countries with more than two environmental NGOs attending UNCED. Asterisks indicate countries in which over 40% of population expressed a "great deal" or "fair amount" of concern about the environment [Dunlap et al. 1993].

Table 13.5 indicates states which are sensitive to specific institutional inducements for compliance with international environmental soft law. It distinguishes states by state capacity, state/society relations, and the extent of domestic environmental concern. State capacity is a measure of a state's technical capacity to formulate and monitor environmental policies. This is a multi-dimensional measure, based on bureaucratic measures of budgets of functional agencies, staffing, resources, administrative influence of functional agencies, and ability to monitor and enforce decisions in civil society.

State capacity is measured by the numbers of scientists and engineers per million population and proportion of GNP devoted to R&D. High capacity is .4% of GNP for R&D or more than 300 scientists & engineers per million population [data is from *UNESCO Statistical Yearbook*, various years]. State–society relations is a measure of the extent to which the state is responsive and accountable to domestic society, and thus is a measure of the extent to which the state is liable to respond to demands for compliance from domestic sources. The obvious measure is whether the government is coded as democratic/non-democratic [data is from Russett 1994]. Public environmental concern is an indicator of the extent to which external pressures will resonate with domestic political forces, and thus the ability of international institutions to amplify domestic pressures for compliance. Environmental concern is measured by a 1992 Gallup survey of public environmental opinions in 24 countries (Dunlap et al. 1993) and the existence of environmental NGOs, indicated by whether more than two environmental NGOs attended UNCED from that country [data from *Who is Who at the Earth Summit* 1992].

Short-term efforts fall in each cell. Longer term efforts, which are more indirect in focus, would be addressed to moving states between cells; improving state capacity and opening up public access. It is for this reason that democratization is widely hailed as a world order strategy with multiple benefits beyond the area of human rights and liberties.

In summary, institutional factors as elaborated by functional institutionalists are likely to vary in their impact on state compliance choices by issue and by country. While verification and monitoring appear to be the most widely applicable institutional factors affecting state compliance, they would operate through different channels in different countries. In democratic or representative societies (pluralistic, parliamentary, presidential, corporatist, or consociational) the information would be converted to leverage through domestic and international channels, while in non-democratic or non-representative societies the information would be converted to pressure on compliance through international channels, either between states, from international institutions, or from MNCs and international banks. NGOs and other transnational actors may exercise an indirect effect through their ability to influence or use influential states, international institutions or firms [Keck and Sikkink 1998]. Institutional factors are probably richest in the areas of trade and economic harmonization, followed by the environment and human rights.

3.4 Social constructivism and knowledge-based compliance

International relations has recently returned to focus on the role of ideas and understanding in shaping choices by goal-seeking states. As international politics becomes increasingly complex and uncertain, it becomes farfetched to assume that states are capable of clearly anticipating how national welfare will be affected by policy choice at home and in conjunction with others. Social constructivism is a recent research program [Ruggie, 1998, Adler 1997, Wendt 1995, Haas 1992, Haas and Haas 1995; see also Franck 1990, Brunnee and Toope 1997] which looks at the process by which collective representations of the world are constructed and diffused. Constructivists assume that states are incapable of searching for new information each time a decision is demanded, and that they satisfice and rely on prior cognitive frames to understand how national interests are likely to be affected by any particular decision. Thus decisions to comply are not based on rational calculations of interest. Rather, compliance is a matter of applying socially generated convictions and understandings about how national interests are likely to be achieved in any particular policy domain.

Collective understandings are thus the source of state choices. Conviction can have two bases: an ethical or moral sense of obligation; or a causal belief in how the world works and how a country's interests will be affected by compliance. *Pacta sunt servanda* (pacts made in good faith are binding) is too general a norm to fit the empirical record of spotty national compliance. States may share moral norms, which precede the choice to comply [Finnemore 1996, Katzenstein 1996, Raymond 1997, Lumsdaine 1993]. Morality may play a stronger role in the case of human rights, and be less likely to apply to other issues, because the very justification of the treaty itself is grounded on normative claims. International institutions which are regarded as legitimate can reinforce the operation of such norms on compliance through monitoring and verification.

The most important sources of influence for social constructivists are the shared causal understandings, or consensual knowledge, which help guide decision makers in making choices in complex and unfamiliar domains. Such domains are increasingly common throughout recent international politics, particularly in the areas of economic and environmental policy. Obligations in these issue areas are largely grounded as well on causal and instrumental warrants rather than normative ones. Following appreciation of new causal factors in the policy environment which affect state interests, states may be said to learn to comply as they learn to recognize undesirable international conditions which will detrimentally affect their national interest, while also learning of new practices by which to alleviate or ameliorate such conditions.

From a constructivist perspective compliance is more likely if there exist relevant widely shared causal beliefs about the operation of the issue to be controlled, and the degree to which the actual rules promote valued ends. The likelihood of compliance with economic regimes is increased if states believe that the regimes accurately reflect the way in which economies behave. Similarly, in environmental areas the potential for compliance may be improved if state officials share beliefs about the operation of ecosystems, and the desirability of their preservation.

The principal mechanism by which such ideas are developed and disseminated is by transnational networks of policy professionals called epistemic communities, who share common values and causal understandings. Members of epistemic communities will seek to introduce national measures consistent with their beliefs, and utilize the enforcement mechanisms of the bureaucratic units in which they operate [Haas 1990, 1992]. Patterns of compliance are thus based on the extent to which epistemic community members are able to acquire influential positions in national administrations and in international institutions from which to encourage compliance. For instance, the UK moved from non-compliance to compliance with EU water quality standards following the penetration of an epistemic community like policy network which convinced the UK that compliance was in the country's interest [Richardson 1994, Bressers et al. 1994]. Mediterranean governments came to embrace marine pollution controls following influence of a regional ecological epistemic community, operating concurrently through national governments and through the United Nations Environment Program [Haas 1990]. Epistemic communities are most likely to gain prompt entry in democratic countries with states which have a high degree of technical competency in the substantive area in question.

International negotiations for particular soft law (and hard law) commitments may then directly affect the prospects for compliance by including new groups in the process of negotiation, and by alerting states to new ideas, and recruiting and institutionalizing groups associated with those ideas. Negotiations may thus contribute to transforming the beliefs and interests of the parties engaged in the negotiations, and thus influencing their subsequent choices regarding compliance [Haas 1989, Haas 1990, Sjostedt 1996].

States may engage in three related types of learning about compliance. States may learn to comply with particular instruments, following the adoption of new consensual knowledge about how state interests are to be achieved. States may also learn to comply with related instruments in the same area of activity. For instance, they may recognize that national interests are protected by arms control or environmental protection, and thus find it easier to comply with other arms control or environmental commitments. States can learn as well about the connection between issues and thus change compliance patterns over time due to the acceptance of new 'policy maps' which identify goals which must be achieved in order to promote national goals.

As perceived causal connections are drawn between issue areas, compliance decisions will come to reflect the broader notion of how national interest is affected by concurrent activity in causally related areas. Of course, in principal, states may learn to reject linkages as well, although most new policy consensus focuses on previously unrecognized connections. Consequently decisions to comply in one area would depend on the nature of the perceived linkages with other functional domains.

Consensus seems to be crystallizing internationally around the linkages between environment and development activities, [Baker, Kousis, Richardson and Young 1997, Nelson and Eglinton 1995, Haas 1996] as policy elites in national governments and international institutions as well as NGO activists increasingly recognize that efforts to promote economic growth are contingent on assuring some degree of environmental protection; although some analysts have stressed the obverse of this connection, that environmental protection may require some minimal economic growth as well. Appreciation of such new linkages would lead states to comply in the newly highlighted policy domains, out of an appreciation of the complementary causal influences between the domains. Conversely, such new consensus might precipitate a backwards movement towards questioning compliance and even non-compliance if stipulated activities in economic areas were deemed to undermine environmental quality.

Such linkages could also be created or contrived; that is, compliance in one area could be made contingent on compliance (or non-compliance) in another area, thereby linking the decisions to comply. The use of trade sanctions as an enforcement device in environmental treaties has been challenged on these grounds.

4 Conclusion

Questions of compliance—to what extent do states comply, which states are likely to comply, what patterns of compliance exist within and across areas of regulation—have not been extensively investigated and remain poorly understood. This chapter argues that compliance is a matter of state choice, and that that choice is often subject to institutional and constructivist forces. While there may be more compliance than skeptics believe, there is also probably less than many international lawyers and international relations scholars would prefer. Future compliance with international soft law is likely to continue to be driven by institutional and constructivist forces so long as globalization and democratization continue. States are likely to retain their legal sovereignty even in the face of global pressures so long as they remain the sole authoritative source for compliance decisions. Further research is called for to combine institutional and constructivist analysis to better understand how institutional design can enhance learning; on prior state beliefs which influence decisions about institutional design; on comparative national studies of state competence, on the degree of autonomy of agencies in different governments; and on the various mechanisms, such as persuasion, recruitment patterns, policy emulation, and third-party inducements, by which shared understandings diffuse more broadly.

Note

1 Copyright © 2000 Oxford University Press. This chapter was first published in Dinah Shelton (ed.) *Commitment and Compliance: The Role of Non-Binding Norms in the International Legal System*. Oxford: Oxford University Press, 2000, Part 1 section 2. By permission of Oxford University Press.

References

Adler, Emanuel (1997) "Seizing the Middle Ground" *European Journal of International Relations* Vol. 3 No. 3 pp. 319–364.

Aggarwal, Vinod (1985) *Liberal Protectionism* Berkeley: University of California Press.

Austin, John (1954) *The Province of Jurisprudence Determined* Lecture I. London: Weidenfeld and Nicholson.

Ausubel, Jesse and David Victor (1992) "Verification of International Environmental Agreements" *Annual Review of Energy and the Environment* Vol. 17 No. 1.

Baker, Susan, Maria Kousis, Dick Richardson and Stephen Young, eds. (1997) *The Politics of Sustainable Development* London: Routledge.

Baldwin, David A. (1985) *Economic Statecraft* Princeton: Princeton University Press.

Bilder, Richard B. (1981)_*Managing the Risks of International Agreement* Madison: University of Wisconsin Press.

Brannum, Katherine (1998) *Compliance with Human Rights Norms* PhD dissertation in political science, University of Massachusetts at Amherst.

Bressers, Hans, Laurence J. O'Toole Jr. and Jeremy Richardson (1994) "Networks as Models of Analysis" *Environmental Politics* Vol. 3 No. 4 pp. 1–24.

Brunnee, Jutta and Stephen J. Toope 1997 "Environmental Security and Freshwater Resources: Ecosystem Regime Building" *American Journal of International Law* Vol. 91 pp. 26–59.

Bull, Hedley (1995) *The Anarchical Society* 2nd edition New York: Columbia University Press.

Burley, Anne-Marie and Walter Mattli (1993) "Europe Before the Court" *International Organization* Vol. 47 No. 1 pp. 41–76.

Caporaso, James A. (1992) "International Relations Theory and Multilateralism" *International Organization* Vol. 46 No. 3 pp. 599–632.

Chayes, Abram and Antonia Handler Chayes (1993) "On Compliance" *International Organization* Vol. 47 No. 2 pp. 175–206.

Chayes, Abram and Antonia Handler Chayes (1995) *The New Sovereignty* Cambridge: Harvard University Press.

Chinkin, Christine M. (1989) "The Challenge of Soft Law" *International and Comparative Law Quarterly* Vol. 38 No. 4 pp. 850–866.

Destler, I. M. and John S. Odell (1987) *Anti-Protection: Changing Forces in United States Trade Politics* Washington, DC: Institute for International Economics.

Donnelly, Jack (1989) *Universal Human Rights in Theory and Practice.* Ithaca: Cornell University Press.

Downing, Paul B. and Kenneth Hanf, eds. (1983) *International Comparisons in Implementing Pollution Laws* Boston: Kluwer-Nijhoff.

Downs, George W., David M. Rocke and Peter N. Barsoom (1996) "Is the Good News About Compliance Good News About Cooperation?" *International Organization* Vol. 50 No. 3 Summer pp. 379–406.

Dunlap, Riley E., George H. Gallup, Jr. and Alec M. Gallup (1993) *Health of the Planet* Princeton: Gallup International Institute.

Evangelista, Matthew (1997) "Domestic Structure and International Change" in Michael Doyle and G. John Ikenberry eds. 1997 *New Thinking in International Relations Theory* Boulder: Westview Press.

Evans, Peter B., Harold K. Jacobson and Robert D. Putnam eds. (1993) *Double-Edged Diplomacy* Berkeley: University of California Press.

Finnemore, Martha, (1996) *National Interests in International Society* Ithaca: Cornell University Press.

Fischer, Wolfgang (1991) "The Verification of a Greenhouse Gas Convention" *Verification Report 1991.*

Franck, Thomas (1990) *The Power of Legitimacy Among Nations* New York: Oxford University Press.

Gallarotti, Giulio (1991) "The Limits of International Organization" *International Organization* Vol. 45 No. 2 pp. 183–220.

Goldstein, Judith (1993) *Ideas, Interests, and American Trade Policy* Ithaca: Cornell University Press.

Gourevitch, Peter (1986) *Politics in Hard Times* Ithaca: Cornell University Press.

Gourevitch, Peter Alexis (1996) "Domestic Sources of International Cooperation" *International Organization* Vol. 50 No. 2 pp. 349–373.

Gross, Andrew C. and Nancy E. Scott (1980) "Comparative Environmental Legislation and Action" *International and Comparative Law Quarterly* Vol. 2.

Haas, Peter M. (1989) "Do Regimes Matter?" *International Organization.*

Haas, Peter M. (1990) *Saving the Mediterranean* New York: Columbia University Press.

Haas, Peter M. (1992) "Introduction: Epistemic Communities and International Policy Coordination" *International Organization* Vol. 46 No. 1 pp. 1–36.

Haas, Peter M. (1993) "Protecting the Baltic and the North Sea" in Peter M. Haas, Robert O. Keohane and Marc A. Levy eds. *Institutions for the Earth* Cambridge: MIT Press.

Haas, Peter M. (1996) "Is 'Sustainable Development' Politically Sustainable?" *The Brown Journal of World Affairs* Vol. III No. 2 pp. 239–248.

Haas, Peter M. and Ernst B. Haas (1995) "Learning to Learn" *Global Governance* Vol. 1 No. 3.

Haas, Peter M., Robert O. Keohane and Marc A. Levy, eds. (1993) *Institutions for the Earth* Cambridge: MIT Press.

Haigh, Nigel et al. (1986) *Comparative Report: Water and Waste in Four Countries* London: Graham & Trotman,

Hanf, Kenneth (1993) "Enforcing Environmental Laws" in Michael Hill ed. *New Agendas in the Study of the Policy Process* New York: Harvester Wheatsheaf.

Hanf, Kenneth and Theo A.J. Toonen, eds. (1985) *Policy Implementation in Federal and Unitary Systems* Dordrecht; Boston: M. Nijhoff.

Hansenclever, Andreas, Peter Mayer, and Volker Rittberger (1996) "Interests, Power, Knowledge: The Study of International Regimes" *Mershon International Studies Review* Vol. 40 No. 2 October pp. 177–228.

Harrison, Kathryn (1995) "Is Cooperation the Answer? Canadian Environment Enforcement in Comparative Context" *Journal of Policy Analysis and Management* Vol. 14 No. 2.

Hudec, Robert E. (1993) *Enforcing International Trade Law: The Evolution of the Modern GATT Legal System* Salem, NH: Butterworth Legal Publishers.

Jackson, John (1989) *The World Trading System* Cambridge: The MIT Press.

Jackson, John H., Jean-Victor Luis, and Mitsuo Matsushita (1984) *Implementing the Tokyo Round* Ann Arbor: University of Michigan Press.

Jacobson, Harold K. and Edith Brown Weiss (1995) "Compliance with International Environmental Accords" *Global Governance* Vol. 1 No. 1 pp. 119–148.

Katzenstein, Peter J. ed. (1977) *Between Power and Plenty* Madison: University of Wisconsin Press.

Katzenstein, Peter J. ed. (1996) *The Culture of National Security* New York: Columbia University Press.

Keck, Margaret and Kathryn Sikkink (1998) *Activists Beyond Borders* Ithaca: Cornell University Press.

Keohane, Robert O. (1989) *International Institutions and State Power* Boulder: Westview Press.

Keohane, Robert O. and Lisa L. Martin (1995) "The Promise of Institutionalist Theory" *International Security* Vol. 20 No. 1 pp. 39–51.

Keohane, Robert O. and Helen Milner eds. (1996) *Internationalization and Domestic Politics* Cambridge: Cambridge University Press.

Kiss, Alexandre Charles (1983) *Selected Multilateral Treaties in the Field of the Environment* Nairobi: UNEP.

Koelble, Thomas A. (1995) "The New Institutionalism in Political Science and Sociology" *Comparative Politics* 27 pp. 231–243.

Krause, Keith and Michael C. Williams (1996) "Broadening the Agenda of Security Studies" *Mershon International Studies Review* Vol. 40 No. 2 pp. 229–254.

Levy, Marc A. (1993) "The Power of Tote-Board Diplomacy" in Haas, Keohane and Levy 1993.

Levy, Marc A., Oran R. Young and Michael Zurn (1995) "The Study of International Regimes" *European Journal of International Relations* Vol. 1 No. 3 pp. 267–330.

Lijphart, Arend (1994) *Electoral Systems and Party Systems* New York: Oxford University Press.

Lipson, Charles (1991) "Why are Some International Agreements Informal?" *International Organization* Autumn, Vol. 45 No. 4 pp. 495–538.

Lumsdaine, David (1993) *Moral Vision in International Politics* Princeton: Princeton University Press.

McCormick, James M. and Neil J. Mitchell (1997) "Human Rights Violations, Umbrella Concepts, and Empirical Analysis" *World Politics* Vol. 49 No. 4 pp. 510–525

Mearsheimer, John J. (1994/1995) "The False Promise of International Institutions" *International Security* Vol. 19 No. 3 pp. 5–49.

Mearsheimer, John J. (1995) "A Realist Reply" *International Security* Vol. 20 No. 1 pp. 82–93.

Michelman, Hans J. and Panayatis Soldatos eds. (1990) *Federalism and International Relations* Oxford: Oxford University Press.

Migdal, Joel S. (1988) *Strong Societies and Weak States* Princeton: Princeton University Press.

Milner, Helen V. (1988) *Resisting Protectionism* Princeton: Princeton University Press.

Milner, Helen V. and David B. Yoffie (1989) "Strategic Trade Policy and Corporate Trade Demands" *International Organization* Vol. 43. No. 2 pp. 239–272.

Moe, Terry M. (1984) "The New Economics of Organization" *American Political Science Review* 28 pp. 739–776.

Muller, Harald (1993) "The Internationalization of Principles, Norms and Rules by Governments" in Volker Rittberger ed. *Regime Theory and International Relations* Oxford: Oxford University Press.

Murphy, Craig N. (1994) *International Organization and Industrial Change* Oxford: Polity Press.

Natural Resources Defense Council and CAPE 2000 (1994) *Four in '94 Assessing National Actions to Implement Agenda 21: A Country-by-Country Progress Report* Washington, D.C.: National Resources Defense Council.

Nelson, Joan and Stephanie J. Eglinton (1995) *Global Goals, Contentious Means* Washington, D.C.: Overseas Development Council.

Nolan, Janne ed. (1994) *Global Engagement: Cooperation and Security in the 21st Century* Washington, D.C.: Brookings Institution.

Noll, Roger ed. (1985) *Regulatory Policy and the Social Sciences* Berkeley: University of California Press.

Nye, Joseph S. (1990) *Bound to Lead* New York: Basic Books.

Oye, Kenneth (1993) *Economic Discrimination and Political Exchange* Princeton: Princeton University Press.

Oye, Kenneth and James H. Maxwell (1995) "Self-Interest and Environmental Management" in Robert O. Keohane and Elinor Ostrom *Local Commons and Global Interdependence* Beverly Hills: Sage Publications pp. 191–222.

Putnam, Robert D. and Nicholas Bayne (1987) *Hanging Together* Cambridge: Harvard University Press.

Raymond, Gregory A. (1997) "Problems and Prospects in the Study of International Norms" *Mershon International Studies Review* Vol. 41 Supplement 2 pp. 205–246.

Richardson, Jeremy (1994) "EU Water Policy" *Environmental Politics* Vol. 3 No. 4 pp. 139–168.

Risse-Kappen, Thomas ed. (1995) *Bringing Transnational Relations Back In* Cambridge: Cambridge University Press.

Ruggie, John Gerard (1995) "The False Premise of Realism" *International Security* Vol. 20 No. 1 pp. 62–70.

Ruggie, John Gerard (1997) *Winning the Peace* New York: Columbia University Press.

Ruggie, J. G. (1998) "What Makes the World Hang Together? Neo-Utilitarianism and the Social Constructivist Challenge." *International Organization* 52(4): 855–885.

Rummel-Bulska, Iwona and Seth Osafo (1991) *Selected Multilateral Treaties in the Field of the Environment* Nairobi: UNEP.

Russell, Clifford S. (1992) "Monitoring and Enforcement of Pollution Control Laws in Europe and the United States" in Tom Tietenberg ed. *Innovation in Environmental Policy* Aldershot: Edward Elgar.

Russett, Bruce (1994) *Grasping the Democratic Peace* Princeton: Princeton University Press.

Sand, Peter ed. (1992) *The Effectiveness of International Environmental Agreements* Oxford: Grotius.

Schelling, Thomas (1960) *The Strategy of Conflict* Cambridge: Harvard University Press.

Schmitter, Philippe and Guillermo O'Donnell (1986) *Transitions from Authoritarian Rule* Baltimore: Johns Hopkins University Press.

Shepsle, Kenneth (1989) "Studying Institutions: Some Lessons from Rational Choice Analysis" *Journal of Theoretical Politics* 1 pp. 131–47.

Sjostedt, Gunnar (1996) "Environmental Aid Negotiations as a Process of Collective Learning" in R.E. Lofstedt and G. Sjostedt eds. *Environmental Aid Programmes to Eastern Europe.*

Spector, Leonard S. (1990) *Nuclear Ambitions* Boulder: Westview Press.

Stone Sweet, Alec and James A. Caporaso (1996) "From Free Trade to Supranational Polity: The European Court and Integration" University of Washington Political Relations and Institutions Group Working Paper 2.45

United Nations Department for Disarmament Affairs (1991) *Verification and the United Nations* New York: United Nations.

United States General Accounting Office (1992) *International Environment International Agreements are Not Well Monitored* GAO/RCED-92-43.

Victor, David, Kal Raustiala and Eugene Skolnikoff eds. (1998) *The Implementation and Effectiveness of International Environmental Commitments* Cambridge: MIT Press.

Ward, Neil, Philip Lowe and Henry Buller (1997) "Implementing European Water Quality Directives" in Susan Baker, Maria Kousis, Dick Richardson and Stephen Young eds. *The Politics of Sustainable Development* London: Routledge.

Weiss, Edith Brown and Harold K. Jacobson eds. (1998) *Engaging Countries* Cambridge: MIT Press.

Wendt, Alexander (1995) "Constructing International Politics" *International Security* Vol. 20 No. 1 pp. 71–81.

Who is Who at the Earth Summit (1992) Waynesville, NC: Visionlink Education Foundation.

Wirth, David A. (1994) "Reexamining decision-making processes in international environmental law." *Iowa Law Review* May, Vol. 79 No. 4, pp. 769–802.

Wirth, David A. (1995) "The Rio Declaration on Environment and Development: two steps forward and one back, or vice versa?" *Georgia Law Review* Spring, Vol. 29 No. 3.

Zacher, Mark W. with Brent A. Sutton (1996) *Governing Global Networks* Cambridge: Cambridge University Press.

14

EVALUATING THE EFFECTIVENESS OF MARINE GOVERNANCE*

Peter M. Haas

1.0 Introduction

Marine governance has become a major focus of policy and scholarly concern over the last 40 years. In the face of increasingly intense human uses of the coastal zones and open oceans, it is now widely accepted that sustainable use of the open oceans and coastal zones requires significant international cooperation to coordinate policies and eliminate externalities, and also requires comprehensive ecosystem management to minimize or eliminate the externalities arising from the interplay of multiple activities in these areas, and the differing needs and interests of multiple users (Sorensen and McCreary 1990; Vallega 1992; Boelaert-Suominen and Cullinan 1994; Cicin-Sain and Knecht 1998).

This chapter focuses on the factors that are associated with managing marine resources in a sustainable manner. What factors contribute to better management, and how have international institutions contributed to better management? This exercise is intended to develop a set of suggestive benchmarks by which specific marine management efforts may be evaluated, and also to identify some best practices by which management may be improved.

Marine governance is a contested and political activity. The functions of effective marine governance serve to induce political will to pursue more sustainable marine governance efforts.

2.0 Evolving understandings about managing marine collective action problems

There has been widespread development in the practice and study of marine governance/management over the last 40 years. Many regimes and MEAs were developed that now cover most of the world's seas, as well as most activities that affect those seas. There is no single global marine governance system. Rather there are

TABLE 14.1 Marine governance efforts

Marine area	*Efforts**
Oil pollution of the open oceans	IMO regimes, MARPOL
Land-based sources	Med Plan, UNEP guidelines, Agenda 21 chapter 17, Ramsar, Global Program of Action for the Protection of the Marine Environment from Land-Based Activities
Marine living resources	ICCAT, NASCO, NAFO, FAO Fisheries arrangements, regional fisheries agreements, UN Fish Stocks Agreement, whaling, CCAMLR, CITES, CBD, CMS, Ramsar Marine mammals guidelines
Marine pollution	UNEP regional seas programs (10), Baltic, North Sea, Black Sea, London Dumping Convention, MARPOL, IMO Funds
POPs and toxics	Stockholm, Rotterdam, Basel, FAO PIC guidelines

* Efforts include global and regional regimes as well as soft law.

Source: http://www.unep.org; http://www.greenyearbook.org/

numerous regional and functional efforts, with varying amounts of experience and study. The following table provides an illustrative list of marine governance efforts.

Yet, as Laurence Boisson de Chazournes observes: "what remains to be fully ascertained is the degree of compliance with them, in particular when such agreements are likely not only to affect the environmental, but also the economic policies of the State parties" (Choucri 1993; Shelton 2000; Chazournes 2005; Zaelke, Kaniaru et al. 2005). There is wide variation in terms of their effectiveness, indeed wide variation in terms of the extent to which information is available for making informed judgment about whether the environment is better, and whether deliberate state policies have had an impact on observed environmental change (for good or bad).

The study of governance of the international commons has also received extensive treatment over the last 30 years, following the emergence of new issue areas in international relations. The findings presented in this chapter are developed deductively from theories about environmental governance and collective action, and inductively from comparative studies of specific environmental governance efforts where the identified factors correlated with more effective governance efforts. Empirical studies of collective governance of shared and transboundary resources have also inferred that more comprehensive substantive governance correlates closely with the provision of specific administrative governance functions (Solkinoff 1972; Ruggie 1975; Kay and Jacobson 1983; Young 1989; Ostrom 1990; Sand 1990; Haas, Keohane et al. 1993; Victor, Raustiala et al. 1998; Weiss and Jacobson 1998; Wettestad 1999; Young 1999; Young 2000; Miles 2002; Sandler 2004).

The social science literature on marine governance rests on broader efforts to study collective action, as they have emerged in political science and economics since the 1960s. These works have helped to clarify the barriers to effective collective action, to describe the mechanisms of collective governance, and how to assess the effectiveness of governance efforts.

2.1 Barriers to collective action

Analysts have identified a wide variety of barriers to collective action at the national and international levels that inhibit effective marine governance for most issues. Administrative design and capacity issues inhibit the ability of states and resource managers to effectively make and enforce comprehensive marine policies (Bell 1997; Bell and Russell 1998; Ascher 2000; Bell 2000; Adeel 2003). Few governments have environmental agencies with sufficient staff, resources, or authority to be able to formulate comprehensive marine policies. Enforcement is also difficult, due to weak political will, few inspectors, minimal enforcement authority, and inadequate environmental monitoring equipment. Moreover, most functional organizations responsible for marine governance lack a sufficiently broad mandate to be able to monitor marine quality, develop management policies that encompass the wide array of activities contributing to environmental risk (in large part because such decisions are the traditional domain of other, older, more powerful and well-entrenched agencies) or to induce such bodies to internalize the externalities that their policies encourage. This problem of institutional myopia exists at the international level as well, where UN and other international institutions generally lack sufficient authority or scope to be able to induce widespread policy change, or to coordinate complex policies effectively with other functional agencies. Another consequence of this organizational incoherence is a paucity of meaningful evaluation studies of environmental regimes, or of environmental conditions. Fisheries management is an extreme problem, due to the sunk investment costs in fisheries fleets which drive over consumption.

Incomplete and poor quality of information and knowledge about the marine environment also impedes the ability to exercise comprehensive management. There is often insufficient information about marine stresses, their causes and effects, or the array of appropriate responses to possible risks.

The common nature of issues means that sovereign governments are unwilling to address them. Effects are displaced in time and space beyond the traditional territorial jurisdiction of most modern states.

Not all governments care about marine governance. Political will is weak for a number of reasons. Public opinion surveys indicate that concern about marine issues are modest, and few publics are eager about committing financial resources to dealing with marine threats, especially if they are outside their coastal waters. Developing country governments are still more worried about resource scarcity issues, whereas industrialized governments are more concerned about pollution matters. Domestic political realities interfere with effective marine management. Typically the costs of marine protection are concentrated in a few politically influential sectors, whereas the benefits of marine management are diffuse. Consequently domestic political pressure tends to only favor delay or modest responses.

2.2 Mechanisms of collective action

Second is the focus on the social mechanisms by which states try to respond collectively to problems of collective action. The study of governance started with a discovery of the literature on public goods in the late 1960s and the appreciation that effective collective action requires that decision makers be made aware of the longer term consequences of national actions, be made accountable to other parties for their actions, and be provided with the material and administrative wherewithal to be able to live up to their collective obligations.

Analysts have grown more sophisticated in their appreciation of the nature of governance for all international issues, not just the environment. International relations initially focused on descriptive studies of international law. In the 1980s attention turned to studies of international regimes: "sets of implicit or explicit principles, norms, rules, and decision-making procedure around which actors' expectations converge in a given area of international relations" (Krasner 1983). By the 1990s the study of regimes expanded to look at international institutions in general: "persistent and connected sets of rules (formal and informal) that prescribe behavioral roles, constrain activity, and shape expectations" (Keohane 1989; North 1990; Ostrom 1990; Young 1997) The virtue of this focus is to sharpen the difference in focus between formal arrangements negotiated between states that guide collective behavior—formal organizations—and shared beliefs that guide understanding of interest and expectations of others' behavior—informal institutions. In this way the study of collective action could take account of both the formal administrative arrangements guiding behavior and also the causal and normative beliefs that would guide substantive management of particular issues.

Later in the1990s this broad research program moved again, to take account of a broader phenomenon of governance. In 1992 James Rosenau and Ernst-Otto Czempiel had launched an analysis of governance without government:

> Governance is ... a system of rule that is as dependent on intersubjective meanings as on formally sanctioned constitutions and charters. Put more emphatically, governance is a system of rule that works only if it is accepted by a majority (or at least, by the most powerful of those it affects), whereas governments can function even in the face of widespread opposition to their policies. In this sense governance is always effective in performing the functions necessary to systemic persistence, else it is not conceived to exist .... Thus it is possible to conceive of governance without government—of regulatory mechanisms in a sphere of activity which function effectively even though they are not endowed with formal authority.
>
> *(Rosenau 1992)*

Kennette Benedict provides a useful definition of the current application of "global governance":

> A purposeful order that emerges from institutions, processes, norms, formal agreements, and informal mechanisms that regulate action for a common good. Global governance encompasses activity at the international, transnational and regional levels, and refers to activities in the public and private sectors that transcend national boundaries.
>
> *(Benedict 2001)*

Governance entails a procedural component of administrative functions diplomats frequently invoke when designing institutional arrangements. Some of these were initially identified as the 3Cs (cooperative environment, capacity building, and building concern) by Haas, Keohane and Levy in 1993, but the list has subsequently been extended and elaborated. These are now regarded as administrative functions that are performed in effective international governance, rather than phases of environmental governance, as in practice they overlap and there is no clear linear sequence with which they were performed (The Social Learning Group 2001). These functions include agenda setting, framing, rule making, enforcement and assessment (Haas 2004b; Kanie and Haas 2004; Speth and Haas 2006). Mark Zacher identifies a similar list of functions performed in United Nations governance of economic issues (Zacher 1999).

Governance is a process, or more accurately a set of social processes that entail contestation. Governance is not purely administrative, and enjoys a number of ancillary effects. Administrative procedures can channel contestation in constructive modes by providing rules for participation and reconciling differences.

Governance entails a shift in analytic focus from the policies that may contribute to the better management of marine resources to the process by which collective action proceeds. While policy analysts argue about the appropriate policies for better marine management, the full array of governance functions helps to understand the conditions under which integrated management is more likely to be widely applied. The United States National Research Council's Committee on the Human Dimensions of Global Change observes:

> Analyzing environmental governance as a problem of institutional design is useful because it reframes the central governance question from one of selecting a single best governance strategy (e.g. choosing between top-down regulation and market oriented policies) to one that considers a full range of governance options and seeks to match institutional forms to specific governance needs.
>
> *(Brewer and Stern 2005)*

In practice, these administrative functions are performed by a number of different groups of actors, including governments (or states), international organizations, MNCs, NGOs, and scientific networks. For the purpose of this chapter I will focus most intensively on the activities of international organizations, although a key criterion for the effectiveness of activities performed by international organizations involves their ability to constructively mobilize participation by other actor groups. Most effective governance in practice utilizes synergies from different groups that utilize institutionalized tensions amongst the parties, such as NGOs and MNCs in joint regulatory arraignments that yield more binding results than MNCs desire and command stronger compliance than NGOs are able to achieve through selective political campaigns.

2.3 Effectiveness

Lastly is effectiveness. That is, do the social institutions actually deliver the desired outcomes? This literature also has evolved over time. Lawyers traditionally investigated compliance, largely counting the number of ratifications and describing domestic enforcement mechanisms converting international obligations to domestic law.

More recently effectiveness itself has been inspected more closely. Questions of whether actions are adequate to achieve environmental protection are important from a policy perspective, as is the questions of causal inference raised by international relations scholars interested in the extent to which institutions are able to induce behavioral change by actors.

'Effectiveness' was widely studied in the 1990s, as academics and policy makers sought to identify factors that contributed to better international environmental cooperation and protection. Effective arrangements are those that entail policy changes by states in accordance with the intentions of negotiated treaties that lead to, or are likely to lead to, improvements in environmental quality (Mitchell, Weiss et al. 1999).

Yet in general it is difficult to measure or obtain good evidence about the extent of environmental change, much less the extent directly attributable to conscious efforts. Accurate, confident long-term data that is regionally specific and is based on the same indicators is hard to come by, as is good data on national compliance. Much easier to rely on proxies associated with effective behavior; that is, what factors have been present in the cases where analysts concur that effective governance occurs. In policy analysis terms this assessment looks at outputs over outcomes.

3.0 Criteria for evaluating marine governance

Studies of marine governance and environmental governance more generally have identified a number of factors that correlate with more effective governance. Examples of effective regimes where the environment is confidently believed to be improved, or on the path to improvement, include the stratospheric ozone regime, European acid rain, the management of Antarctic living resources, and efforts to protect the North Sea and Baltic Sea. Some mid-level successes in which environmental decline has been reversed include the Mediterranean, South Pacific and southeast Pacific regions.

Three causal mechanisms relate these governance functions to more effective marine governance. The principal causal mechanisms operating in marine governance are coercion, inducements, and persuasion. Some functions, particularly related to compliance, rely on coercion to compel states to behave in new ways and to pursue more active marine governance. Some functions enable states to get something else they value from engaging in marine governance: inducement. Other functions persuade states of the value of marine governance; either of the desirability from their own perceptions of national interests that marine governance is desirable, or that they should engage in more vigorous marine governance: persuasion. Particular governance functions, such as environmental monitoring, policy analysis, and public education, among others, influence effectiveness through multiple causal mechanisms, both inducements and persuasion. These forces operate directly on government administrations (the state) and on the population (the public) to whom they are accountable. Thus some capacity building efforts are addressed to building administrative capabilities, while other efforts are addressed to building and mobilizing public concern and pressure on their governments.

Thus capacity building may have multiple effects (Sagar and Vandeveer 2005). Capacity building can build administrative capacity for managing problems, as well as inducing longer term and unpredictable learning by government officials and policy elites.

Governance factors will not be equally influential on all countries. For instance, administrative capacity building may be most important for less developed countries. Public education may be more influential in richer democracies.

Table 14.2 summarizes the factors associated with effective marine governance. These factors and some best practices in performing those factors are discussed below.

3.1 Agenda setting

Agenda setting is necessary for a problem to enjoy sustained attention from the international community. Agenda setting entails monitoring, publicity and framing. Ongoing scientific monitoring is necessary to generate meaningful data about possible marine risks. Publicity is necessary to relay such findings to decision makers and the public. Frames are necessary so that responses and policies are made possible in a way that can generate responsible action.

TABLE 14.2 Summary table of effectiveness factors

	Assessment factors	*Best practices*
Agenda setting	Providing accurate environmental data through monitoring Publicizing findings Framing the issues broadly	Selective use of NGOs, IOs, states and scientific networks Nurture media contacts Diverse participation
Norm development	High-profile individuals associated with marine governance Collectively endorsed statements/ declarations Soft law	

Rule making	Venue Numbers (small number of actors whose behavior needs changing, relatively small number of actors involved in developed policy responses) Voting rules Frequency Perceived fairness Profile Financial transfers Technology transfers Knowledge Transfers Treaty design: Formal enforcement provisions (sanctions, arbitration procedures);Verification, monitoring Institutional density	Rely on legitimate IOs with autonomy and resources Technology panels with lead countries to identify BEPs and BATs UNEP Industry & Environment office, Tokyo Technology Center
Science-based rule making	Epistemic community Standing science panels	Interdisciplinary standing international panels Research precedes policy, or at least independent Info should be timely, authoritative, and useful
Compliance	Verification Sanctions Arbitration Capacity building • Administrative resources • Financial transfers • Technology transfer • Public education • Elite education • Knowledge Transfers	High-quality information through 3rd party impartial evaluations Shaming by NGOs LEAD training programs
Financing	Support for collective activities Support for national compliance Ability to mobilize funding for marine governance	Technology transfer Financial transfers Lead countries on panels
Synergies	Horizontal linkages Vertical linkages Joint activities	 Pool monitoring, training

3.1.1 Stable environmental monitoring programs are necessary to provide an accurate picture of marine conditions

Effective monitoring must provide accurate long-term data on compatible indicators over time. Accurate time series data is vital for providing an early warning sign of environmental threats, as well as assessing the impact of governance efforts. Good monitoring programs should be capable of providing timely data on environmental quality that is capable of ensuring that potential alarms reject false positives and false negatives.

Monitoring may be done by networks organized by international institutions—such as UNESCO's IOC for the open oceans, UNEP's monitoring networks undertaken in concert with FAO, WHO and WMO, and the Mediterranean Action Plan's monitoring network—or by networks organized by governments or through networks of scientific institutions. In all cases national laboratories are coordinated through an international network through the guidance of international organizations. Some other bodies collect and publicize monitoring data, including GESAMP, UNEP's Global Environment Outlook, and ICES.

Satellite monitoring is surely helpful for assessing organic threats, and oil spills, although they are not adequate for assessing inorganic contaminants (UNEP 2013). Moreover, reliable remote sensing requires periodic interaction with ground stations and direct observations to validate observations and data.

NGO campaigns can be highly effective, so long as they are based on accurate monitoring. For instance, the shifting Baselines Ocean Media Campaign issues Rotten Jellyfish awards for media stories documenting the worst instances of marine decline (Shifting Baselines 2013).

Few monitoring programs pursue long-term synoptic measurements. In part because they are funded by governments and intended to serve as early warning signals, such monitoring schemes are abandoned if they do not yield worrisome results. Yet these then fail to provide a meaningful baseline, nor are they available should environmental stresses grow. A current temptation to monitor only hot-spots, in part for reasons of cost-cutting, runs the risk then of failing to provide true warnings of threats from substances whose study is ignored.

There is untapped policy potential in considering non-traditional indicators of environmental disruption, such as focusing on the behavior of benchmark species or of social indicators that would be reflective of environmental disruptions (such as prices of species as a sign of scarcity, or migratory patterns of people highly dependent upon access to marine resources).

3.1.2 Environmental alarms must receive publicity in order to attract widespread attention

Decision makers seldom learn directly of results from monitoring activities. Monitoring results must be publicized broadly for the mass public to become concerned and for governments to react. Publicity often occurs from widely publicized disasters, such as Jacques Cousteau's public announcements of the impending death of the Mediterranean in the early 1970s, or oil spills in the open oceans for oil spill regimes. Thus governance arises in response to an unusual event, rather than from observations of changes in natural systems. NGO publicity campaigns and NGOs' media contacts also play a role in disseminating monitoring findings. Thus effective governance institutions may do well to foster relationships with the media to be able to relay new findings.

3.1.3 Marine governance issues are more likely to be addressed in an integrated manner when they are framed comprehensively

The existence of accepted norms or frames can provide a roadmap to governance for a particular issue by identifying presumptive institutions for governance and doctrines or policies for governance. In order for information to be regarded as meaningful or authoritative it must be presented in a way that implies human responsibility and also that such observations are not ambiguous or may be a fluke or attributable to non-anthropogenic sources. Participating organizations must be seen as legitimate, so that their information will be processed quickly and taken seriously. Membership of the organizations must be sufficiently broad and diverse to ensure that information is sufficiently broad, encouraging an integrated view of marine governance. The mandate of the framing institution must be sufficiently broad and comprehensive so as to confer the multiple ingredients for a comprehensive marine governance approach.

For instance, in the early days of the Mediterranean Action Plan FAO provided monitoring reports of Mediterranean water quality which emphasized the relatively depleted organic material in the Sea, consistent with its institutional mandate to focus on factors relating to fisheries productivity. UNEP's monitoring at the time included a far broader array of indicators which also related to chemical contaminants in the Sea. Similarly studies of IMO in the 1980s criticized its agenda and framing for representing the interests of its principal constituency, tank owners, and not a more comprehensive or environmental policy agenda (M'Gonigle and Zacher 1979). Sponsoring organizations also have policy orientations which may affect the presumptive substance of negotiated arrangements which they will help guide. For instance, Bretton Woods institutions tend to favor economic policy levers, whereas UNEP has been more inclined towards adopting environmental quality standards.

3.2 Norms: widely accepted norms can stiffen states' resolve to participate in marine governance

Widely accepted norms of behavior (or principles) can set states' expectations about marine governance, as well as providing legitimate guidelines against which they can be held politically accountable. In the international realm functionally specific universal norms have been developed in the areas of human rights and refugees protection. Such norms typically take the form of UN declarations and soft law. They are developed through the actions of transnational activist networks. Independent International Commissions are often associated with the development and popularization of such ideas. High-profile international figures are then associated with the adoption and diffusion of such norms, using their international positions (such as the High Commissioner for Human Rights and the High Commissioner for Refugees) to further promote the norms and to cajole other parties to adhere to the norms. In the environmental arena writ large the Executive Director of UNEP has aspired to such a role, although UNEP lacks sufficient authority in the international system to provide a firm foundation for such action.

Ongoing efforts have been promoted to establish principles of and for marine governance by such luminary international policy advocates as Elizabeth Mann Borgese and Arvid Pardo, through the Pacem in Maribus conferences, and at the 1998 UN Year of the Oceans annual global coastal zone management conferences (Costanza 1998; Independent World Commission on the Oceans 1998). A number of norms have been asserted for environmental and marine governance, although they do not appear to command widespread support, and many are probably incompatible in practice Some current governance norms in wide circulation include common but differentiated responsibility: the precautionary principle, the polluter pays principle, and sustainable development, to name a few.

Corporate guidelines can provide a normative basis for marine governance behavior as well. UNEP's Industry and Environment office in Paris has promoted voluntary guidelines documenting best environmental practices and identifying best environmental technologies for use. The Exxon Valdez principles provide a corporate set of norms. Industry groups, particularly when operating in conjunction with NGOs and scientists, can provide powerful guidelines for effective marine management, such as are emerging through the Marine Stewardship Council.

3.3 Rule making and negotiated settlements

A number of factors combine to strengthen the bargaining context within which environmental negotiations are conducted and thus make it easier for states to reach meaningful agreements. An array of factors are identified that improve the prospects of reaching binding obligations. In the absence of such factors negotiated arrangements are only likely to take the form of least common denominator type efforts, such as occur in most international fisheries.

3.3.1 Venue

At the very least a venue is necessary for discussions to occur. A venue is typically provided by a host government, international organization, or, after entry into force, the COP.

The venue must be regarded as legitimate for member governments to willingly participate. Not just any venue will work to provide the necessary logistical and administrative functions. Potentially contentious issues can be far more expeditiously addressed in institutions that command legitimacy in the eyes of stakeholders. Such institutions are increasingly those with representation open to elements of civil society. In addition legitimacy rests on a widespread sense or perception of the fairness by which decisions are taken, including the fairness of the process by which negotiated outcomes are pursued as well as the distributional equity of the outcomes themselves.

3.3.2 Participation

The number of participants is important. A relatively small number of parties make it easier to negotiate and to develop meaningful policies to mitigate shared problems. Negotiations are more efficient with smaller numbers. Consequently, efforts that have a minimal number of parties, organized within regional bodies rather than global functional bodies, are more likely to be able to efficiently resolve political disagreements and reach meaningful negotiated settlements expeditiously. In practice bloc-diplomacy and caucusing often reduces the number of parties that have to participate in negotiations.

Access by NGOs, scientists and civil society are also likely to make for more robust and legitimate governance arrangements. Widespread involvement of stakeholders in negotiations makes the arrangements more effective.

3.3.3 Frequency of meetings

The more interaction between parties the more likely it is to see meaningful compromises and more integrated policies adopted, and for parties to learn from one another. Continuing negotiations rather than one-shot negotiating sessions are better at generating meaningful compromises, as over time states grow more familiar with one another's positions and are also willing to make concessions at one session in the anticipation that those concessions may be reciprocated at future meetings. Single high-level conferences are unlikely to yield the same results as ongoing preparatory meetings leading to treaty adoption followed by ongoing COPs.

3.3.4 Voting rules

Voting rules can also be important. Consensus makes it most difficult to achieve agreement, whereas a simple majority facilitates agreement. Weighted majority schemes have been devised to assure a better trade-off between legitimate outcomes and the potential for deadlock or bullying

3.3.5 Political profile of negotiations

A high-level profile for negotiations also encourages breakthroughs and meaningful commitments that mid-level bureaucrats lack the authority to make. Thus, many COPs and meetings now include brief high-level Ministerial meetings before or after the longer sessions, at which such deals may be presented. Also, the North Sea and Baltic countries intersperse their annual meetings with periodic Ministerial meetings every 3–5 years at which major commitments and new agendas are developed for the regimes.

3.3.6 Institutional strength: the more resources institutions can provide the more likely states are to support marine governance

The institutional sponsor of marine governance negotiations must have independent resources in order to induce recalcitrant parties to accept compromises and move beyond a least common denominator negotiated outcome. Such resources typically include the ability to provide material inducements for negotiating parties to accept meaningful marine governance arrangement (note that the causal mechanism by which such resources induce state compromise may have either to do with their administrative utility for improving a state's ability to actually enforce its obligations, or because these are generic resources of value to the state for which they will make compromises).

Three broad sets of factors have been identified that are associated with inducing states to support and comply with environmental treaties. These are various forms of transfers or rewards that states may enjoy as a consequence of cooperating: financial transfers, technology transfers, and knowledge transfers. Financial transfers provide money for improving compliance with international obligations, and are particularly attractive to poorer developing countries. Technology transfers include sales and gifts of environmental clean-up technology and technical equipment for environmental monitoring. Knowledge transfers consist of training programs for government officials in environmental management, monitoring and verification activities. Knowledge transfers also include environmental training programs for national scientists and even NGOs.

3.3.7 Treaty design

Ensuring effective governance requires that participants are able to develop policies in which they have faith, and also have the political and administrative ability to enforce their wishes. Without this other countries are reluctant to enter into joint arrangements, so that the resources are under-protected, and also efforts are not effective if states cannot enforce their obligations.

International environmental lawyers believe the most important legal design provisions include policy verification elements, environmental monitoring, arbitration and adjudication procedures, and sanctioning mechanisms for non-compliance.

3.3.8 Building public concern

Heightened national concern can pressure governments to take stronger action on the environment. Effective regimes and organizations have included programmatic elements for building national concern. These efforts include public education campaigns; sponsoring TV and radio and media shows; and also expanding the participation of national NGOs and scientists in international discussion. Building environmental norms at the international level can heighten national concern.

Similarly, publicizing monitoring results can build national environmental concern as well as contributing to improved compliance.

UNEP has engaged in a number of public education efforts, including radio and television shows, posters, popular print materials, and developing rotating museum exhibitions. NGOs have much experience with public education campaigns that may be informative to study to yield best practices. Labeling and certification campaigns also serve an educational role.

3.4 Science-based rule making

With the involvement of organized scientific inputs to the policy process negotiated outcomes are much more likely to yield integrated management efforts rather than mere political compromises. For instance, in the North Sea marine governance efforts take the form of across the board reductions in emissions of a variety of substances that were chosen by a process of political compromise. Conversely, in the Mediterranean the specific substances and the environmental standards chosen for those substances reflect scientific consensus about the nature of the threats to the Mediterranean environment.

Much experience has been amassed in the last 20 years about mobilizing environmental science for science-based rule making. Global environmental assessments have been completed for ecosystems, climate change, biodiversity, and other topics. Global marine assessments have been conducted for fisheries-related issues, ecosystems assessments, and other topics. While generating usable information, global assessments tend to be expensive (estimates of the Millennium Ecosystem Assessment run at 20 million dollars, with similar values for each IPCC assessment of global warming) and may monopolize the available scientific community.

A number of standing commissions also operate, such as GESAMP. GESAMP suffers from its lack of close connection to the member governments and decision makers. Some IOs have their own science network resources that can be deployed, such as ICES. While ICES has a comparative advantage for fisheries knowledge, it lacks a specialized network for pollution issues, and does not have the ear of decision makers. Some regimes have standing research bodies that are responsible for developing independent policy advice about the marine resource, identifying problems, and suggesting environmental or resource standards for adoption. These include the Mediterranean, North Sea and Baltic, CCAMLR, and Whaling.

Studies of environmental assessments that were intended to identify usable policy knowledge for environmental governance have identified a number of salient points about assessments that successfully informed negotiated rule making. The core group developing and disseminating policy-relevant science must be part of a network of like-minded scientists—an epistemic community. In addition the knowledge deployment process must be seen as legitimate and be accessible to decision makers. Assessments must be credible, salient and legitimate (Haas 2004a; Haas 2004c; Farrell and Jaeger 2005). Credibility means that the key knowledge producers and their consumers believe their product is true. Legitimate means that the claims are believed to be legitimate, that is, developed through a process that minimizes the potential for bias and is more equitable in terms of participation by those who are dependent upon the information. Finally saliency means that such information is provided in a timely manner and contains information that is useful for making public policy by decision makers: that is, in practice that it arrives in conjuncture with the policy process and provides advice which can be converted into laws or decisions by decision makers.

Criteria of these points have been elaborated by UNEP (UNEP and UNESCO 2003):

> Legitimacy: Governance efforts are undertaken at country request or in response to international/regional convention; national stakeholders involved in all phases.
>
> Credibility: QA (quality assurance) mechanisms are in place, external peer review method guidelines adopted with regular review, assessment is based on empirical data, assessment involves partners, and assessment uses an indicator framework.
>
> Saliency: assessment responds to a convention or a national request, is regular, provides policy advice, has provision for review, identifies policy makers as end-users, has stakeholder involvement, outputs are oriented to user, information freely available.

Best practices on mobilizing scientific networks to provide policy-relevant information include (Haas 2004):

- Create standing international interdisciplinary scientific panels or committees organized around specific topics.
- Create subcommittees responsible for different functions of governance, such as basic research, environmental monitoring, policy analysis, and policy verification and evaluation.
- Carefully survey the scientific population to identify individuals who share causal understandings.
- Ensure that networks and international panels have interdisciplinary representation, including the social sciences. Individuals should have high regard in their own disciplines as well as being able to talk to experts from other disciplines.
- Funding should come from multiple sources.
- Avoid governmental designation of scientists to meetings and to participate on panels.
- Assure timely submission of scientific reports before political meetings.
- Models and assessments should be conducted at geographic scales that are meaningful to policy makers.

Broader considerations of the proper institutional design of science policy entails timing: When consensus has been achieved before an issue reaches the agenda and policy discussions begun, then scientists can merely be introduced as experts, following the lessons above. However, at times it is necessary to simultaneously develop scientific consensus and advance policy debates. For such issues, such as was the case in the Mediterranean and Ozone regimes, the parallel development of science and policy must be kept insulated from ongoing policy debates, with the two streams united only when consensus has been achieved. In other cases, where consensus remains elusive and policy debates have already attained their own momentum, as in climate change and biodiversity, it may be best if the two activities can be kept as separate as possible.

3.5 Enforcement and compliance

Not all parties, even well-intentioned ones, necessarily live up to their international obligations. Enforcement and compliance factors help induce or teach countries to integrate integrated management and ecosystems management techniques into national marine practices. Several factors have been identified as critical: verification measures, arbitration and adjudication measures, clearly written commitments, sanctions for non-compliance, resources to achieve compliance, and environmental monitoring.

3.5.1 Verification measures make it easier to identify and deter non-compliance

Verification is important to assure that parties are living up to their obligations. States are more likely to comply with their international obligations if infractions are promptly and accurately reported. Most treaties and regimes include verification measures that collect information about parties' compliance with their obligations. In practice it appears that compliance is poorly verified, and presumed to vary widely by country and by regime (Haas 1998; UNCSD 1999).

Treaties vary widely in terms of who collects this information and how frequently. This is best done by impartial third parties (IOs, NGOs, or contracted private firms) collecting disseminating information about the activities of states. Self-reporting runs the risk of allowing states to lie or misrepresent their records. For instance, after the Cold War ended it was discovered that the Soviets had been routinely lying about their whale catches and the amount of radioactive wastes they were dumping in the ocean. The Montreal Protocol encourages self-reporting by states that are incapable of reducing CFC emissions, but the Montreal protocol is unusual in that it also provides financial transfers for states to construct non-CFC based chemicals. In practice most treaties rely on a complicated mix of verification arrangements.

Verification is also performed through general reports about national environmental performance, as conducted by the OECD and the EEA for European countries. A review of these efforts finds that they are often weak in terms of comprehensive evaluations, because they do not rely on the same indicators over time, and may not be published in a timely fashion. They do have an effect on building public concern (Rosenstrom and Lyytmaki 2006). EU annual reporting on beach quality is also an effective means of verification, so long as the information is widely available, the assessments are believed accurate, and tourists are aware of the scheme.

A key lesson about verification drawn from arms control studies is that the most accurate and credible verification activities are those conducted by impartial third parties. The Baltic Sea parties have discussed the prospects of surprise inspections by third parties, but have not adopted these procedures.

3.5.2 Arbitration and adjudication can mitigate disputes about compliance

Arbitration and adjudication arrangements are necessary for reconciling disputes about compliance and interpretation of obligations. There are widespread provisions for the creation of arbitration bodies in most marine governance regimes, but none have ever been convened.

3.5.3 Clearly documented obligations make parties more willing to commit to negotiated rules

Clearly documented obligations make parties more willing to commit to negotiated rules because they make signatory parties more confident about the expectations and behavior of other parties using the marine resource on which they depend. Not only do they contribute to stronger negotiated outcomes, they can make compliance easier for states. Once written, most stringent rules become public knowledge as potential BEPs (Environmental Data Services 1992).

3.5.4 Sanctions for non-compliance can deter non-compliance

Clear cut penalties for non-compliance can deter non-compliance. Formal enforcement provisions include economic sanctions against parties that are in violation, and legal provisions for arbitration over disputes in interpretation or for enforcing sanctions. Curiously, few international environmental treaties contain strong sanctions or compliance mechanisms.

In practice sanctioning mechanisms take indirect forms. Governments can chose to penalize other countries who are seen to be violating marine governance standards, such as the US did to Mexico in the instance of tuna/dolphins. IOs can apply conditionality terms to countries seen to be violating their environmental norms. The World Bank can suspend loans, as it did in with Brazil in the 1980s over Amazonian forestry policies. Less acute means of pressure can also be applied, through parallel institutional arrangements (see synergies below).

A major mechanism of recent sanctioning occurs through NGO shaming campaigns. If accurate and timely information about infractions is available, NGOs may launch shaming campaigns against governments, firms or municipalities that have failed to meet their international obligations.

3.5.5 Providing administrative and material resources can promote compliance

Sometimes states do not achieve compliance for reasons of inadequate capacity. If mechanisms are available to generate administrative- and resource-based resources for member governments then they may help promote compliance.

Enhancing access to providing material resources can help states and other actors enforce their marine governance commitments. Marine monitoring technologies, clean-up technology, training in its use, can all help. Direct outreach to include the private sector through identifying market opportunities for firms has helped achieved compliance in the Mediterranean, Baltic and North Seas. Certification schemes can also help encourage private investment in activities that can contribute to compliance.

Many governments require administrative assistance in complying with marine governance arrangements as well. Officials may benefit from training in ICZM techniques, as well as in verification and enforcement techniques. Training activities have been provided by UNEP, the World Bank, through national ODA, and also by NGOs supported by foundations, such as LEAD. Some states may also need direct financial transfers to pay for the costs of inspectors and monitoring.

3.5.6 Building social capacity can improve compliance

Building political will may make governments more willing to commit resources to marine governance. Public education efforts can mobilize public pressure and demands for more comprehensive marine management (Weinthal 2003). If the USA is any example, popular movies on marine subjects may have potential.[1] Governance activities that mobilize NGOs and domestic scientific groups can contribute to broadening public awareness and concern at the domestic level. Democratization is a means of capacity building.

3.5.7 Environmental monitoring provides an accurate picture of environmental quality

Scientific certainty about the causes of marine problems and the extent of the threats enhance the response by which governments will mobilize resources to promote marine governance. Robust monitoring programs generate the necessary information about problems to galvanize concern and commitment for compliance, as well as providing ongoing information for enabling verification and also feedback for ongoing assessments, as states can ascertain if their efforts are having an impact, and can also develop new policies if new threats are identified, or if earlier concerns are shown to be exaggerated. Enables resilient responses to new threats, and provides baseline for self assessments.

3.6 Financing

Adequate financing is necessary to pay for collective governance activities, as well as to pay for national responses. The total sum required for marine governance is unknown, although estimates in 1992 of what it would cost to achieve the overall Sustainable Development goals elaborated in Agenda 21 were in the range of 65–70 billion dollars per year.

The national contributions to the annual budgets of international organizations (such as FAO, UNESCO, and UNEP) and to the dedicated trust funds for the support of specific regimes tend to be modest at best. Such financing is supplemented by project grants to COPs, and at times IOs, from the GEF, UNDP, World Bank, and regional development banks. National financial support tends to come from these sources as well as from private DFI and foreign aid (ODA) from governments. DFI now vastly eclipses the annual flows of ODA for most countries other than the least developed.

Effective marine governance thus relies on the ability of the key actors to access and mobilize such financial transfers.

Private technology transfers have been encouraged by creative institutional design. In the Mediterranean UNEP and the IMO sponsored trade shows at which information about oil spill technology was made available to government officials. In the North Sea and Baltic technical technology panels operate to highlight best environmental technologies for marine clean-ups. The panels are chaired by countries with a specialization in the specific technologies being investigated by the panel, thus providing the chairs with a strong incentive to publicize new technologies.

NGOs have developed some novel financial instruments through partnerships with local firms, developing ecotourism projects and debt for nature swaps (Miles 2005).

Economic growth from trade liberalization is often lauded as a source of increased revenue sources for environmental protection. In addition, as economies develop they shift away from pollution-intensive industries towards more benign service sectors, along with presumably increasing public environmental concern and willingness to pay for environmental and resource protection (Cole and Neumauyer 2004). But such economic benefits are contingent on political will to commit to activities that will contribute to better marine governance, rather than accentuate the conflicts between economic activities that affect marine quality and resources. Thus, the greater the rate of economic growth, the stronger the need for institutional mechanisms and efforts to mobilize public concern.

3.7 Synergies

Governance efforts do not operate in a vacuum. Indeed, given the international matrix of states' social obligations, isolation inhibits effectiveness in most instances.

While there has been much recent attention to synergies between governance arrangements, much less effort has been directed towards clearly identifying the dimensions or mechanisms by which synergies occur, and how synergies affect marine governance.

It would seem that the relative poor performance of fisheries governance vis-à-vis other marine governance efforts has to do, in part, with the relative isolation of fisheries regimes from the broader institutions of global environmental governance.

3.7.1 A large number of horizontal linkages will increase the robustness of marine governance efforts over time

Horizontal linkages have to do with the overlapping memberships of marine governance arrangements (COPs, IOs, regimes). Overlapping memberships offers the potential for states to pursue policy harmonization across the institutions of which they are a member, in order to rationalize policies and ensure that their firms face uniform regulations (Stokke and Coffrey 2004).

A few examples can indicate the potential from horizontal linkages that contributed to efforts to strengthen standards and broaden governance coverage. TBT was first regulated in the North Sea by OSPAR, and then the standards were transmitted to the Mediterranean by the IMO (Ship Paint Pollutes Mediterranean). The overlapping memberships of the EU, North Sea and Baltic regimes led environmental leader countries to seek to transfer commitments made within one regime to the others. LRTAP even adopted a cooperative arrangement with PARCOM and the Med Plan to take account of airborne emissions into the ocean. Discharge of PCBs into the ocean was first regulated at North Sea Ministerial Conferences in1984 and 1987. The OECD then adopted a decision on PCBs, followed by the Med Plan in 1995, the EU in 1996, UNEP's Governing Council in 1996, and ultimately in the Stockholm Convention on POPs (Koppe and Keys 2001; Santillo, Johnston et al. 2001).

Forum shopping may be a mechanism for transmitting obligations. Anti-whaling advocates have tried to add whales to the CITES list of protected species.

3.7.2 Tight vertical linkages may increase the robustness of marine governance efforts over time, if those linkages are tied to integrated frames

Vertical linkages have to do with the extent to which issue-specific governance efforts are subordinate to universally shared obligations or principled commitments or to a legally binding formal authority. In this regard higher level notions are legally binding obligations or powerful international organizations whose enforcement powers reach down to marine management issues.

Tight connections between formal authorities operating at different physical scales may improve the potential for mitigating international externalities. For instance, if local authorities are involved in decision making with national authorities, governance efforts may be better able to capture cross-scale effects, as well as improving compliance prospects though including stakeholders in decision making (Young 2002).

While tight nesting may relate marine governance to more comprehensive universal norms or prospects for legal enforcement, not all higher level international frames encourage integrated or comprehensive planning. Many international economic frames encourage market-based policy approaches and tools. For vertical nesting to promote integrated marine management the marine management efforts must be subordinate to complementary higher level integrated frames. In this regard the trade and environment conflict in the WTO is emphasized by environmentalists, because of a fear that trade liberalization concerns (or frames) will prevail over environmental regulations.

3.7.3 Pooling resources may be a source of efficiency

Many marine governance arrangements require similar governance functions. Administrative efficiency gains can be achieved by joint provision of many overlapping functions. In addition to the administrative benefits from pooling efforts it is possible to build cross-region networks capable of building political will and capacity within their own governments. Many monitoring activities can be done across regions. Much training—of government officials in ICZM techniques as well as laboratory scientists in research, monitoring and verification techniques—can easily be done to include multiple regimes. Public education activities can easily be pooled. Other functions are region specific and can't be as easily pooled.

It is important to keep these efforts focused on the core elements of marine governance, still. During the 1990s UNEP sought to hook many regional training efforts to the more popular and better funded climate change and biodiversity projects, which actually seemed to have the effect of diverting attention from marine governance to the more lucrative issue areas.

3.7.4 Opportunities for sharing experiences

Many governments and actors can learn about integrated management and its application through indirect links with one another. These can occur in several ways. Actors can share best practices and experiences at seminars or meetings, or also through indirect networks of policy specialists. Thus marine governance can be strengthened through convening such meetings, and also by building and strengthening networks of expertise.

Prominent actors, including the World Bank, and the United States, provide a demonstration effect for other countries to emulate. ICZM, environmental impact assessments, and particular environmental standards are often imitated from authoritative sources. European governments also look to one another for lessons. Joint workshops at which national experiences are presented also provide a mechanism for lesson drawing and sharing (Janicke and Carius 1992; Janicke and Weidner 1995; Janicke and Weidner 1995; Weidner and Janicke 2002).

4.0 Conclusion

Much paper and many bytes have been devoted to writing and evaluating marine governance. In this paper I have presented some of the broad outlines of the challenges presented by marine governance, and some evocative lessons that can be applied to assess and improve marine governance in the future. Further work needs to be conducted to carefully assess individual efforts to appraise progress to date, to ascertain the potential for improvements, as well as to gauge the potential contributions to effective marine governance by multiple actors active in global governance. Such assessments would focus on the criteria presented in this paper, as well as surveying participant stakeholders regarding their satisfaction, regime successes, regime failures, and lessons about best practices.

Marine governance will face new challenges, beyond that of merely appraising and reinforcing existing arrangements. New threats to marine governance are sure to arise, and will have to be addressed. Functional gaps in marine governance should be systematically investigated and addressed. Evolving governance highlights the growing role of non-state actors, and marine governance efforts will have to be reformed and developed to accommodate the involvement of new non-state actors, and to constructively deploy their contributions to improve marine governance efforts. Lessons drawn from appraisals of ongoing efforts can contribute to institutional reforms and policy lesson drawing for improved marine governance in the future.

Notes

* Copyright © 2008 PEMSEA. This chapter was first published in Chua Thia-Eng, Gunnar Kullenberg, and Danilo Bonga (eds) *Securing the Oceans* Quezon City, Philippines: Global Environment Facility/United Nations Development Programme/International Maritime Organization Regional Programme on Building Partnerships in Environmental Management for the Seas of East Asia (PEMSEA) and the Nippon Foundation, 2008. Reprinted with permission of PEMSEA.

1 Such as Al Gore's *An Inconvenient Truth* on global warming, but possibly the *Free Willy* movies for marine conservation issues.

References

Adeel, Z. (2003). *East Asian Experience in Environmental Governance*. Tokyo, United Nations University Press.

Ascher, W. (2000). "Understanding Why Governmens in Developing Countries Waste Natural Resources." *Environment* 42(2): 8–18.

Bell, R. G. (1997). "Developing a Culture on Compliance in the International Environmental Regime." *The Environmental Law Reporter* 27(8): 10402–10412.

Bell, R. G. (2000). "Building Trust." *Environment* 42(2): 20–32.

Bell, R. G. and C. Russell (1998). "Environmental Policy for Developing Countries". *Issues in Science and Technology*. R. G. Bell.

Benedict, K. (2001). "Global Governance". *International Encyclopedia of the Social and Behavioral Sciences.*

Boelaert-Suominen, S. and C. Cullinan (1994). Legal and Institutional Aspects of Integrated Coastal Area Management in National Legislation. Rome, Food and Agriculture Organization of the United Nations: 118.

Brewer, G. and P. C. Stern, Eds. (2005). *Decision Making for the Environment.* Washington D.C., National Academies Press.

Chazournes, L. B. d. (2005). "The Global Environment Facility: A Unique and Crucial Institution." *RECEIL* 14(3): 193.

Choucri, N. (1993). *Global Accord.* Cambridge, MA, MIT Press.

Cicin-Sain, B. and R. W. Knecht (1998). *Integrated Coastal and Ocean Management: Concepts and Practices.* Washington, D.C., Island Press.

Cole, M. A. and E. Neumauyer (2004). "Environmental Policy and the Environmental Kuznets Curve: Can Developing Countries Escape the Detrimental Consequences of Economic Growth?". *Handbook of Global Environmental Politics.* P. Dauvergne. Cheltenham, UK, Edward Elgar.

Costanza, R., Francisco Andrade, Paula Antunes, et al. (1998). "Principles for Sustainable Governance of the Oceans." *Science* 281: 198–199.

Environmental Data Services (1992). *Dangerous Substances in Water: A Practical Guide.* London.

Farrell, A. and J. Jaeger, Eds. (2005). *Assessments of Regional and Global Environmental Risks: Designing Processes for the Effective Use of Science in Decisionmaking.* Washington, D.C., Resources for the Future.

Haas, P. (1998). "Compliance with EU Directives: Insights from International Relations and Comparative Politics." *Journal of European Public Policy* 5(1): 38–65.

Haas, P. (2004a). "Addressing the Global Governance Deficit." *Global Environmental Politics* 11(4): 1–19.

Haas, P. (2004b). "Science Policy for Multilateral Environmental Governance". *Emerging Forces in Environmental Governance.* N. Kanie and P. Haas. Tokyo, UNU Press: 115–136.

Haas, P. M. (2004c). "When Does Power Listen to Truth? A Constructivist Approach to the Policy Process." *Journal of European Public Policy* 11(4): 569–592.

Haas, P. M., R. O. Keohane, et al., Eds. (1993). *Institutions for the Earth: Sources of Effective International Environmental Protection.* Global Environmental Accords Series. Cambridge, The MIT Press.

Independent World Commission on the Oceans (1998). *The Ocean Our Future.* New York, Cambridge University Press.

Janicke, M. and A. Carius (1992). Framework Conditions for Environmental Policy Success: An International Comparison. *Annual Meeting of the American Political Science Association.* Chicago.

Kanie, N. and P. M. Haas, Eds. (2004). *Emerging Forces in Environmental Governance.* Tokyo, UNU Press.

Kay, D. and H. Jacobson, Eds. (1983). *Environmental Protection: The International Dimension.* Totowa, Allanheld Osmun.

Keohane, R. O. (1989). *International Institutions and State Power.* Boulder, Westview Press.

Koppe, J. G. and J. Keys (2001). "PCBs and the Precautionary Principle". *Late Lessons From Early Warnings: The Precautionary Principle 1896–2000.* E. E. Agency.

Krasner, S. D., Ed. (1983). "Structural Causes and Regime Consequences: Regimes as Intervening Variables". *International regimes.* Ithaca, NY, Cornell University Press.

M'Gonigle, R. M. and Mark W. Zacher (1979). *Pollution, Politics, and International Law: Tankers at Sea*. Berkeley, University of California Press.

Miles, E. L. (2002). *Environmental Regime Effectiveness: Confronting Theory with Evidence*. Cambridge, MA, MIT Press. Environmental policy International Cooperation Case studies.

Miles, K. (2005). "Innovative Financing: Filling in the Gaps on the Road to Sustainable Environmental Funding." *RECEIL* 14(3): 202–211.

Mitchell, Weiss, et al. (1999). "The Effect of Global Environmental Regimes." *International Political Science Review* 20.

North, D. C. (1990). *Institutions, Institutional Change and Economic Performance*. Cambridge, Cambridge University Press.

Ostrom, E. (1990). *Governing the Commons: The Evolution of Institutions for Collective Action*. Cambridge, Cambridge University Press.

Rosenau, J. N. (1992). "Governance, Order, and Change in World Politics". *Governance without Government*. J. N. Rosenau and E.-O. Czempiel. Cambridge, Cambridge University Press.

Rosenstrom, U. and Jari Lyytmaki (2006). "The Role of Indicators in Improving Timeliness of International Environmental Reports." *European Environment* 14: 32–44.

Ruggie, J. G. (1975). "International Responses to Technology." *International Organization* 29(3): 557–584.

Sagar, A. D. and S. D. Vandeveer (2005). "Capacity Building for the Environment." *Global Environmental Politics* 5(3).

Sand, P. H. (1990). *Lessons Learning in Global Environmental Governance*. Washington, D.C., World Resource Institute.

Sandler, T. (2004). *Global Collective Action*. Cambridge, Cambridge University Press.

Santillo, D., P. Johnston, et al. (2001). "Tributyltin (TBT) Antifoulants: A Tale of Ships, Snails and Imposex". *Late Lessons from Early Warnings: The Precautionary Principle 1896–2000*. E. E. Agency.

Shelton, D. (2000). *Commitment and Compliance*. Oxford, Oxford University Press.

Shifting Baselines (2013). from http://www.shiftingbaselines.org/news/jellyfish_awards.htm.

Ship Paint Pollutes Mediterranean. "Toxic ship paint pollutes Mediterranean sea." from http://ens.lycos.com/ens/sep2w000/2000L-09-05-10.html.

Solkinoff, E. B. (1972). *The International Imperatives of Technology*. Berkeley, Institute of International Studies of California.

Sorensen, J. C. and S. T. McCreary (1990). *Coasts: Institutional Arrangements for Managing Coastal Resources and Environments*. Washington, D.C., U.S. Agency for International Development.

Speth, J. G. and P. M. Haas (2006). *Global Environmental Governance*. Washington, D.C., Island Press.

Stokke, O. S. and C. Coffrey (2004). "Precaution, ICES and the Common Fisheries Policy: A Study of Regime Interplay." *Marine Policy* 28: 117–126.

The Social Learning Group (2001). *Social Learning and the Management of Global Environmental Risks*. Cambridge, MIT Press.

UNCSD (1999). *Oceans and Seas Report of the Secretary General Trends in National Implementation*.

UNEP (2013). "Satellite Systems for land cover monitoring." from http://www.na.unep.net/.

UNEP and UNESCO (2003). *Global Marine Assessments: A Survey of Global and Regional Marine Environmental Assessments and Related Scientific Activities*, UNEP.

Vallega, A. (1992). *Sea Management: A Theoretical Approach.* London, Elsevier Applied Science.

Victor, D. G., K. Raustiala, et al. (1998). *The Implementation and Effectiveness of International Environmental Commitments: Theory and Practice.* Laxenburg, Austria; Cambridge, MA, International Institute for Applied Systems Analysis; MIT Press.

Weidner, H. and M. Janicke, Eds. (2002). *Capacity Building in National Environmental Policy.* Berlin, Springer.

Weinthal, E. a. Y. P. (2003). "Two Steps Forward, One Step Backward: Societal Capacity and Israel's Implementation of the Barcelona Convention and the Mediterranean Action Plan." *Global Environmental Politics* 3(1): 51–71.

Weiss, E. B. and H. K. Jacobson (1998). *Engaging Countries: Strengthening Compliance With International Accords.* Cambridge, MA; London, MIT Press.

Wettestad, J. (1999). *Designing Effective Environmental Regimes: The Key Conditions.* Cheltenham, UK, Edward Elgar Publishing Company.

Young, O. (1989). *International Cooperation.* Ithaca, Cornell University Press.

Young, O. R. (1997). "Global Governance: Toward a Theory of Decentralized World Order". *Global Governance: Drawing Insights From the Environmental Experience.* O. R. Young. Cambridge, MA, MIT Press: 273–299.

Young, O. R. (1999). *The Effectiveness of International Environmental Regimes: Causal Connections and Behavioral Mechanisms.* Cambridge, MA, MIT Press.

Young, O. R. (2000). *Governance in World Affairs.* Cornell University Press.

Young, O. R. (2002). *The Institutional Dimensions of Environmental Change.* Cambridge, MA, MIT Press.

Zacher, M. (1999). *The United Nations and Global Commerce.* New York, United Nations.

Zaelke, D., D. Kaniaru, et al., Eds. (2005). *Making Law Work: Environmental Compliance & Sustainable Development, 2 volumes.* London, Cameron May.

PART V

Science policy

It is now clear that science and expert advice have an impact on collective governance and policy effectiveness. But which knowledge claims and epistemic communities are likely to be heard and which aren't? In these pieces I seek to develop a better understanding of the conditions under which epistemic communities influence international decisions in the environmental realm.

In general their advice is channeled to international negotiations through international science panels and through domestic channels. The pieces in this section look at international science panels. What is key is the nature of the consensual knowledge, and the organizational channels by which the epistemic communities are able to transmit their knowledge to governments.

"When does power listen to truth?" (Chapter 15) focuses in detail on the relative failure of the Intergovernmental Panel on Climate Change (IPCC) to significantly influence global negotiations on climate change. While the international scientific community agrees on the existence of anthropogenic climate change and the urgency of addressing it, the political pace of negotiations is seemingly slower than the glacial melting from climate change. It argues that epistemic communities must command "usable knowledge" and enjoy regular channels of access to governments. Moreover, the science must come from groups which enjoy social authority, so that delegation or deference to their information is more widely justifiable. In this regard epistemic communities return as the viable information conduits to decision makers. The chapter also summarizes best practices for recruiting and organizing international science panels.

In "Organized science" (Chapter 16) Casey Stevens and I investigate more systematically the organizational features of more influential science panels. In order to meaningfully impart their advice to target audiences they need to command usable knowledge and be able to maintain an ongoing connection with the target audience.

15

WHEN DOES POWER LISTEN TO TRUTH?*

A constructivist approach to the policy process

Peter M. Haas

While speaking truth to power has long been a major theme in political science and policy studies (Wildavsky 1979), commentators are increasingly skeptical about whether modelers and scientists are capable of developing truth, and whether power ever listens to them anyhow. Indeed, at the international level international relations (IR) scholars tend to be surprised by the occasions when it does. This article applies the political science literature to the related question of when power listens to science, particularly with regard to the management of complex environmental issues associated with sustainable development.

Knowledge and sustainability science

Sustainable development is now one of the major mantras invoked in the area of international environmental governance. Sustainable development was popularized in the seminal 1987 World Commission on Environment and Development report, *Our Common Future*. The report served as the justificatory document for the 1992 World Conference on Environment and Development, and put forward a new doctrinal approach to economic development that 'meets the needs of the present without compromising the ability of future generations to meet their own needs'. Sustainable development requires a reorientation of collective understanding and of formal institutions to focus on the key intersecting and interacting elements of complex problems. Technically, efforts to cope with environmental threats must be comprehensive if they are to address the complex array of causal factors associated with them. Yet comprehensiveness is difficult to achieve, because few governments or international institutions are organized to cope with the multiple dimensions of environmental problems, and many states lack the technical resources to develop and apply such efforts. Sustainable development urges a simultaneous assault on pollution, economic development, unequal distribution of economic resources, and poverty reduction. It argues that most social ills are nondecomposable, and that environmental degradation cannot be addressed without confronting the human activities that give rise to it. Thus sustainable development dramatically expanded the international agenda by arguing that these issues needed to be simultaneously addressed, and that policies should seek to focus on the interactive effects between them.

This new policy doctrine – or consensual wisdom within the international community of environmental policy analysts (and at times advocates) – rests on two key foundations (Ravetz 1986; National Research Council 1999; Clark 2001; Kates, Clark et al. 2001). One is that procedurally policy should be participatory and transparent: in part to include new perspectives in the knowledge base brought to bear on understanding a particular problem, as well as to promote buy-in and the inclusion of stakeholders in the subsequent application of these policies, and thus to improve the prospects for more effective enforcement and compliance.

The second foundation is substantive: and involves more comprehensive and systematic approaches to planning and policy formulation, often through the technique of sustainability impact assessment and forms of interdisciplinary integrated modeling intended to bring together partial insights to sustainable development from different academic disciplines.

Communicating scientific consensus to power

While sustainability science has been percolating in the various laboratories of environmental analysts for nearly twenty years, there remain few, clearly observable occasions on which there has been effective technology transfer from the laboratory to a government agency or international regime. Even when scientists think they have developed truths for power, power appears disinterested at best, and possibly even uninterested. Yet sustainable development has gradually encroached on environmental policy-making worldwide, both in international regimes and in national legislation and practices. This article analyzes the conditions under which power does listen to knowledge, the characteristic forms of collective action that emerge when sustainable development is applied to addressing shared transboundary and global environmental threats, lessons for improving the process, and an application of these inductive insights to efforts to apply scientific understanding to combating global climate change. Analyzing sustainable development and environmental protection provides a sense of the circumstances under which knowledge and power can effectively interact. While this is a relatively rare phenomenon, it is too often entirely dismissed by science policy analysts and political scientists.

Many arguments are harnessed to account for why science seldom influences policy-making. This is because science is seldom directly converted to policy. The path from truth to power is a circuitous route at best. Let me cluster the reasons for this under a set of convenient categories (for a recent summary of arguments about science policy and an argument for a pragmatic procedural approach to generating scientific knowledge for public policy, see Ziman 2000).

1 Science isn't wisdom (or true). Science studies scholars dismiss the prospects of objective knowledge about the world, and stress the political dimensions to science and science policy.
2 Science is politically tainted and suspect. Organized modern science embodies implicit values of control, so that decisions made with scientific warrants may unconsciously reflect such hidden values. In addition the distributional consequences of science-based advice are themselves political (Jasanoff, Markle et al. 1995; Miller and Edwards 2001). Science is political in its consequences, because some benefit and others suffer as a consequence of policy options that are supported by the application of scientific understanding. To the extent that those affected by the use of science in formulating policy are not consulted in its development and application, the use of science is potentially regarded by those affected as an illegitimate and exploitative set of discursive practices (Lidskog and Sundqvist 2002).
3 Power doesn't care about truth anyhow. Politicians don't want science; they want a justification for pre-existing political programs which are driven principally by political anticipations of gain (Nelkin 1979; Miles 1998).

Before discussing these reasons in greater detail, let me first dismiss the standard rationalist account that major problems create the incentives for their resolution, and thus modern bureaucracies (or bureaucratic institutions) either develop effective responses almost automatically, or are so powerfully constrained by the strategic interests of powerful member states or participants that institutional applications of knowledge are little different than the aggregate wishes of the more powerful members (for international institutions, see Koremenos, Lipson et al. 2001; for domestic, see Moe 1984). I reject this proposition for two reasons.

One is that, particularly for sustainable development, the material incentives and the nature of the presumptive useful information are not in synch. Few organizations, if any, have the available mission and resources to be able to address the full integrative range of issues encompassed by sustainable development. Yet there is an empirical record of shifts beyond what formal organizational assignments would predict (Haas and Haas 1995).

The second reason can be addressed by way of the historical analogy of scurvy. That is, we shouldn't assume that all organizations are rational and will automatically recognize and adopt what prove to be the appropriate policy responses with the virtue of hindsight or reflect the material needs of their most powerful constituencies. Arguably scurvy was the single most important limiting factor to the expansion of trade and geopolitical influence in the fifteenth and sixteenth centuries. Trade and exploration was significantly hampered because the mortality rate of sailors on long-distance expeditions was often in excess of 90 percent. Expeditions would return with far fewer ships not because of storms but because too many crew members died to enable the entire fleet which had left to be served. And yet in the early 1600s scurvy was solved by Captain James Lancaster – bringing citrus trees along on ships – and the solution was forgotten for nearly 150 years, until Captain

Cook rediscovered it in the 1760s (Milton 1999)! So much for rational societies responding effectively to important issues. So we should not expect a fully formed sustainable development science to be developed and applied in response to the presumptive need for such a view.

There is a well-developed literature that lays out a variety of arguments for the limitations of science for policy because many policy-makers do not view it as essential for policy-making or discredit science's impartiality. Ironically, those who do accept science's claims to impartiality may be particularly suspicious because science may undermine their political agendas.

Scientific consensus is often suspect because the scientists themselves are part of a broader cultural discourse, and thus lack autonomy or independent stature: in part, scientific findings may reflect the bias of sponsors, but more deeply they may reflect the broader culture of the society from which they emerge and about which they may not be fully conscious. The universe of what is known or deemed knowable may be biased by the availability of funding resources for research, and thus reflect the conscious or unconscious bias of major public and private funding bodies. Public sources of research support tend to reflect the broad political mission of the funding agency, whereas private sources reflect short-term commercial concerns, and philanthropic funding generally tends to cluster around a small number of topics and shift in ways that are seen as capricious and cyclical by recipients.

Science has become extremely politicized. The use of science is mediated and thus possibly distorted by the political goals of potential users. Truth claims are politically suspect because of potential funding bias or participation exclusion. Science isn't pure in the area of sustainable development because scientists' contributions exceed their technical skills. In 1972 Alvin Weinberg observed that scientists were often asked to provide advice that exceeded their formal disciplinary training (Weinberg 1972).

Science may not be sufficiently simple for the needs of policy-makers. Harry Truman is supposed to have complained about his economist advisers who would say 'on the one hand' and 'on the other hand'. Truman said, just give me a one-handed economist.

Science may provide advice that is out of synch with the political plans of decision-makers or parliaments, and thus be dismissed. While in principle parliamentarians seek to pursue the goals which they think the constituency that elected them wants. Thus they will selectively cull advice to find material that will either help them to identify what their constituency wants, if there is a dominant constituency on an issue that could potentially unseat them if displeased, or to achieve what the constituency already wants. Similarly, executive branch bodies solicit knowledge. But they are guided by a quest for information that will either help to pursue the traditional foreign policy goals of advancing material wealth and power, or the goals that will satisfy their parliamentary oversight committees. Either way the information heard by power is not the same as the truths that scientists think they are delivering.

Usable knowledge: when knowledge speaks to power

I take a late modern view of truth – the domain of science and its ability to confer truth is bounded, but we can talk of better and worse science (for an application to the study of international institutions, see Haas and Haas 2002). In the context of this discussion we can speak of usable knowledge. Several different schools of thought exist about usable knowledge, although the core insights are quite similar and complementary. In short, usable knowledge encompasses a substantive core that makes it usable for policy-makers, and a procedural dimension that provides a mechanism for transmitting knowledge from the scientific community to the policy world and provides for agency when theorizing about broader patterns of social learning, policy-making, and international relations.

Current research from comparative politics, IR, policy studies, and democratic theory suggests that science remains influential if its expertise and claims are developed behind a politically insulated wall (Andresen, Skodvin et al. 2000; Botcheva 2001; Social Learning Group 2001). Moreover, efficiency gains by relying on one single source of policy advice are more than offset by the loss of legitimacy, analytic blinders imposed by relying on just one institutional source for usable knowledge, and the political doubts of bias that are raised by narrowing the source of information. Studies of international environmental assessments and science panels suggest the need for fluid bodies that can bring together multiple sources of information and are not beholden to one single funder or political sponsor (Clark 2001; Siebenhuner 2002; Siebenhuner 2003; Haas 2004; Jasanoff and Martello 2004; Farrell and Jager 2005). Studies of national-level environmental policy processes have convincingly argued against relying on individual institutions for research and policy advice, because they may bias the information flow, and control resources (Brown 1997; Skoie 2001). There is much national-level experience with establishing standing scientific panels, such as in the United States the National Academy of Sciences, the Environment Protection Agency Science Advisory Board, and the now defunct Office of Technology Assessment, among others. Each is regarded as generating usable knowledge for the government, and enjoyed sufficient autonomy to identify research questions and to convene panels to develop reports. Comparative studies with other countries would be valuable. There are also experiences with ad hoc bodies created to develop usable knowledge for particular issues, such as the German Enquette Commissions for Ozone and Climate Change (Smith 1990; Carnegie Commission 1992; Smith 1992; Morgan and Peha 2003).

I call the relevant body of scientific knowledge 'usable knowledge' (see Clark 1990) for an earlier, and slightly different, usage; Dimitrov 2003 also has a slightly different usage). Usable knowledge is accurate information that is of use to politicians and policy-makers. It must be accurate and politically tractable for its users. It frequently exceeds the mastery of any individual disciplinary approach. Recent studies of global environmental assessments apply the criteria of credibility, legitimacy and saliency (Siebenhuner 2002; Siebenhuner 2003; Mitchell, Clark et al. 2006). Credibility means that the key knowledge producers and their consumers believe that their product is true. Legitimate means that the claims are believed to be legitimate; that is, developed through a process that minimizes the potential for bias and is more equitable in terms of participation by those who are dependent upon the information. Finally, saliency means that such information is provided in a timely manner and contains information that is useful for making public policy by decision-makers: that is, that in practice it arrives in conjuncture with the policy process and provides advice which can be converted into laws or decisions by decision-makers. In practice credibility and legitimacy are mutually reinforcing, as a procedural approach to developing consensual knowledge is likely to generate both accurate and acceptable knowledge. Yet saliency and credibility may be at odds, as the long time often necessary for developing credible knowledge may interfere with the short-term needs for applying the knowledge to making policy. In practice, then, existing knowledge is more likely to play a role in usable knowledge than is knowledge being developed concurrently with the policy process.

This articulation of usable knowledge builds from prior efforts to formulate a sense of what kind of technical information is likely to be useful for policy-making relating to matters of complexity, which is also likely to be used by decision-makers.

Clark and Majone offer four criteria of usable knowledge: its adequacy, value, legitimacy, and effectiveness. Adequacy relates to including all the relevant knowledge or facts germane to the matter at hand. Value has to do with contributing to further understanding and meaningful policy. Legitimacy relates to its acceptance by others outside the community that developed it. Effectiveness relates to its ability to shape the agenda or advance the state of the debate, and, ultimately, improve the quality of the environment (Clark and Majone 1985; Social Learning Group 2001).

The Center for International Climate and Environmental Research in Oslo applies three requirements for a solution design model to be considered adequate (CICERO 1998):

- It must be capable of mobilizing sufficient political support to produce agreement.
- It must be capable of generating solutions that can be implemented.
- It must be capable of generating solutions that are instrumental towards solving the problems for which they were designed.

In short usable knowledge, even, or particularly, when it is expressed in the form of a model, must be seen as accurate and accessible, and contribute to the achievement of collective goals. It must represent consensus, and be provided through a medium that is politically palatable.

A new consensus is emerging amongst social scientists who study the use of science in international regimes that a procedural or discursive model of the policy process is normatively superior and is growing in frequency of its application, as against the former policy analytic approach associated with cost–benefit analysis and analytic efforts to identify the best policy solution. Policy analysis is a process of exchange and mutual learning between policy-makers and policy analysts. Policies themselves are experiments that participants monitor and about which they reflect in order to improve them over time (Ascher 1986; Funtowicz and Ravetz 1991; Lee 1993; Lindner and Peters 1995; Funtowicz and Ravetz 2001). Consequently, policy-making is a process rather than a fixed set of analytic techniques. Language may play a role (Fischer and Forester 1993; Hajer and Wagenaar 2003) but the constructivist argument pursued here stresses the political parameters within which debate occurs rather than the connotations generated by the language that is used.

Constructivists argue that under conditions of uncertainty – such as are associated with contemporary globalization and highly technical issues – it is impossible to create ex-ante sufficient information to follow the policy analytic model (Checkel 1998; Hopf 1998; Kubalkova, Onuf et al. 1998; Ruggie 1998; Christiansen, Jorgensen et al. 1999; Guzzini 2000; Haas 2001; Adler 2002). Alternatively, the key is to design policy analytic processes from which actors learn about the world and about each other. This view is anti-rationalist in the sense that the process model entails path dependency and uncertainty assumptions that presume that outcomes, and indeed preferences, are often underspecified or indeterminate, and thus that regimes and political interactions can best be appraised in terms of process rather than outcome. All outcomes will be sub-optimal in some sense as compared to the ex-post ideal outcome; but one can just hope that by an open discursive process better outcomes may result (Dryzek 1997).

Constructivist approaches to policy analysis suggest that science must be developed authoritatively, and then delivered by responsible carriers to politicians. Doctors, scientists and engineers remain the most esteemed professions in Europe, and thus command the greatest social legitimacy and deference when providing policy advice (Jasanoff and Wynne 1998; Drori, Meyer et al. 2003).

The transmission belt of like-minded scientists is called an 'epistemic community' (Haas 2001). The more autonomous and independent science is from policy the greater its potential influence (Andresen, Skodvin et al. 2000; Botcheva 2001; Haas 2001). Consensus in isolation builds value and integrity, and then its consequences should be discussed publicly. Measures of autonomy and integrity include the selection and funding of scientists by international organizations rather than by governments, their recruitment by merit on important panels, and reliance on individuals whose reputation and authority rest on their role as active researchers rather than policy advocates or science administrators. Accuracy can be achieved via peer review, interdisciplinary research teams, and independence from sponsoring sources. Increasingly, sustainability scientists, themselves an epistemic community, argue for the need to include local knowledge with the more formally technical understanding of traditional disciplinary elites associated with formulating sustainability policies (Haas and Haas 2002; Jasanoff and Martello 2004). However, the criteria for participation remain loosely defined, but perhaps no more so than the broad injunctions for multidisciplinary participation that do not clearly identify which disciplines need to be consulted for which types of questions.

Political legitimacy rests on a process of knowledge development and diffusion that is scrupulously free of political interference. International institutions can help to foster and disseminate information, and sanitize it so that it is not seen as compromised by potential users who may fear that the information is controlled by one country.

Usable knowledge is developed by international and transnational networks of scientists. It is heeded, to the extent that it is, after widely publicized shocks or crises. While usable knowledge contributes to broader patterns of social learning, the delivery of knowledge and its application are often by different communities. I am not talking of reflective learning by decision-makers. Rather, I am talking about the recognition by decision-makers of the limits of their abilities to master new issues and the need to defer or delegate to authoritative actors with a reputation for expertise. In the aggregate, social learning and human betterment emerge when the experts have been able to develop usable knowledge, and the decision-makers feel compelled to apply it (Haas and Haas 2002).

Empirical record of usable knowledge in multilateral environmental governance

Over the last thirty years there has been a massive surge in the adoption of international treaties and regimes to address many aspects of transboundary and global environmental threats (Tolba and Rummel-Bulska 1998; Haas 2001; Miles, Underdal et al. 2002).

When regimes are negotiated with the involvement of epistemic communities and strong international institutions they develop through a process of 'social learning'. Negotiations occur within a scientific discourse, in which political debate and compromise reflect expert consensus on the behavior of ecosystems and their ability to sustain stress. The substance of regimes reflects scientific consensus about the most important environmental threats, and negotiated standards reflect consensus about the degree of environmental stress the target environment can sustain. Social learning generates treaties with differentiated national obligations and substantive commitments, based on expert consensus on causes and environmental effects. For instance, the 1980 Land-Based Sources Protocol for the Mediterranean requires more stringent emission controls on the industrialized countries than on the developing countries because the magnitude of degradation of the northern coast of the Mediterranean was much more severe than it was on the southern coast (Haas 1990).

Other regimes developed through social learning include the stratospheric ozone protection regime, the 1979 Geneva Convention on Long-Range Transboundary Air Pollution (LRTAP), and subsequent treaties addressing European acid rain, and pollution control efforts for the Mediterranean, Persian Gulf, South Pacific, and South East Pacific.

Maurice Strong, Secretary General of the United Nations Conference on the Human Environment (UNCHE) and the United Nations Conference on

Environment and Development (UNCED), and the United Nations Environment Programme's (UNEP's) first executive director, helped to design the outlines of this process of social learning. Strong believed that 'the policy is the process': that is, by generating an open political process in which states are exposed to consensual science, government officials may be persuaded to adopt more sustainable policies, and individual scientists may gain heightened political profiles at home which may ultimately increase their effectiveness as well. Most social learning treaties have standing environmental monitoring and research committees, to provide timely warnings of new problems, monitor achievements of regime goals, and educate politicians and policy-makers on environmental issues.

However, social learning takes time. Substantively the regimes are more comprehensive, and attuned to the emergent sustainable development doctrine's sensibility and injunctions. Comprehensive treaties are slower to negotiate than others, because they require persuasion and consensus rather than mere compromise. From a policy perspective, though, comprehensive regimes are likely to be superior in their ability to protect the environment in a cost-effective and politically acceptable manner. Moreover, treaties developed with help from the scientific community typically enter into force more rapidly than without it, presumably because of the weight that involvement of scientists carries in the ratification process.

Regimes that were built with usable knowledge appear to be more effective at inducing states to achieve their intended goals of improving environmental quality. For instance, stratospheric ozone, and European acid rain efforts are widely hailed as some of the more successful and effective international environmental governance efforts of the contemporary era (Andresen, Skodvin et al. 2000; Haas 2001; Miles, Underdal et al. 2002).

Epistemic communities often work in conjunction with broader policy networks, functional bureaucrats, transnational scientific organizations, nongovernmental organizations (NGOs), and international civil servants. A small number of international institutions have supported the development and transmission of usable knowledge. UNEP has played a powerful role in environmental protection over the last thirty years. With a staff of less than 200 professionals and a budget now in the order of $150 million a year, UNEP has led global environmental monitoring efforts, catalyzed environmental protection activities in other UN bodies, served as the environmental conscience of the UN system, and sponsored the conclusion of dozens of international environmental treaties. The UN Economic Commission for Europe (UNECE) and the International Institute for Applied Systems Analysis (IIASA) helped to develop and circulate usable knowledge for the effective management of European acid rain.

Lessons for generating and mobilizing usable knowledge

Given our thirty-year experience with addressing transboundary and global environmental threats, what lessons are available about developing and mobilizing usable knowledge for sustainable development? The following lessons can be drawn, based on comparative case studies of multilateral environmental regimes with and without usable knowledge, and based on process-tracing studies of efforts to institutionalize usable knowledge (Haas 1991; Victor, Raustiala et al. 1998; Weiss and Jacobson 1998; Young 1999; Andresen, Skodvin et al. 2000; Haas 2000; Haas 2000; Reinicke and Deng 2000; Miles, Underdal et al. 2002; Haas 2004). Thus, the conditions are found in effective regimes and are absent in ineffective regimes.

Lessons about mobilizing networks of scientific expertise for sustainable development

1. Create standing international interdisciplinary scientific panels or committees to address specific topics.
2. Multilateral Environmental Agreements should create and rely on separate subcommittees responsible for different functions of governance, such as basic research, environmental monitoring, policy analysis and policy verification and evaluation.
3. Carefully survey the population of scientists to identify a core group sharing values and causal beliefs. For instance, in the Mediterranean a UNEP consultant spent nine months visiting national laboratories to inventory national capabilities and to personally build the scientific network.
4. Ensure that networks and international panels have interdisciplinary representation, including the social sciences. Individuals should have high regard in their own disciplines as well as be able to talk to experts from other disciplines.
5. Recruit carefully for national and regional institutions. Base judgments on professional credentials and networking ability.
6. Avoid relying on one national institution to provide or sponsor research and training.
7. Provide professional outlets for members through conferences and publications in refereed professional journals. This also elevates the domestic profile of individual scientists in the community of expertise who may then be recruited to fill positions in national administrations.
8. Promote scientific discussions on topics that are likely to lead to consensus, i.e. ripe research topics.
9. Avoid government designation of scientists to international meetings.
10. Try to make use of joint international panels for environmental risk assessment rather than relying on national assessments. Avoid capture by one scientific discipline or school of expert analysis.
11. Assure the timely submissions of scientific advice in advance of meetings. The timely submission of reports according to the legislative cycle in the major countries is also key.
12. Arrange for focused interactions between scientists and policy-makers to discuss the technical substance of the issues. For instance, in LRTAP the IIASA

arranged for two-day sessions to familiarize policy-makers with acid rain transfer and deposition models developed by scientists.

13 Maintain momentum within the community by continuing to have projects and research opportunities so those members don't drift away. This avoids having to reconstitute the community each time a new problem emerges.
14 Seek funding for studies from multiple sources in order to avoid budgetary shocks if money is withheld from a principal funder. Thus economically inefficient redundancy is politically warranted.
15 Construct models so that effects are calculated at meaningful political scales, i.e. corresponding to significant political divisions that are relevant in developing policy applications. For instance, at the international level this means that models should explicitly identify effects by country, and even, if possible, by domestic districts (at least in countries without proportional representation). So far climate change models have only yielded effects at a scale of resolution sufficient to demonstrate to countries able to vote in the General Assembly that they are likely to suffer, and thus they keep the issue on the international agenda.
16 Train or recruit scientists who have a high profile within their own discipline and who are able to effectively communicate with counterparts from other disciplines, as well as with the media, politicians and popular audiences.
17 Try to recruit networks from as broad a national basis as possible, as governments are more likely to rely on experts who share their nationality than on foreign experts.

Broader considerations of the proper institutional design of science policy entail timing: when consensus has been achieved before an issue reaches the agenda and policy discussions begin, scientists can merely be introduced as experts, following the lessons above. However, at times it is necessary to simultaneously develop scientific consensus and advance policy debates. For such issues, such as was the case in the Mediterranean and ozone regimes, the parallel development of science and policy must be kept insulated from ongoing policy debates, with the two streams united only when consensus has been achieved. In other cases, where consensus remains elusive and policy debates have already attained their own momentum, as in climate change and biodiversity, it may be best if the two activities can be kept as separate as possible.

Climate change, usable knowledge and the limits to social learning

Internationally, states are increasingly relying on this procedural approach to the development and application of truth to power. The European Commission developed a set of guidelines and proposals for the collection and use of expertise by the Commission that is very similar in orientation (European Commission 2001; Commission of the European Communities 2002).

The climate change regime, the focus of the majority of global environmental diplomatic efforts during the 1990s, has also developed through a deliberate effort by diplomats to base policy on strong scientific foundations through its Intergovernmental Panel on Climate Change (IPCC). The IPCC is of interest because it highlights the way in which states may choose to shape the science advisory process. A closer look at the interplay of science and power in the IPCC reveals empirically how this dynamic interaction operates in this key contemporary issue, as well as analytically establishing the political limits to autonomous science and social learning. The scientific consensus is not yet strong, and thus the available scientific knowledge is not fully usable. Yet, in the case of climate change, the fact that usable knowledge is not yet available for climate change has much to do with the political choices associated with the design of the IPCC, and thus suggests the political limits to states' willingness to confer some degree of autonomy to scientific institutions and to defer to their guidance.

The IPCC is one of the most concerted efforts by the international community to harness usable knowledge for addressing transboundary and global environmental threats (Agrawala 1998; Agrawala 1998; Skodvin 2000; Skodvin 2000; Siebenhuner 2002; Siebenhuner 2003; Kameyama 2004). The IPCC was established in 1988 as the principal international science policy advisory body for global warming, but is widely believed to have also been formed politically in order for governments to reassert control over the science process in an issue which was accelerating on the policy agenda more rapidly than most leaders in the North were comfortable with. The IPCC was the consequence of a General Assembly initiative for a climate change regime, and did not rely on UNEP as had most previous international environmental regime initiatives. The IPCC consists of three working groups, with members chosen by governments subject to the scientific reputation of the candidates. Working Group One addresses questions of atmospheric science; Working Group Two assesses social and economic impacts and adaptation measures, and Panel Three looks at mitigation alternatives. Each working group is administered by a bureau composed of the IPCC bureau members, working group co-chairs, and vice-chairs and a technical advisory unit drawn from the country of the working group's chair, except for the Technical Support Unit of the scientific Working Group One, which has always remained in the UK. To date the IPCC has produced three three-volume assessment reports (1990 with 1992 revisions, 1995, and 2001, with a fourth assessment underway), each with a summary for policy-makers, as well as various ad hoc special reports and technical papers. The Working Groups report to a plenary composed of government delegates, who review the Summary Reports on a line-by-line basis, and also approve the Working Group Assessments and Special Reports in more block-like fashion. All reports other than the Summary Reports rely on extensive peer reviews, and the material presented must come from peer-reviewed journals or be in the process of publication in a peer-reviewed journal. The Summary Reports, which command the greatest publicity and hence public attention, are written by the Working Group leaders along with the lead authors and specially invited experts (Skodvin

2000; Swart, Mitchell et al. 2002). The bureaus are responsible for drafting an initial table of contents and topics for each chapter. This agenda is then approved by the political plenary of the IPCC. The reports are drafted by the scientific committees, and are then approved subject to careful scrutiny by government representatives in the Plenary.

Thus governments have sought to exercise control over the scientific process, while also allowing for some degree of scientific latitude in order to generate accurate advice, even if the process is designed in such a way that the advice is unlikely to be particularly salient. Governments appoint the IPCC chair, and in 2002 the United States vetoed the appointment of the climatologist Robert Watson, a well-regarded American candidate, in favor of Rajendra Pachuari, an Indian engineer, based on the belief that Watson was too independent of the US administration. Politically charged language in the Third Assessment Report (2001) was criticized by the US government for containing language which the US claimed was stronger than had been approved by the Plenary, although the authors were able to subsequently prove that their draft was consistent with the IPCC's rules of procedure (Edwards and Schneider 2001; Siebenhuner 2002). All individuals are nominated and chosen by governments, although there is little evidence of direct government manipulation in recruitment or the inclusion of material. The procedures are carefully designed to maximize the seemingly procedural scientific legitimacy and accuracy of the work, by stressing peer review and reliance on peer-reviewed material.

The degree to which the IPCC is capable of generating usable knowledge is largely politically circumscribed. The state of scientific understanding of the key global systems that affect global warming remains relatively immature. It is unclear to what extent funding and research choices have been shaped by political factors, but there is no strong evidence to assert that the state of knowledge about the phenomenon is directly biased or controlled by political influences. The accuracy of the IPCC science remains limited. The estimates of global warming and their effects are crude, and global carbon models are unable to account fully for the global carbon cycle.

The IPCC does not enjoy a high degree of legitimacy in the eyes of many science policy consumers (Biermann 2002). It suffers from the appearance of governmental control, because governments appoint the scientists and also vote on the reports. The distribution of scientific participants comes overwhelmingly from the North, despite the best efforts of the bureaus to increase participation from developing countries since the release of the First Assessment Report (Biermann 2002).

Table 15.1 shows the crude national breakdown of scientific involvement in the Third Assessment Report by country of origin of the scientific participants.

The IPCC is limited in its legitimacy through its seeming lack of equity in participation. While the IPCC has helped to pay for the participation of scientists from developing countries, it faces the deeper structural science policy issue that the overwhelming majority of climate change research is conducted in the North by northern scientists.

TABLE 15.1 Third Assessment Report: distribution of nationality of scientific participants

Nationality	*WG 1 lead authors (98)*	*WG 1 contributing authors (708)*	*WG 2 authors and expert reviewers (626)*	*WG 3 authors and reviewers (22)*	*Synthesis report authors and expert reviewers (375)*
Industrialized countries	80%	97%	79%	75%	81%
Developing countries	20%	3%	21%	25%	18%
Technical support unit	UK		USA	Netherlands	

Source: http://www.ipcc.ch (retrieved January 7, 2004).

Note: Industrialized and developing country categories follow accounting scheme of the 2003 Human Development Report.

Its saliency is particularly poor. For one thing it has been unable to develop policy advice that resonates domestically in any of the major countries. To date the scientific work has narrowed the range of likely warming that will occur in 2100, and generated scenarios of what the global environmental consequences may be of such effects. However, the scenarios are sufficiently crude that they do not engage any significant political interests in any of the member countries, other than reinforcing the prior knowledge of Egypt, Bangladesh and the small island countries that they may well be submerged. Thus the political effect of the new knowledge is only adequate for the aggrieved parties to continue to keep the issue on the General Assembly agenda, where they have sufficient votes, but not to directly influence policy outcomes. More salient science would present data at a level of resolution that corresponded to domestic political divisions (i.e. states or congressional districts in the US, and parliamentary districts in parliamentary systems). Some observers have criticized the IPCC Assessment Reports for focusing excessively on adaptive policies, which reflect the preferences of the industrialized countries historically responsible for GHG emissions, rather than extensively addressing preventative or mitigating policies which would favor the developing countries.

The publication of the Assessment Reports did not fit in well with the negotiating cycle of the Kyoto Protocol and efforts to follow up the 1992 United Nations Framework Convention on Climate Change. A few of the special reports, such as the 2000 Report on Land Use, Land Use Changes and Forestry, corresponded to meetings of the Conference of the Parties, and thus were able to provide a timely submission of scientific policy advice to the actual negotiators. Otherwise, though, the reports have not coincided closely with the negotiation process, so have had little impact other than periodically reminding mass publics of the potential severity of global warming. IPCC scientists have not fared well at communicating their findings to the public and engaging in elite or public education.

In short, the IPCC is designed to keep science on a tight leash by controlling the selection and autonomy of individual scientists engaged in the assessment process. Consequently, the degree of usable knowledge generated by the IPCC has been limited. And not surprisingly IPCC scientists have been unable to exercise sufficient discretion to develop more politically tractable advice. In turn, negotiated treaties within the climate change regime have not reflected a strong degree of scientific basis, despite the ongoing IPCC efforts. The 1992 United Nations Framework Convention on Climate Change and the 1997 Kyoto Protocol (not yet in force) reflect broad rhetorical commitments or political compromises about target emissions, but do not reflect any scientifically grounded targets for specific atmospheric concentrations of greenhouse gases. Few governments outside the European Union (EU) have proven willing to make national decisions based on this collective enterprise, and EU greenhouse gas targets reflect political and energy policy realities rather than scientific justifications.

The limits of the IPCC's ability to speak truth to power can be explained by principal–agent theory (Moe 1984; Koremenos, Lipson et al. 2001; see Guston 1998; Guston 2003 for an effort to capture science policy issues more generally within a principal–agent framework). The principals strategically design the institution on which they rely for information and advice (the agent) in order to minimize the surprises that the agent will provide and to ensure that the agent will provide advice that will run counter to the principals' ex-ante interests and preferences. The principals here are the governments of the industrialized countries. In 1988 they were tired of being lambasted by UNEP in multilateral environmental negotiations, and were concerned that uncontrolled scientific panels may give rise to policies that they did not deem warranted. The Villach meetings from 1985 to 1988 had forcefully put the issue of global warming on the international agenda, and the 1988 Toronto conference, convened by non-state actors, had called for 20 percent reductions in greenhouse gas emissions; thus alerting foreign ministry officials of the political threats posed by non-state actors' potential control of climate change agenda. The recent lesson of the powerful influence of the Ozone Trends Panel on the rapid conclusion of the Montreal Ozone Protocol was also fresh in the minds of foreign ministries. Moreover, the short-term costs of addressing the problem for the industrialized countries appeared prohibitive (or even, one could say, the net discounted value of efforts for mitigation adaptation) so they wished to rein in any independent political pressure that would be generated from an organized scientific involvement in collective discussions on climate change.

The agents in this regard are the scientific community, who are recruited and organized in a way through the IPCC that assures governments that they will be able to exercise maximum control over individual scientists, as well as remaining able to shape the political agenda for climate change negotiations. Admittedly, by relying on extensive peer review some degree of control was lost, but in the end governments appoint the individuals and the reports are approved on a line-by-line basis by foreign ministry officials. Thus the principals establish and enforce the parameters within which the agents have to act, ensuring cautious outcomes by the agents.

But the IPCC is an extreme case. It is extreme in terms of the clarity with which the powerful industrialized countries recognized their material interests in the matter, and it is extreme in terms of the widespread transformations in industrial and individual behaviors that effective preventive policies were presumed to entail. In addition, the concurrent negotiations and science advisory process made governments extremely sensitive to the strategic implications that scientific assessments may have for national policies. In most other transboundary and global environmental issues the scientific consensus preceded active political discussions, so that governments were less acutely aware of the implications of their choice to delegate rule-making to scientists. Most other environmental and sustainable development issues do not share these characteristics, and hence there is greater potential for governments to create more legitimate and porous scientific processes to guide their policy work.

Conclusion

Thirty years of speaking truth to power have yielded tangible beneficial outcomes. Sometimes power listens. There now exist far more extensive legal commitments on environmental protection than ever existed before, and most areas of human activity in most countries of the world are now regulated as a consequence. In the aggregate, major changes have occurred internationally in environmental governance. Many multilateral environmental regimes are more comprehensive than before science was systematically applied to multilateral environmental diplomacy, stronger national environmental protection commitments have been undertaken in compliance with international obligations, and green markets for cleaner technologies have grown as a consequence. Markets estimated at 600 billion dollars a year now exist for green technologies (US OTA 1994; Hoffman 2001). These markets were created in response to the regulatory obligations resulting from international treaties negotiated on the basis of the application of new scientific understandings of the behavior of ecosystems (Garcia-Johnson 2000; Haufler 2001). The protection of ecological integrity has become elevated to a state interest on a par with the pursuit of national wealth and power. Sustainable development is now the doctrine by which states seek to reconcile (or optimize) the pursuit of those three objectives.

This broad process of increasing reliance on sustainable development for environmental policy-making over the last thirty years can be explained by constructivism. Constructivism in the United States became a popular activity in the late 1980s. Constructivists soon responded to initial theoretical challenges calling for a more fully elaborated theory of the state and of their rationality assumptions (Keohane 1989; Milner 1992) with a richer, deductively derived set of propositions about social construction and the role of different forms of shared understandings (ideas) on IR and policy-making. A core set of assumptions emerged that most constructivist research now implicitly reflects (Hopf 1998; Ruggie 1998; Christiansen, Jorgensen et al. 1999; Haas 2001; Checkel 2004). States are functionally differentiated actors; states respond to threats under conditions of uncertainty; knowledge is asymmetrically distributed within and between states; and the most important independent variables that shape state behavior are distributions of power, knowledge, and formal institutional properties (although material capabilities play a very small role in multilateral environmental governance). The nature of responses is subject to the particular configuration of these variables in place at the time of negotiations. Behavior is path-dependent and reflects the decisions taken under the influence of the initial configuration of factors. States' perceptions of problems to be addressed, the nature of appropriate responses, and their willingness to adhere to commitments are all a function of the configuration of these factors. Non-state actors are important players in policy-making, and pressure states. The core insight is that the nature of external incentives for actors to behave is not obvious, and requires articulation for them. Many constructivist variables and actors have been put forward over the years.

Constructivist theorizing requires interactive theories that can look at institutional and knowledge-based influences at different levels of political analysis (Zurn 1998; Botcheva 2001). Most constructivists emphasize the role of agency and seek to develop causal theories about the role of ideas. Constructivist theories about patterns of policy-making have generated a number of provocative propositions or hypotheses, including: (1) whether actors follow a logic of consequences or appropriateness (March and Olson 1988; Searle 1995) is a function of how the issue is framed for decision-makers by knowledgeable advisers; (2) policies vary with issue-area because different arrays of policy communities and different degrees of technical understanding exist in different issues; (3) distinctive patterns of social learning are associated with configurations of strong institutions and usable knowledge; (4) over time institutionalized new understandings can lead to the transformation of state identities and patterns of practices; and (5) multiple ideational factors are involved in constructivism, including generative and constitutive norms, operational and aspirational norms, discourses, and causal beliefs. This article has focused exclusively on the role of causal beliefs.

A number of important empirical results have been discovered as a consequence of the application of constructivist theory. In the environmental realm, new patterns of more comprehensive environmental governance have occurred at the international and national levels as a consequence of the dramatic transformation in the understanding of the behavior of natural ecosystems and their impact on human life. Studies of international political economy show, for instance, how Keynesianism spread internationally through a similar process of epistemic persuasion in a core set of countries, followed by the creation of international institutions to legitimate and create international rules based on Keynesian dictates. Keynesianism first caught on in England, then traveled to the US through academic networks of economists, and after its adoption at the highest levels in the two countries, provided the intellectual foundations for the Bretton Woods system, which provided for coordinated and institutionalized national trade, monetary and financial policies that persisted for decades (Colander and Coats 1989; Hall 1989; Ruggie 1991; Ikenberry 1993; Helleiner 1994).

Recently, many constructivist claims have been confirmed by constructivists and analysts from other research programs. There is growing appreciation of the role of ideas as informing formal institutions by rational choice theorists (Mantzavinos, North et al. 2004), and of the transformative influence of social communication leading to new political identities and communities extending the Deutschian liberal communitarian tradition (Adler and Barnett 1998). Studies suggest the analytic primacy of constructivism before the application of other approaches to understanding political behavior and collective action. Comparative policy studies have emphasized the role of institutionalized policy networks in transferring policy lessons from one country to another, and their contribution to policy convergence (Rose 1993; Janicke 1995; Coleman and Perl 1999; Weidner and Janicke 2002). At the international level comparative environmental regime studies have confirmed that distinctive patterns of comprehensive environmental policy-making occur through a process of social learning (Andresen, Skodvin et al. 2000; Miles, Underdal et al. 2002). Edward Miles observes: 'we may, for all practical purposes, consider a base of consensual knowledge about the basic characteristics of the problem to be a necessary, though by no means sufficient, condition for achieving effective solutions to truly malign problems' (Miles, Underdal et al. 2002: 470). Thus it is clear that knowledge plays a distinctive role in shaping multilateral environmental protection and sustainable development, and that knowledge in conjunction with strong international institutions yields the distinctive pattern of social learning. Mainstream traditions of IR scholarship that focus exclusively on material incentives, the exercise of power or of institutional incentives are incapable of accounting for this variation in regulatory forms, the variation in national behavior, or for the occasional autonomy and agency enjoyed by state-created scientific advisory bodies. With the presence of usable knowledge, regimes adopt policy instruments and rules that apply environmental standards involving command and control approaches with very few market mechanisms, contrary to the expectations of non-constructivist approaches. Arild Underdal writes: 'all cases where the predictions derived from [assumptions] ... based on rational anticipations of political difficulty ... proved too pessimistic are instances of institutional growth or improvement in the knowledge base' (Miles, Underdal et al. 2002: 440).

Knowledge can speak volumes to power. Current research from comparative politics, IR, policy studies, and democratic theory suggests that science remains influential if its expertise and claims are developed behind a politically insulated wall. Epistemic communities are the transmission belts by which new knowledge is developed and transmitted to decision-makers. Knowledge must also possess the substantive characteristics of usable knowledge: credibility, legitimacy, and saliency. Practical administrative lessons about generating and mobilizing epistemic communities and usable knowledge were presented in this paper. Real limits to learning persist. National administrative design often inhibits learning and the diffusion of environmental lessons. Most learning remains highly localized: lessons are limited to changes in national policy with regard to one particular environmental medium, for instance air pollution or marine pollution. This limitation is due to administrative design in most countries. Epistemic community members' expertise is highly specialized, so their influence and involvement in agency choices are limited to a specific regime and environmental issue-area. Moreover, environmental responsibilities within governments vary by environmental issue, as different agencies are responsible for coordination of national activities in different areas. Thus the influence of epistemic community members in the short term is highly restricted. Few governments have the domestic potential for these lessons spreading within the governments, because the functional agencies have little influence over other agencies.

The prospects for improving national sensitivity to epistemic community influence rest on recruitment and retention patterns within national administrations. Environmental agencies often lack the ability to spread their influence within government administrations, since most environmental agencies are so specialized and the staff remain in position for limited periods so there is limited potential for individuals educating each other. Scientific consensus is often poorly represented to decision-makers. There is a need for eloquent and articulate specialists from within scientific disciplines who are able to communicate across technical and cultural divides to policy analysts and politicians (Koehn and Rosenau 2002).

The IPCC made it clear that it is much harder to mobilize knowledge to address politicized and high-stakes issues, particularly when knowledge is being developed at the same time as policy debates are occurring. The future of effective scientific policy advice may also suffer from declining public faith in science as an institution that can contribute to public welfare. This is an increasingly common theme in the public discussions of science use in the US and in Europe, and leads public officials to become disillusioned with the technical and political utility of relying on science following from a decline in trust in scientists (Topf 1993: 109).

While in general constructivists are reluctant to rely on the rationalist assumptions of theories such as principal–agent theory, in the particular issue of global warming the material incentives to the principals are so clear that the principal–agent theory's rationality assumptions are warranted. It is only in cases with greater uncertainty about material national interests and national effects from environmental harm that more constructivist approaches are warranted, although some dimensions of the global warming issues are subject to fruitful constructivist analysis (Miller and Edwards 2001; Jasanoff and Martello 2004).

In conclusion, institutionalized knowledge has contributed to more effectiveness in multilateral environmental management. Further work can advance the understanding of the interplay between knowledge and power by studying when states are willing to create autonomous institutions and defer to their authority; the interplay between different ideational factors in shaping collective action; and the interplay between various non-state actors at and across multiple scales of multi-level global governance (Annan 1998; Camilleri, Malhotra et al. 2000; Kanie and Haas 2004).

Acknowledgements

Much of this paper is based on two lectures at the Universite´ Catholique de Louvain-la-Neuve: a keynote address to the Workshop on Sustainability Impact Assessment of Trade Agreements and New Approaches to Governance, 28 March 2003, and to the Workshop on Science, Legitimacy and the IPCC, 8 November 2003. Versions were also presented at the University of California at Berkeley, 20 April 2004, and at Ohio Wesleyan University, 22 April 2004. My thanks to Frank Laird for his comments and to Rebecca Root for research assistance, to Bernd Siebenhuner and Robert Watson for comments on my interpretation of the IPCC, and to an anonymous *JEPP* referee.

Note

References

Adler, E. (2002). *Changing Identities: The Road to Peace.* Toronto, University of Toronto.

Adler, E. and M. Barnett, Eds. (1998). *Security Communities.* Cambridge, Cambridge University Press.

Agrawala, S. (1998). "Context and Early Origins of the Intergovernmental Panel on Climate Change." *Climatic Change* 39: 605–620.

Agrawala, S. (1998). "Structural and Process History of the Intergovernmental Panel on Climate Change." *Climatic Change* 39: 621–642.

Andresen, S., T. Skodvin, et al. (2000). *Science and Politics in International Environmental Regimes.* Manchester, Manchester University Press.

Annan, K. (1998). "The Quiet Revolution." *Global Governance* 4: 123–138.

Ascher, W. (1986). "The Evolution of the Policy Sciences." *Journal of Policy Analysis and Management* 5(2): 365–389.

Biermann, F. (2002). "Institutions for Scientific Advice." *Global Governance* 8(2): 195–219.

Botcheva, L. (2001). "Expertise and International Governance." *Global Governance* 7(2): 197.

Brown, G. (1997). "Environmental Science Under Siege in the US Congress." *Environment* 39(2): 12–20, 29–31.

Camilleri, J. A., K. Malhotra, et al. (2000). *Reimagining the Future.* Bandoora, Victoria, Australia, Department of Politics, La Trobe University.

Carnegie Commission (1992). *Science and Technology in U.S. International Affairs.* New York, Carnegie Commission on Science, Technology and Government.

Checkel, J. (1998). "The Constructivist Turn in IR Theory." *World Politics* 50(2): 324–348.

Checkel, J. T. (2004). "Social Constructivisms in Global and European Politics." *Review of International Studies* 30(3): 229–244.

Christiansen, T., K. E. Jorgensen, et al. (1999). "The Social Construction of Europe." *Journal of European Public Policy* 6(4): 528–544.

CICERO (1998). *The Intergovernmental Panel on Climate Change (IPCC) and scientific consensus: how scientists come to say what they say about climate change.* Oslo, Oslo Center for International Climate and Environmental Research.

Clark, W. C., Ed. (1990). *Usable Knowledge for Managing Global Climate Change*. Stockholm, Stockholm Environmental Institute.

Clark, W. C. (2001). "Knowledge Systems for Sustainable Development." *Environment* 43(8): 1.

Clark, W. C. and G. Majone (1985). "The Critical Appraisal of Scientific Inquiries with Policy Implications." *Science, Technology and Human Values* 10(3): 6–19.

Colander, D. C. and A. W. Coats, Eds. (1989). *The Spread of Economic Ideas*. Cambridge, Cambridge University Press.

Coleman, W. D. and A. Perl (1999). "Institutionalized Policy Environments and Policy Network Analysis." *Political Studies* XLVII: 691–709.

Commission of the European Communities (2002). *Improving the Knowledge Base for Better Policies*. Luxembourg, Communication from the Commission on the Collection and Use of Expertise by the Commission: Principles and Guidelines.

Dimitrov, R. S. (2003). "Knowledge, Power and Interests in Environmental Regime Formation." *International Studies Quarterly* 47(1): 123–151.

Drori, G. S. M., John W. Meyer, Francisco O. Ramirez, and Evan Schofer (2003). *Science in the Modern World Polity: Institutionalization and Globalization*. Stanford, Stanford University Press.

Dryzek, J. S. (1997). *The Politics of the Earth*. Oxford, Oxford University Press.

Edwards, P. N. and S. H. Schneider (2001). "Self-Governance and Peer Review in Science-for-Policy: The Case of the IPCC Second Assessment Report". *Changing the Atmosphere*. C. A. Miller and P. N. Edwards. Cambridge, MA, MIT Press.

European Commission (2001). Report of Working Group on Broadening and Enriching the Public Debate on European Matters: Democratising Expertise and Establishing Scientific Reference Systems. Brussels, European Commission.

Farrell, A. E. and J. Jager, Eds. (2005). *Assessments of Regional and Global Environmental Risks: Designing Processes for the Effective Use of Science in Decisionmaking*. Washington, D.C., Resources for the Future.

Fischer, F. and J. Forester, Eds. (1993). *The Argumentative Turn in Policy Analysis and Planning*. Durham, NC, Duke University Press.

Funtowicz, S. O. and J. R. Ravetz (1991). "A New Scientific Methodology for Global Environmental Issues". *Ecological Economics*. R. Costanza. New York, Columbia University Press: 137–152.

Funtowicz, S. O. and J. R. Ravetz (2001). "Global Risk, Uncertainty, and Ignorance". *Global Environmental Risk*. J. X. Kasperson and R. Kasperson. London, Earthscan.

Garcia-Johnson, R. (2000). *Exporting Environmentalism: U.S. Multinational Chemical Corporations in Brazil and Mexico*. Cambridge, MA., MIT Press.

Guston, D. H. (1998). "Principal-Agent Theory and the Structure of Science Policy." *Science and Public Policy* 23(4): 229–240.

Guston, D. H. (2003). "Principal-Agent Theory and the Structure of Science Policy, Revisited." *Science and Public Policy* 30(5): 347–357.

Guzzini, S. (2000). "A Reconstruction of Constructivism in International Relations." *European Journal of International Relations* 6(2): 147–182.

Haas, P. M. (1990). *Saving the Mediterranean: The Politics of International Environmental Cooperation*. New York, Columbia University Press.

Haas, P. M. (1991). "Save the Seas". *Ocean Yearbook 9*. E. M. Borgese, N. Ginsburg and J. R. Morgan. Chicago, University of Chicago Press: 188–212.

Haas, P. M. (2000). "Prospects for Effective Marine Governance in the NW Pacific Region." *Marine Policy* 24(4): 341–348.

Haas, P. (2000). "Letter to the Editor." *Marine Policy* 24(6): 499–500.

Haas, P. (2001). "Environment: Pollution." *Managing a Globalized World*. Washington, D.C., Carnegie Foundation.

Haas, P. M. (2001). "Epistemic Communities and Policy Knowledge." *International Encyclopedia of Social and Behavioral Sciences*: 11578–11586.

Haas, P. (2004). "Science Policy for Multilateral Environmental Governance". *Emerging Forces in Environmental Governance*. N. Kanie and P. Haas. Tokyo, UNU Press: 115–136.

Haas, P. and E. B. Haas (1995). "Learning to Learn: Improving Global Governance." *Global Governance* 1(3): 255–284.

Haas, P. M. and E. B. Haas (2002). "Pragmatic Constructivism and the Study of International Relations." *Millennium: Journal of International Studies* 31(3): 573–601.

Hajer, M. A. W. and H. Wagenaar, Eds. (2003). *Deliberative Policy Analysis: Understanding Governance in the Network Society*. Cambridge, Cambridge University Press.

Hall, P. A., Ed. (1989). *The Political Power of Economic Ideas*. Princeton, Princeton University Press.

Haufler, V. (2001). *A Public Role for the Private Sector*. Washington, D.C., Carnegie Endowment for International Peace.

Helleiner, E. (1994). *States and the Emergence of Global Finance*. Ithaca, Cornell University Press.

Hoffman, A. J. (2001). *From Heresy to Dogma, An Institutional History of Corporate Environmentalism*. Stanford, CA, Stanford University Press.

Hopf, T. (1998). "The Promise of Constructivism in International Relations Theory." *International Security* 23(1): 171–200.

Ikenberry, G. J. (1993). "Creating Yesterday's New World Order". *Ideas and Foreign Policy*. J. Goldstein and R. O. Keohane. Ithaca, Cornell University Press: 57–86.

Janicke, M. W., H., Ed. (1995). *Successful Environmental Policy: A Critical Evaluation of 24 Cases*. Berlin, Edition Sigma.

Jasanoff, S., G. E. Markle, et al. (1995). *Handbook of Science and Technology Studies*. Thousand Oaks, CA, Sage Publications.

Jasanoff, S. and M. B. Martello, Eds. (2004). *Localizing and Globalizing: Knowledge Cultures of Environment and Development*. Cambridge, MA, MIT Press.

Jasanoff, S. and B. Wynne (1998). "Science and Decisionmaking". *Human Choice and Climate Change*. S. Rayner and E. L. Malone. Columbus, OH, Batelle Press. 1: 1–87.

Kameyama, Y. (2004). "The IPCC". *Emerging Forces in Environmental Governance*. N. Kanie and P. M. Haas. Tokyo, United Nations University Press: 137–154.

Kanie, N. and P. M. Haas, Eds. (2004). *Emerging Forces in Environmental Governance*. Tokyo, UNU Press.

Kates, R. W., W. C. Clark, et al. (2001). "Sustainability Science." *Science* 292(5517): 641–642.

Keohane, R. O. (1989). *International Institutions and State Power*. Boulder, Westview Press

Koehn, P. and J. N. Rosenau (2002). "Transnational Competence in an Emergent Epoch." *International Studies Perspectives* 3(2): 105–127.

Koremenos, B., C. Lipson, et al. (2001). "The Rational Design of International Institutions." *International Organization* 55(4).

Kubalkova, V., N. Onuf, et al. Eds. (1998). *International Relations in a Constructed World*. Armonk, NY, M.E. Sharpe.

Lee, K. N. (1993). *Compass and Gyroscope*. Washington, D.C., Island Press.

Lidskog, R. and G. Sundqvist (2002). "The Role of Science in Environmental Regimes: The Case of LRTAP." *European Journal of International Relations* 8(1): 77–101.

Lindner, S. H. and B. G. Peters (1995). "The Two Traditions of Institutional Designing: Dialogue versus Decision". *Institutional Design*. D. L. Weimer. Boston, Kluwer Academic Publishers: 133–160.

Mantzavinos, C., D. C. North, et al. (2004). "Learning, Institutions and Economic Performance." *Perspectives on Politics* 2(1): 75–84.

March, J. G. and J. P. Olson (1988). "The Institutional Dynamics of International Political Orders." *International Organization* 52(4): 943–970.

Miles, E. L. (1998). "Personal Reflections on an Unfinished Journey through Global Environmental Problems of Long Timescale." *Policy Sciences* 31(1): 1–33.

Miles, E. L., A. Underdal, et al., Eds. (2002). *Environmental Regime Effectiveness: Confronting Theory with Evidence*. Cambridge, MA, MIT Press.

Miller, C. A. and P. N. Edwards, Eds. (2001). *Changing the Atmosphere: Expert Knowledge and Environmental Governance*. Cambridge, MA, The MIT Press.

Milner, H. (1992). "International Theories of Cooperation Among Nations." *World Politics* 44(3): 466–496.

Milton, G. (1999). *Nathaniel's Nutmeg*. New York, Farrar, Straus and Giroux.

Mitchell, R. B., W. C. Clark, et al., Eds. (2006). *Global Environmental Assessments: Information and Influence*. Cambridge, MA, MIT Press.

Moe, T. M. (1984). "The New Economics of Organization." *American Journal of Political Science* 28(4): 739–777.

Morgan, M. G. and J. M. Peha, Eds. (2003). *Science and Technology Advice for Congress*. Washington, D.C., Resources for the Future.

National Research Council (1999). *Our Common Journey: A Transition Toward Sustainability*. Washington, D.C., National Academy Press.

Nelkin, D., Ed. (1979). *Controversy: Politics of Technical Decisions*. London, Sage Publications.

Ravetz, J. R. (1986). "Usable Knowledge, Usable Ignorance: Incomplete Science with Policy Implications". *Sustainable Development of the Biosphere*. W. C. Clark and R. E. Munn. New York, Cambridge University Press: 415–432.

Reinicke, W. H. and F. Deng (2000). *Critical Choices*. Cambridge, Cambridge University Press.

Rose, R. (1993). *Lesson Drawing in Public Policy*. Ottawa, IDRC.

Ruggie, J. G. (1991). "Embedded Liberalism Revisited". *Progress in Postwar International Relations*. E. Adler and B. Crawford. New York, Columbia University Press: 201–234.

Ruggie, J. G. (1998). "What Makes the World Hang Together?" *International Organization* 52(4): 855–886.

Searle, J. (1995). *The Construction of Social Reality*. New York, The Free Press.

Siebenhuner, B. (2002). "How do Scientific Assessments Learn?" *Environmental Science & Policy* 5(5): 411–427.

Siebenhuner, B. (2003). "The Changing Role of Nation States in International Environmental Assessments." *Global Environmental Change* 13: 113–123.

Skodvin, T. (2000). "The Intergovernmental Panel on Climate Change". *Science and Politics in International Environmental Regimes*. S. Andresen, T. Skodvin, A. Underdal and J. Wettestad. Manchester, Manchester University Press: 146–180.

Skodvin, T. (2000). *Structures and Agents in the Scientific Diplomacy of Climate Change*. Dordrecht, Kluwer.

Skoie, H. (2001). The Research Councils in the Nordic Countries – Developments and Some Challenges, Rapport 10/2001 NIFU: Norsk Institutt for Studier av Forskning og utdanning.

Smith, B. L. R. (1990). *American Science Policy Since World War II*. Washington, D.C., The Brookings Institution Press.

Smith, B. L. R. (1992). *The Advisers.* Washington, D.C., The Brookings Institution Press.
Social Learning Group (2001). *Social Learning and the Management of Global Environmental Risks.* Cambridge, MA, MIT Press.
Swart, R , J. Mitchell, et al. (2002). "Stabilisation Scenarios for Climate Impact Assessment." *Global Environmental Change* 12(3): 155–165.
Tolba, M. and I. Rummel-Bulska (1998). *Global Environmental Diplomacy: Negotiating Environmental Agreements for the World, 1973–1992.* Cambridge, MA, MIT Press.
Topf, R. (1993). "Conclusion". *The Politics of Expert Advice.* A. Barker and B. G. Peters. Edinburgh, Edinburgh University Press: 103–117.
US OTA (1994). *Industry, Technology and the Environment: Competitive Challenges and Business Opportunities, OTA-ITE-586.* Washington, D.C., US Government Printing Office.
Victor, D. G., K. Raustiala, et al. (1998). *The Implementation and Effectiveness of International Environmental Commitments: Theory and Practice.* Laxenburg, Austria; Cambridge, MA, International Institute for Applied Systems Analysis; MIT Press.
Weidner, H. and M. Janicke, Eds. (2002). *Capacity Building in National Environmental Policy.* Berlin, Springer.
Weinberg, A. (1972). "Science and Trans-Science." *Minerva* 10: 2009–2222.
Weiss, E. B. and H. K. Jacobson, Eds. (1998). *Engaging Countries.* Cambridge, MA, MIT Press.
Wildavsky, A. B., Ed. (1979). *Speaking Truth to Power.* Boston, Little Brown.
Young, O. R., Ed. (1999). *The Effectiveness of International Environmental Regimes.* Cambridge, MA, MIT Press.
Ziman, J. (2000). *Real Science: What It is, and What It Means.* Cambridge, Cambridge University Press.
Zurn, M. (1998). "The Rise of International Environmental Politics." *World Politics* 50: 617–649.

16

ORGANIZED SCIENCE, USABLE KNOWLEDGE, AND MULTILATERAL ENVIRONMENTAL GOVERNANCE*

Peter M. Haas and Casey Stevens

There is a widespread acceptance that effective multilateral environmental governance requires the involvement of environmental scientists (Gore 1996). Although speaking truth to power has long been a major theme in political science and policy studies (Wildavsky 1979), commentators are increasingly skeptical about whether modelers and scientists are capable of developing truth and whether power ever listens to them anyhow. Indeed, international relations scholars tend to be surprised by the occasions when it does. This chapter applies the political science literature to the related question of when power does listen to science, particularly with regard to the management of complex environmental issues.

The Convention on Long-Range Transboundary Air Pollution (CLRTAP) can assist in drawing larger lessons about the situations when science can speak to power and create effective multilateral environmental governance. This chapter places the analysis of CLRTAP in the wider empirical record of effective mobilization of usable knowledge. We begin by focusing on the development of sustainability science, which requires the substantial involvement of scientific expertise in policymaking. We then move to consider the typical arguments about the limits that knowledge places on power and the situations in which knowledge may actually contribute to power. Next, we look at the empirical examples of knowledge speaking to power in three sections: first, an overview of 30 different multilateral environmental agreements and their construction of procedural science–policy connections; next, the lessons from looking at the empirical record; and third, the limiting case of the Intergovernmental Panel on Climate Change (IPCC). The conclusion reflects on these empirical sections and on their implications for constructivism in international relations theory.

Knowledge, sustainability science, and the social construction of consensus

Sustainable development is now one of the major mantras invoked in the area of international environmental governance. Sustainable development was popularized in the seminal 1987 World Commission on Environment and Development report *Our Common Future*. This new policy doctrine—or consensual wisdom within the international community of environmental policy analysts (and at times advocates)—rests on three key foundations (Ravetz 1986; National Research Council 1999; Clark 2001; Kates, Clark et al. 2001; Clark 2007): (1) a normative belief in systematic intervention in human activities to make those activities sustainable; (2) a procedural belief that policy should be participatory and transparent—in part to include new perspectives in the knowledge base brought to bear on understanding a particular problem as well as to promote buy-in and the inclusion of stakeholders in the subsequent application of these policies and thus to improve the prospects for more effective enforcement and compliance; (3) a substantive and more comprehensive belief that systematic approaches to planning and policy formulation are necessary, often through the technique of sustainability impact assessment and forms of interdisciplinary integrated modeling intended to bring together partial insights to sustainable development from different academic disciplines.

Although sustainability science has been percolating in the various laboratories of environmental analysts for nearly 20 years, there has been effective technology transfer from the laboratory to a government agency or international regime in few observable occasions and perhaps not in any. Even when scientists think they have developed truths for power, power appears disinterested at best and possibly even uninterested. Yet sustainable development has gradually encroached on environmental policymaking worldwide, both in international regimes and in national legislation and practices. Power listening to truth is rare, and science policy analysts and political scientists too often dismiss truth entirely.

Many arguments are harnessed to account for why science seldom influences policymaking, as indicated by the fact that science is seldom directly converted to policy. The coproduction of science and social institutions is a leitmotif of the social studies of science literature (Jasanoff and Wynne 1998; Jasanoff 2004). Although the two activities do coevolve and may be mutually constitutive, in this chapter we seek to elaborate some of the causal mechanisms by which the two interact. For convenience, we keep scientific institutions and broader social institutions discrete for the purpose of closer analysis. The path from truth to power is circuitous at best. Let us cluster the reasons for the skepticism under a set of convenient categories (for a summary of arguments about science policy and an argument for a pragmatic procedural approach to generating scientific knowledge for public policy, see Ziman 2000).

1 Science isn't wisdom (or true). Science studies dismiss the prospects of objective knowledge about the world and stress the political dimensions to science and science policy. Scientific findings may in part reflect the bias of sponsors, but they may more deeply reflect the broader culture of the society from which they emerge and about which they may not be fully conscious (Litfin 1994).
2 Science is politically tainted and suspect. Organized modern science embodies implicit values of control, so that decisions made with scientific warrants may unconsciously reflect such hidden values. In addition, the distributional consequences of science-based advice are themselves political (Jasanoff, Markle et al. 1995). Science is political in its consequences because some people and areas benefit and others suffer as a consequence of policy options that are supported by the application of scientific understanding. To the extent that those affected by the use of science in formulating policy are not consulted in its development and application, they potentially regard it as an illegitimate and exploitative set of discursive practices (Miller and Edwards 2001; Lidskog and Sundqvist 2002).
3 Power doesn't care about truth anyhow. Power doesn't want science; it wants a justification for preexisting political programs, which are driven principally by political anticipations of gain (Nelkin 1979; Miles 1998; Andresen, Walloe et al. 2005). In addition, science may not be sufficiently simple for policymakers. Harry Truman is supposed to have complained about economic advisers who would say "on the one hand" and "on the other hand," "Just give me a one-handed economist."

Before discussing situations in which power does listen to truth, let us first dismiss the standard rationalist account that major problems create the incentives for their resolution, and thus modern bureaucracies (or bureaucratic institutions) either develop effective responses almost automatically or are so powerfully constrained by the strategic interests of powerful member states or participants that institutional applications of knowledge are little different than the aggregate wishes of the more powerful members (for international institutions, see (Koremenos, Lipson et al. 2001; for domestic, see (Moe 1984). We reject this proposition for two reasons.

One is that, particularly for sustainable development, the material incentives and the nature of the presumptive useful information are not in synch. Few organizations, if any, have the available mission and resources to be able to address the full integrative range of issues encompassed by sustainable development. Yet there is an empirical record of shifts in policies and institutions beyond what formal organizational assignments would predict (Haas and Haas 1995). The second reason can be addressed by way of a historical analogy. Scurvy was arguably the single most important limiting factor to the expansion of trade and geopolitical influence in the fifteenth and sixteenth centuries. Trade and exploration was significantly hampered because the mortality rate of sailors on long-distance expeditions was often in excess of 90 percent. Expeditions would return with far fewer ships not because of storms, but because too few crew members were left to be able to serve the entire fleet. And yet in the early 1600s the scurvy problem was solved by Captain James Lancaster, who suggested that citrus trees be brought along on ships, but the solution was forgotten for nearly 150 years, until Captain Cook rediscovered it in the 1760s (Milton 1999)! So much for rational societies responding effectively to important issues. We should not expect a fully formed sustainable development science to be developed and applied in response to the presumptive need for such a view. We shouldn't assume that all organizations are rational and will automatically recognize and adopt what prove to be the appropriate policy responses with the virtue of hindsight or reflect the material needs of their most powerful constituencies.

Usable knowledge: when knowledge speaks to power

We take a late-modern view of truth: the domain of science and its ability to confer truth is bounded, but we can talk of better and worse science (for an application of this view to the study of international institutions, see Haas and Haas 2002). In the context of this discussion, we can think of the domain of science that is likely to be adopted by decision makers as *usable knowledge* (the literature has multiple uses of the term: for earlier and slightly different usages, see Clark 1990 and Jasanoff 1990; Dimitrov 2003 calls a similar concept "serviceable truths"). Several different schools of thought exist about usable knowledge, although the core insights are quite similar and complementary. In short, usable knowledge encompasses a substantive core that makes it usable for policymakers and a procedural dimension that provides a mechanism for its transmission from the scientific community to the policy world, and it allows for agency when theorizing about broader patterns of social learning, policymaking, and international relations.

William Clark and Giandomenico Majone (Clark and Majone 1985) define the core ingredients to usable knowledge as adequacy, value, legitimacy, and effectiveness. Adequacy relates to including all the relevant knowledge or facts germane to the matter at hand. Value has to do with contributing to further understanding and meaningful policy. Legitimacy relates to the acceptance of the knowledge by others outside the community that developed it. Effectiveness relates to its ability to shape the agenda or advance the state of the debate and ultimately to improve the quality of the environment (see also Social Learning Group 2001). The substantive core of usable knowledge can be augmented in a number of ways.

Current research from comparative politics, international relations, policy studies, and democratic theory suggests that science remains influential if its expertise and claims are developed behind a politically insulated wall (Andresen 2000; Botcheva 2001; Social Learning Group 2001). Moreover, any efficiency gains by relying on one single source of policy advice are more than offset by the loss of legitimacy, analytic blinders imposed by relying on just one institutional source for usable knowledge, and the political doubts of bias that are raised by narrowing the source of information. Studies of international environmental assessments and science panels suggest the need for fluid bodies that can bring together multiple sources of information and are not beholden to one single funder or political sponsor (Siebenhuner 2002; Siebenhuner 2003; Haas 2004; Martello and Jasanoff 2004; Mitchell, Clark et al. 2006). Studies of national-level environmental policy processes have convincingly argued against relying on individual institutions for research and policy advice because they may bias the information flow and control resources (Brown 1997; Skoie 2001). There is much national-level experience with establishing standing scientific panels, such as in the US National Academy of Sciences, the Environmental Protection Agency Science Advisory Board, and the now defunct Office of Technology Assessment, among others. Each panel is regarded as generating usable knowledge for the government and enjoys sufficient autonomy to identify research questions and to convene panels to develop reports. In CLRTAP, the European Monitoring and Evaluation Program (EMEP), the working groups, and the Network on Air Pollution and Health (AIRNET) all serve to identify research questions and guide science with some degree of autonomy from the political process.

Based on these insights, the general factors that influence the likelihood that decision makers will apply scientific knowledge are credibility, legitimacy, and saliency (Schroeder, King et al. 2008). Credibility means that the knowledge claims are believed to be accurate: within a consensus theory of truth, this means that they are publically created through a deliberative process by people widely regarded as experts. Legitimacy means that knowledge is developed by people who have social authority and that it is accepted by people outside the community that developed it. In practice, the assignment of legitimacy often rests on peer review and scholarly reputation. Salience means that the information is timely and is organized on a politically meaningful timescale and scale of resolution.

In addition to the substantive core of usable knowledge, there is a focus on the processes of transmitting that usable knowledge from scientists to policymakers. A new consensus is emerging among social scientists who study the use of science in international regimes that a procedural or discursive model of the policy process is normatively superior and that the frequency of its application is growing, as against the former policy-analytic approach associated with cost–benefit analysis and analytic efforts to identify the best policy solution. Policy analysis is a process of exchange and mutual learning between policymakers and policy analysts. Policies themselves are experiments that participants monitor and about which they reflect in order to improve them over time (Asher 1986; Funtowicz and Ravetz 1991; Lee 1993; Funtowicz and Ravetz 2001; Lindner 2003; Siebenhüner, chap. 4 in this volume). International states are increasingly relying on this procedural approach in the development and application of truth to power. The Commission of the European Communities (European Commission) developed a set of guidelines and proposals for the collection and use of expertise by the commission that is very similar in orientation to expert groups assembled and utilized for specific, procedural tasks (European Commission 2001; European Commission 2002; Gornitzka and Sverdrup 2008). As a consequence, policymaking is a process rather than a fixed set of analytic techniques. Language may play a role (Fischer and Forester 1993; Hajer and Wagenaar 2003), but the constructivist argument pursued here stresses the political parameters within which debate occurs rather than the connotations generated by the language that is used.

Constructivists argue that under conditions of uncertainty—such as those associated with contemporary globalization and highly technical issues—it is impossible to ascertain ex ante sufficient information to follow the policy-analytic model (Checkel 1998; Hopf 1998; Kubalkova, Onuf et al. 1998; Ruggie 1998; Christiansen, Jorgensen et al. 1999; Guzzini 2000; Haas 2001; Adler 2002). Alternatively, the key is to design policy-analytic processes from which actors learn about the world and about each other. This view is antirationalist in the sense that the process model entails path dependency and uncertainty assumptions that presume that outcomes and, indeed, preferences are often underspecified or indeterminate and thus that regimes and political interactions can best be appraised in terms of process rather than outcome. All outcomes will be suboptimal in some sense as compared to the ex post ideal outcome, but one can just hope that better outcomes may result from an open discursive process (Dryzek 1997).

Constructivist approaches to policy analysis suggest that science must be developed authoritatively and then delivered by responsible carriers to politicians. Doctors, scientists, and engineers remain the most esteemed professionals in Europe and thus command the greatest social legitimacy and deference when providing policy advice (Jasanoff and Wynne 1998; Drori, Meyer et al. 2003; Drori and Meyer 2006). The transmission belt of like-minded scientists is called an "epistemic community" (Haas 2001; Haas 2007). The more autonomous and independent science is from policy, the greater its potential influence (Andresen, Skodvin et al. 2000; Botcheva 2001; Haas 2001). Consensus in isolation builds value and integrity, and then its consequences should be discussed publicly. Measures of autonomy and integrity include the selection and funding of scientists by intergovernmental organizations rather than by governments, the recruitment of scientists by merit on important panels, and reliance on individuals whose reputation and authority rest on their role as active researchers rather than on their role as policy advocates or science administrators. Accuracy can be achieved via peer review, interdisciplinary research, and independence from sponsoring sources. Sustainability scientists, themselves an epistemic community, increasingly argue for the need to include local knowledge along with the more formally technical understanding of traditional disciplinary elites associated with formulating sustainability policies (Haas and Haas 2002; Martello and Jasanoff 2004). However, the criteria for participation remain loosely defined, but perhaps no more so than the broad injunctions for multidisciplinary participation that do not clearly identify which disciplines need to be consulted for which types of questions.

International and transnational networks of scientists develop usable knowledge. That knowledge is heeded, to the extent that it is, after widely publicized shocks or crises. Although usable knowledge contributes to broader patterns of social learning, different communities often deliver and apply that knowledge. We are not talking of reflective learning by decision makers. Rather, we are talking about decision makers' recognition of the limits of their abilities to master new issues and the need to defer or delegate to authoritative actors with a reputation for expertise. In the aggregate, social learning and human betterment emerge (Levy, Cavender-Bares et al. 2001) when the experts have been able to develop usable knowledge, and the decision makers feel compelled to apply it (Haas and Haas 2002).

Empirical record of usable knowledge in multilateral environmental governance

Over the past 30 years, there has been a dramatic adoption of international treaties and regimes to address many aspects of transboundary and global environmental threats ((Tolba and Rummel-Bulska 1998; Haas 2001; Miles, Underdal et al. 2002; Mitchell 2002; and see Mitchell's Web site at http://pages.uoregon.edu/rmitchel/data.html). When usable knowledge is successfully constructed and transmitted, it yields distinctive results: regimes are developed by a process of social learning; the regime rules reflect scientific consensus about environmental sustainability; and the regimes tend to be effective. Maurice Strong, secretary-general of the United Nations (UN) Conference on the Human Environment (in Stockholm in 1972) and of the UN Conference on Environment and Development (in Rio de Janeiro in 1992) and the first executive director of the UN Environment Program (UNEP), helped design the outlines of this process of social learning. Strong believed that "the policy is the process": that is, by generating an open political process in which states are exposed to consensual science, government officials may be persuaded to adopt more sustainable policies, and individual scientists may gain heightened political profiles at home, which may ultimately increase their effectiveness as well. In addition, Marc Levy and his colleagues explain that social learning "is characterized by acceptance of a core set of common values, mutual recognition of the legitimacy of each actor, and a high degree of trust" (Levy, Cavender-Bares et al. 2001 p. 110). International environmental negotiations develop through scientific discourse in which political debate and compromise reflect expert consensus on the behavior of ecosystems and their ability to sustain stress. The substance of regimes reflects scientific consensus about the most important environmental threats, and negotiated standards reflect consensus about the degree of environmental stress the target environment can sustain. For instance, the 1980 Land-Based Sources Protocol for the Mediterranean required more stringent emission controls on the industrialized countries than on the developing countries because the magnitude of degradation of the northern coast of the Mediterranean was much more severe than it was on the southern coast (Haas 1990).

The CLRTAP offers one case that demonstrates the permissive conditions for successful scientific guidance. Within CLRTAP, the EMEP serves as a body within the convention that organized much of the science work on issues of air pollution. The EMEP and the various working groups under CLRTAP are composed of experts nominated by parties to the treaty. Most important, EMEP manages a series of centers in many member countries that monitored a wide variety of atmospheric conditions. Through discussions between EMEP and the working groups, a consensus emerged regarding the amount of pollutants that needed to be reduced below the critical-loads level. This dialogue between different groups of scientists with defined focuses, coupled with their ability to set their own agenda with guidance from the Conference of the Parties (COP) and the secretariat of the treaty, results in the construction of a politically usable scientific consensus. This consensus that emerged and the monitoring procedures that were established have resulted in a series of protocols leading to emission cuts and a system to supervise compliance. The construction of bodies that were largely insulated and respected by the members allowed the development of usable knowledge and the resultant social learning by the parties to the treaty (Tuinstra, Hordijk et al. 1999; Tuinstra 2008; Wettestad, chap. 2 in this volume). It is through these systems that CLRTAP was able to grow from a treaty that required little from members in terms of cutting emissions into a significant environmental regime that both lowered emissions from members and reduced the environmental pollution caused by transboundary air pollution. The chapter's final section investigates a limiting case where science was unable to influence policymakers.

Social learning regimes are substantively more comprehensive and attuned to the emergent sustainable development doctrine's sensibility and injunctions. Regimes built with usable knowledge appear to be more effective at inducing states to achieve their intended goals of improving environmental quality. For instance, the stratospheric ozone and European acid rain efforts are widely hailed as among the more successful and effective international environmental governance efforts of the contemporary era (Andresen 2000; Haas 2001; Miles, Underdal et al. 2002). Social learning is aided through processes and organizational regime design that both keep the construction of science insulated and construct procedures for the connection of science to policy decisions. For example, most social learning treaties have standing environmental monitoring and research committees to provide timely warnings of new problems, monitor achievements of regime goals, and educate politicians and policymakers on environmental issues.

Regimes vary widely in how they organize scientific participation. There are detailed case studies about many of these regimes (see, e.g., Farrell and Jager 2005; Mitchell, Clark et al. 2006). Regimes that delegate some responsibilities to organized science include many multilateral environmental treaties and organizations (for a partial list of some organizations that maintain some form of science panel, see appendix 16.1). Some institutions have effectively constructed connections between science and policymakers that assist the building of usable knowledge.

The maintenance and support of science bodies within multilateral environmental governance arrangements is absolutely vital for the construction of usable knowledge within the regime. Standing scientific panels allow the constant construction and transmission of accurate and timely information. Three types of standing-committee structure are typical in multilateral environmental regimes.

The first type includes scientific groups that set their own schedules and research agendas. These groups meet as they see fit between meetings of policymakers and according to their own determined needs. The second type is made up of those groups whose meeting times are set by the COP. For example, most fishery agreements require their scientific bodies to meet a month or two months prior to the COP meeting. The UN's Joint Inspections Unit found that the Convention Concerning Desertification largely organized the science body meetings directly in line with COP meetings, thereby creating a polarized and political atmosphere in the meetings (Ortiz and Tang 2005). And the third type includes ad hoc science panels, which, especially if called for by member states, introduce a high level of political involvement in the reporting of science. In addition, long intervals between reports impede the timely involvement of science in policy discussions. Table 16.1 shows the distribution among the various types of science committee organization in multilateral environmental agreements. Standing committees that set their own agenda and schedule appear to be the most politically insulated, whereas ad hoc science panels appear to be the most vulnerable to political involvement.

In addition to the timing of scientific bodies' meetings, the process of selecting scientists to serve on various bodies varies, with different effects on the legitimacy of the science created. The COP and the states involved in political negotiations sometimes directly select authors and experts. For some Multilateral Environmental Agreements' (MEAs) science bodies, the selection of experts is begun by the secretariat

TABLE 16.1 Committee organization in multilateral environmental regimes

Standing Committee: sets its own schedule	*Standing Committee: schedule set by COP*	*Ad hoc Committee*
Stratospheric ozone	Climate change	CCMPRCBS
CLRTAP	Ramsar Convention on Wetlands	IATTC
CBD	Bonn Convention on Migratory Species	ICCAT
CCAMLR	IWC	IPHC
CITES	IMO	NASCO
MEDPOL	CCSBT	South Pacific
North Sea	IOTC	Persian Gulf (pre-1990)
Stockholm POPs	UNCCD	PERSGA
Polar Bears	IOTC	
	NAFO	
	SEAFO	
	Baltic Sea	
	Basel Convention on Hazardous Wastes	
	Rotterdam PIC	
	Southeast Pacific	

Source: UNEP 1998; Fritz 2000; UNEP and UNSECO 2003.

Note: Abbreviations are spelled out in the appendix.

TABLE 16.2 Selection procedures for scientific bodies

Selected by states/COPs	*Selected by secretariat*	*Selected by other international organizations*
IPCC (climate change)	Ozone	STAP GEF
CLRTAP	Baltic Sea	Polar Bears
CBD (biodiversity)	MEDPOL (Mediterranean Sea)	IMO
CCAMLR (Antarctica)	South Pacific	
Ramsar (wetlands)	Southeast Pacific	
Bonn (migratory species)	North Sea	
CITES (endangered species)		
IWC (whaling)		
IOTC (fishery)		
Basel (hazardous wastes)		
Stockholm (POPs)		
Rotterdam (PICs)		
Persian Gulf (pre-1990)		
PERSGA (Red Sea)		

Sources: UNEP 1998; Fritz 2000.

Note: Abbreviations are spelled out in the appendix.

(sometimes in addition to selection by the parties) or even by secretariats of other international organizations in order to insulate the process (Fritz 2000). Table 16.2 illustrates the different mechanisms for selecting scientific bodies. The most insulated involves selection of authors from secretariats or even scientific bodies of other intergovernmental organizations. Most MEAs, however, select the authors and peer reviewers of scientific reports using either the secretariat or the COP.

The actual design of a scientific body in multilateral environmental governance reflects one of three different design types (see Table 16.3). The most popular institutional design (50 percent of regimes) is an open-access design that allows each member country to appoint a representative to a scientific body. Some of these open-membership bodies require the representatives from states to meet minimal and often vague expert qualifications in order to serve (e.g., the Bonn Convention on Hazardous Wastes). This design does not typically create proper insulation between the scientific bodies and the policymakers. It is popular in many regimes and institutions dealing with biodiversity and fishery management. A second type of design is the small expert group. For example, the Ramsar Convention on Wetlands uses a small expert group of 12 members nominated and accepted by the COP and distributed geographically based on the convention's membership. The trade-off in these panels is that although they are efficient, the complex environmental–social linkages and large amount of information necessary often prevents the development of adequate knowledge. A third and relatively recent design is a roster of experts that can either stand alone or be coupled to either of the previous forms. Jan-Steffan Fritz identifies these rosters as being increasingly

TABLE 16.3 Different forms of scientific and expert bodies in MEAs

Open-membership body	*Small expert panel*	*Roster of experts*
CBD (biodiversity)	CLRTAP	STAP GEF
CCAMLR (Antarctica)	Ozone	IPCC (climate change)
Bonn (migratory species)	Ramsar (wetlands)	CBD (biodiversity)
CCMPRCBS (fishery)	CITES (endangered species)	UNCCD (desertification)
IMO (fishery)	IWC (whaling)	Basel (hazardous wastes)
CCSBT (fishery)	IPHC (fishery)	Stockholm POPs
IATTC (fishery)	Stockholm POPs	
ICCAT (fishery)	Rotterdam PIC	
IOTC (fishery)	MEDPOL (Mediterranean)	
NAFO (fishery)	Polar Bears	
NASCO (fishery)	Persian Gulf (pre-1990)	
SEAFO (fishery)	Baltic Sea	
Basel (hazardous wastes)	Southeast Pacific	
UNCCD (desertification)	South Pacific	
PERSGA (Red Sea)		

Sources: UNEP 1998; Fritz 2000; UNEP and UNSECO 2003.

Note: Abbreviations are spelled out in the appendix.

important in filling the need for specialists to provide cutting-edge information but within broader contexts (Fritz 2000). Although this form has been most extensively used in the Scientific and Technical Advisory Panel (STAP) of the Global Environment Facility (GEF) (along with a small expert panel), other organizations are moving toward creating a roster of experts and some unique ways to use them (Fritz 2000). These rosters are most applicable in monitoring functions—for example, in the Convention on Biological Diversity (CBD) biosafety protocol (Hill and Sendashonga 2006). The STAP of the GEF uses the roster of experts for a variety of different functions, from providing peer review of reports to serving as regional science hubs to providing funding oversight. The sheer number of scientists and the evaluation system developed in the STAP ensures insulation between the science and the policymakers.

Despite all these constructions that aim to facilitate the creation of usable knowledge, social learning takes time. Comprehensive treaties are slower to negotiate than others are because they require persuasion and consensus rather than mere compromise. From a policy perspective, though, comprehensive regimes are likely to be superior in their ability to protect the environment in a cost-effective and politically acceptable manner. Moreover, treaties developed with help from the scientific community typically enter into force more rapidly than without it, presumably because of the weight that scientists' involvement carries in the ratification process.

A small number of international institutions have supported the development and transmission of usable knowledge. Epistemic communities often work in conjunction with broader policy networks, functional bureaucrats, transnational scientific organizations, nongovernmental organizations, and international civil servants. UNEP has played a powerful role in environmental protection the past 30 years. With a staff of less than 200 professionals and a budget now on the order of $150 million a year, UNEP has led global environmental monitoring efforts, catalyzed environmental protection activities in other UN bodies, served as the environmental conscience of the UN system, and sponsored the conclusion of dozens of international environmental treaties. The UN Economic Commission for Europe and the International Institute for Applied Systems Analysis helped develop and circulate usable knowledge for the effective management of European acid rain. Since 1991, the GEF has been a major supporter of the transmission of usable knowledge.

Lessons for generating and mobilizing usable knowledge

Given the 30-year experience with addressing transboundary and global environmental threats, what lessons are available about developing and mobilizing usable knowledge for multilateral environmental governance? The lessons listed here seem evident, based on comparative case studies of multilateral environmental regimes with and without usable knowledge and based on process-tracing studies of efforts to institutionalize usable knowledge (Victor, Raustiala et al. 1988; Haas 1990; Weiss and Jacobson 1998; Young 1999; Andresen 2000; Haas 2000; Haas 2000; Reinicke, Deng et al. 2000; Miles, Underdal et al. 2002; Haas 2004; Haas 2004; Kanie and Haas 2004; Oberthur and Gehring 2006). Thus, these conditions are found in effective regimes and are absent in ineffective regimes. Effective regimes:

1. Create standing international interdisciplinary scientific panels or committees to address specific topics.
2. Create and rely on separate subcommittees responsible for different functions of governance, such as basic research, environmental monitoring, policy analysis, and policy verification and evaluation.
3. Carefully survey the population of scientists to identify a core group that shares values and causal beliefs. For instance, in the Mediterranean a UNEP consultant spent nine months visiting national laboratories to inventory national capabilities and to build the scientific network personally.
4. Ensure that network and international panels have interdisciplinary representation, including social sciences. Ensure that individuals are highly regarded in their own disciplines and are able to talk to experts from other disciplines.
5. Recruit carefully for national and regional institutions. Base judgments on professional credentials and networking ability.
6. Avoid relying on one national institution to provide or sponsor research and training.
7. Provide professional outlets for members through conferences and publications in refereed professional journals. By this means, elevate the domestic

profile of individual scientists in the community of experts, who may then be recruited to fill positions in national administrations.

8 Promote scientific discussions on topics that are likely to lead to consensus—that is, ripe research topics. Critical loads and the Regional Acidification Information System are excellent examples of ripe research topics.
9 Avoid government assignment of scientists to international meetings.
10 Try to make use of joint international panels for environmental risk assessment rather than relying on national assessments. Avoid capture by one scientific discipline or school of expert analysis.
11 Assure the timely submission of scientific advice in advance of meetings. Submit reports in a timely way according to the legislative cycle in the major countries.
12 Arrange for focused interactions between scientists and policymakers to discuss the technical substance of the issues. For instance, in CLRTAP the International Institute of Applied Systems Analysis arranged for two-day sessions to familiarize policymakers with acid rain transfer and deposition models developed by scientists.
13 Maintain momentum within the community by continuing to have projects and research opportunities so members don't drift away, which avoids having to reconstitute the community each time a new problem emerges.
14 Seek funding for studies from multiple sources in order to avoid budgetary shocks if money is withheld from a principal funder. Thus, economically inefficient redundancy is politically warranted.
15 Formally institutionalize a roster of experts within the organization so that it can operate more effectively. The risk is that without institutionalizing and embedding the roster of experts the group will float freely and largely fail to affect policymaking.
16 Construct models so that effects are calculated at meaningful political scales—that is, corresponding to significant political divisions that are relevant in developing policy applications. For instance, at the international level this means that models should explicitly identify effects by country and even, if possible, by domestic districts (at least in countries without proportional representation). Climate-change models have so far yielded effects only at a scale of resolution sufficient to demonstrate to countries able to vote in the UN General Assembly that they are likely to suffer, and thus they keep the issue on the international agenda.
17 Train or recruit scientists capable of communicating with the media, politicians, and popular audiences.
18 Try to recruit networks from as broad a national basis as possible because governments are more likely to rely on experts who share their nationality than on experts who are not.

Broader consideration of the proper institutional design of science policy entails timing. When consensus is achieved before an issue reaches the agenda and policy discussions have begun, scientists can merely be introduced as experts, following these lessons. However, at times it is necessary to develop scientific consensus and to advance policy debates simultaneously. For these issues, such as in the Mediterranean and ozone regimes, the parallel development of science and policy must be kept insulated from ongoing policy debates, with the two streams united only when consensus has been achieved. In other cases, where consensus remains elusive and policy debates have already attained their own momentum, as in climate change and biodiversity, it may be best if the two activities are kept as separate as possible.

Overall, there has been extensive social learning in governance of Mediterranean pollution (Program for the Assessment and Control of Pollution in the Mediterranean Region, MEDPOL), European acid rain (CLRTAP), and stratospheric ozone. Thus, for example, epistemic communities were vital in the creation of MEDPOL and then for the degree of social learning that occurred within the organization. As a consequence, independent panels helped to develop standards for achieving reduction of land-based pollutants (Haas 1990). With respect to MEDPOL, CLRTAP, and stratospheric ozone, epistemic communities have been institutionalized generally in the regimes and played key roles in legal negotiations. Social learning has occurred in the creation of many environmental regimes (e.g., the International Whaling Convention [IWC] and the North Sea), but these regimes have mostly stalled, and little social learning has happened within them since they were established. In the IWC, the scientific committee developed a revised management procedure to allow whaling of nonendangered species. Years later this revised management scheme has still not been adopted, and Japan, Iceland, and Norway's continued hunting of nonendangered species threatens the legitimacy of the regime (Andresen 2002). There has been moderate social learning in some biodiversity and chemical conventions (Rotterdam Convention on Prior Informed Consent, Bonn Convention on Migratory Species, Ramsar, and Convention on International Trade of Endangered Species [CITES]). For example, the establishment of scientific committees in the 1987 CITES allowed a continued scientific appraisal of the regime's effectiveness and the expansion or reduction of the list of endangered species. The scientific committees have been able to update and monitor endangered species status effectively, but these functions have translated minimally into changing the organization's policy significantly. There has been little to no social learning in a number of regimes regarding fisheries, biodiversity (namely, the CBD), desertification, and climate change (see relevant chapters in Social Learning Group 2001; Miles, Underdal et al. 2002). Table 16.4 gives a breakdown of various MEAs.

The insulation of the science body within the MEA is key to the social learning that can occur within the organization (see appendix 16.1 for a combination of the information given in all the tables). For example, in MEDPOL, CLRTAP, and the ozone regime there is an insulated and separate science panel. Most cases in which there are small groups of experts that are selected by the secretariat and that can set their own agenda lead to some degree of social learning (this is true in the

TABLE 16.4 Degrees of social learning in MEAs

No significant social learning	*Some social learning within regime*	*Social learning*
Climate change	Bonn (migratory species)	MEDPOL (Mediterranean)
CCMPRCBS (fishery)	Ramsar (wetlands)	CLRTAP (air pollution)
IMO (fishery)	Rotterdam PIC	Ozone
CCSBT (fishery)	CITES (endangered species)	CCAMLR (Antarctica)
IATTC (fishery)	IWC (whaling)	Polar Bears
ICCAT (fishery)	North Sea	Stockholm POPs
IOTC (fishery)	Baltic	
IPHC (fishery)	Persian Gulf (pre-1990)	
NAFO (fishery)	South Pacific	
NASCO (fishery)	Southeast Pacific	
SEAFO (fishery)		
UNCCD (desertification)		
Basel (hazardous wastes)		
PERSGA (Red Sea)		

Sources: see Young 1999; Social Learning Group 2001; Miles, Underdal et al. 2002.

Note: Abbreviations are spelled out in the appendix.

ozone, MEDPOL, CLRTAP, Rotterdam, and Baltic Sea regimes). For example, with regard to CLRTAP, Bernd Siebenhüner (chap. 4 in this volume) discusses various *reflexive mechanisms*, including bringing international experts into the panels of the CLRTAP negotiations and external reviews as forms of insulation to increase the legitimacy of the science. There are several cases in which the use of some of these procedures has produced varied results in terms of social learning. The Bonn Convention on Migratory Species is the only institution with an open advisory system (each nation appoints one member to the science committee) that has demonstrated any social learning. Bonn, however, has several other safeguards and mechanisms that insulate science from political control. Fisheries management in general has failed to insulate science bodies from direct or indirect political control. The design of the science committees in fisheries reflects an institutional pathology that limits the ability of science to speak to power. Insulation is vital procedurally if usable knowledge is to lead to social learning; however, many organizations that aim toward deliberation and transparency are still limited in their level of social learning. Thus, the mere existence of a roster of experts or a small expert group does not guarantee that social learning will occur. Issues of comprehensiveness and adequacy are vital if the regime is to maintain social learning.

Climate change, usable knowledge, and the limits to social learning

The climate-change regime, the focus of the majority of global environmental diplomatic efforts during the 1990s, has also developed through a deliberate effort by diplomats to base policy on strong scientific foundations through the IPCC. The IPCC is of interest because it highlights the way in which states may choose to shape the science advisory process. A closer look at the interplay of science and power in the IPCC reveals empirically how this dynamic interaction operates in this key contemporary issue and establishes analytically the political limits to autonomous science and social learning. The knowledge base developed so far has not proven particularly usable largely because the construction of knowledge in the IPCC is politically suspect and because the results are not presented in a way that connects to individual experiences (Haas 2008).

The IPCC is one of the most concerted efforts by the international community to harness usable knowledge for addressing transboundary and global environmental threats (Agrawala 1998; Agrawala 1998; Skodvin 2000; Skodvin 2000; Siebenhuner 2002; Siebenhuner 2003). The IPCC was established in 1988 as an international science policy advisory body for global warming, but the general belief is that it was formed politically so that governments would be able to reassert control over the science process in an issue that was accelerating on the policy agenda more rapidly than most leaders in the North were comfortable with. The IPCC was the consequence of a UN General Assembly initiative for a climate-change regime and did not rely on UNEP, as had most previous international environmental regime initiatives. The IPCC consists of three working groups, with members chosen by governments subject to the scientific reputation of the candidates. Working Group One addresses questions of atmospheric science; Working Group Two assesses social and economic impacts and adaptation measures; and Working Group Three focuses on mitigation alternatives. Each working group is administered by a bureau composed of the IPCC Bureau members, working group co-chairs and vice-chairs, and a technical advisory unit drawn from the country of the work group's chair. To date, the IPCC has produced 4 three-volume assessment reports (1990 with 1992 revisions, 1995, 2001, and 2007), each with a summary for policymakers as well as various ad hoc special reports and technical papers. The working groups report to a plenary composed of government delegates, who review the summary reports on a line-by-line basis and approve the working group assessments and special reports in more block-like fashion. All reports other than the summary reports rely on extensive peer reviews, and the material presented must come from peer-reviewed journals or be in the process of publication in a peer-reviewed journal. The summary reports, which command the greatest publicity and hence public attention, are written by the working group leaders along with the lead authors and specially invited experts (Skodvin 2000; Swart, Mitchell et al. 2002). The bureaus are responsible for drafting an initial table of contents and topics for each chapter. The IPCC political plenary then approves this agenda. The scientific committees draft reports, and then government representatives in the plenary approve them subject to careful scrutiny.

Thus, governments have sought to exercise control over the scientific process while also allowing for some degree of scientific latitude in order to generate accurate advice, even if the process is designed in such a way that the advice is unlikely to be particularly salient. Governments appoint the IPCC chair, and in

2002 the United States vetoed the appointment of the climatologist Robert Watson, a well-regarded American candidate, in favor of Rajendra Pachauri, an Indian engineer, based on the belief that Watson was too independent of the US administration. The United States criticized the *Third Assessment Report* (2001) for containing language that the United States claimed was stronger than had been approved by the plenary, although the authors were able to prove subsequently that their draft was consistent with the IPCC's rules of procedure (Miller and Edwards 2001; Siebenhuner 2002). Diplomats in the plenary evaluate the summaries for policymakers on a line-by-line basis. All individuals are nominated and chosen by governments, although there is little evidence of direct government manipulation in recruitment or the inclusion of material. The procedures are carefully designed to maximize the seeming procedural scientific legitimacy and accuracy of the work by stressing peer review and reliance on peer-reviewed material.

The degree to which the IPCC is capable of generating usable knowledge is largely politically circumscribed. The state of scientific understanding of the key global systems that affect global warming remains relatively immature. It is unclear to what extent funding and research choices have been shaped by political factors, but there is no strong evidence to assert that the state of knowledge about the phenomenon is directly biased or controlled by political influences. The saliency of the IPCC science remains limited. The estimates of global warming and their effects are crude, and global carbon models are unable to account fully for the global carbon cycle.

The IPCC does not enjoy a high degree of legitimacy in the eyes of many consumers of science policy (Biermann 2002). It suffers from the appearance of governmental control because governments appoint the scientists and vote on the reports. The distribution of scientific participants comes overwhelming from the North, despite the bureaus' best efforts to increase participation from developing countries since the release of the *First Assessment Report* (Biermann 2002). Table 16.5 shows the crude national breakdown of scientific involvement in the *Fourth Assessment Report* by the scientific participants' country of origin. The IPCC is limited in its legitimacy through its seeming lack of equity in participation. Although

TABLE 16.5 IPCC Fourth Assessment Report: distribution of nationality of scientific participants

Nationality	*WG1 authors (n = 625)*	*WG2 authors and expert reviewers (n = 1,503)*	*WG3 authors (n = 272)*	*Synthesis lead authors (n = 39)*
Industrialized countries	88%	75%	68%	62%
Developing countries	12%	25%	32%	38%
Technical support unit	United States	United Kingdom	Netherlands	

Source: see http://www.ipcc.ch (retrieved 10 December 2007).

Note: Industrialized countries are considered those that are members of the Organization of Economic Cooperation and Development (OECD), which includes those countries at high levels of industrialization and development. Developing countries are those that are not members of OECD. See http://www.oecd.org.

the IPCC has helped pay for the participation of scientists from developing countries, it faces the deeper structural science policy issue that the overwhelming majority of climate-change research is conducted in the North by northern scientists.

The IPCC's saliency is particularly poor. For one thing, it has been unable to develop policy advice that resonates domestically in any of the major countries. To date, its scientific work has narrowed the range of likely warming that will occur in 2100 and generated scenarios of what the global environmental consequences may be of such effects. However, the scenarios are sufficiently crude that they do not engage any significant political interest in any of the member countries, other than reinforcing the prior knowledge that Egypt, Bangladesh, and the small island countries may well be submerged. Thus, the political effect of the new knowledge is adequate only for the aggrieved parties to continue to keep the issue on the General Assembly agenda, where they have sufficient votes, but not to influence policy outcomes directly. More salient science would present data at a level of resolution that corresponded to domestic political divisions (i.e., states or congressional districts in the United States and parliamentary districts in parliamentary systems). In addition, in attempting to be politically neutral, the IPCC aims to provide "policy-relevant, not policy-prescriptive" information, which results in information that is too broad to provide even basic policy assistance (Hasselman and Barker 2008).

The publication of the assessment reports did not fit in well with the negotiating cycle of the Kyoto Protocol and efforts to follow up the 1992 UN Framework Convention on Climate Change. A few of the special reports—such as the 2000 *Report on Land Use, Land Use Changes, and Forestry*—corresponded to COP meetings and thus were able to provide timely scientific policy advice to the actual negotiators. The reports have more often not coincided closely with the negotiation process, so they have had little impact other than periodically reminding the mass public of the potential severity of global warming. IPCC scientists have not fared well at communicating their findings to the public or in engaging in either elite or public education. This failure can be contrasted with the timely issuing of reports by other organizations to precede major political conferences on the issue. For example, both the *Scientific Assessment of Ozone Depletion* in the 1990s and the *World Water Report* in the 2000s preceded major negotiating cycles internationally. However, many reports from scientists are timed on cycles (e.g., every two years) that seem to have little or nothing to do with the political negotiations. The IPCC seems to have altered this pattern slightly with the *Fourth Assessment Report*, which was released in early 2007 following a momentous year that saw Al Gore's *An Inconvenient Truth* receiving every possible award for a movie, book, and book on tape (with an opera in the planning stages), the IPCC's receiving the 2007 Nobel Peace Prize, and then major international climate-change negotiations. However, this slight alteration in timing failed to result in effective negotiations among the COP at the Copenhagen UN Framework Convention on Climate Change (COP 15) in 2009, where, in the measured words of one analyst, "the conference was a failure whose magnitude exceeded our worst fears, and the resulting Copenhagen Accord was a desperate attempt to mask that failure" (Dimitrov 2010 p. 18).

In short, the IPCC is designed to keep science on a tight leash by controlling the selection and autonomy of individual scientists engaged in the assessment process. As a consequence, the degree of usable knowledge generated by the IPCC has been limited. And IPCC scientists not surprisingly have been unable to exercise sufficient discretion to develop more politically tractable advice. In turn, negotiated treaties within the climate-change regime have not reflected a strong degree of scientific basis despite the ongoing IPCC efforts. The 1992 UN Framework Convention on Climate Change and the 1997 Kyoto Protocol reflect broad rhetorical commitments or political compromises about target emissions but do not point to any scientifically grounded targets for specific atmospheric concentrations of greenhouse gases. Few governments outside the European Union have proven willing to make national decisions based on this collective enterprise, and European Union greenhouse gas targets reflect political and energy policy realities rather than scientific justifications (Haas 2008).

The limits of the IPCC's ability to speak truth to power can be explained by principal–agent theory (Moe 1984; Koremenos, Lipson et al. 2001; see also Guston 1998; Guston 2003 for an effort to capture science policy issues more generally within a principal–agent framework). According to this theory, the principals strategically design the institution (the agent) on which they rely for information and advice in order to minimize surprises from the agent and to ensure that the agent will not provide advice that runs counter to the principals' ex ante interests and preferences. The principals in the IPCC case are the governments of the industrialized countries. In 1988, they were tired of being lambasted by UNEP in multilateral environmental negotiations and were concerned that uncontrolled scientific panels might give rise to policies that they did not deem warranted. The Villach meetings from 1985 to 1988 had forcefully put the issue of global warming on the international agenda, and the 1988 Toronto conference convened by nonstate actors had called for 20 percent reductions in greenhouse gas reductions, thus alerting foreign ministry officials of the political threats posed by nonstate actors' potential control of the climate-change agenda. The recent lesson of the Ozone Trends Panel's powerful influence on the rapid conclusion of the Montreal Ozone Protocol was also fresh in the minds of foreign ministries. Moreover, the short-term costs of addressing the problem for the industrialized countries (or, put another way, the net discounted value of efforts for mitigation adaptation) appeared prohibitive so that they wished to rein in any independent political pressure that would be generated from an organized scientific involvement in collective discussions on climate change.

The agents in this regard are the scientific community, who are recruited and organized through the IPCC in a way that assures governments that they will be able to exercise maximum control over individual scientists and will remain able to shape the political agenda for climate-change negotiations. Through the reliance on extensive peer review, some degree of control was admittedly lost, but at the end of the day governments appoint the individuals, and foreign-ministry officials approve reports on a line-by-line basis. Thus, the principals establish and enforce the parameters within which the agents have to act, thus ensuring cautious outcomes by the agents.

But the IPCC is an extreme case in terms of the clarity with which the powerful industrialized countries recognized their material interests in the matter and in terms of the widespread transformations in industrial and individual behaviors that effective preventive policies were presumed to entail. In addition, the concurrent negotiations and science advisory process made governments extremely sensitive to the strategic implications that scientific assessments may have for national policies. In most other transboundary and global environmental issues, the scientific consensus preceded active political discussions so that governments were less acutely aware of the implications of their choice to delegate rule making to scientists. Most other environmental and sustainable development issues do not share the characteristics of the IPCC case, so there is greater potential for governments to create more legitimate and porous scientific processes to guide their policy work.

Conclusion

Thirty years of speaking truth to power has yielded tangible beneficial outcomes. Power listens sometimes. There now exist far more extensive legal commitments on environmental protection than ever existed before, and most areas of human activity in most countries of the world are now regulated as a consequence. In the aggregate, major changes have occurred internationally in environmental governance. Many multilateral environmental regimes are more comprehensive than before science was systematically applied to multilateral environmental diplomacy; stronger national environmental protection commitments have been undertaken in compliance with international obligations; and green markets for cleaner technologies have grown as a consequence. Markets estimated at more than $650 billion a year now exist for green technologies (*Environmental Business Journal* 2006). These markets were created in response to the regulatory obligations resulting from international treaties negotiated based on the application of new scientific understandings of the behavior of ecosystems (Garcia-Johnson 2000; Haufler 2001). The protection of ecological integrity has become elevated to the position of a state interest on a par with the pursuit of national wealth and power. Sustainable development is now the doctrine by which states seek to reconcile (or optimize) the pursuit of those three objectives.

Knowledge can speak volumes to power. Much current work suggests that science remains influential if its expertise and claims are developed behind a politically insulated wall. Epistemic communities are the transmission belts by which new knowledge is developed and transmitted to decision makers. Knowledge must also possess the substantive characteristics of usable knowledge: credibility, legitimacy, and saliency. And successful regimes demonstrate practical administrative lessons about generating and mobilizing epistemic communities and usable knowledge.

Real limits to learning persist. National administrative design often inhibits learning and the diffusion of environmental lessons. Most learning remains highly localized: lessons are limited to changes in national policy with regard to one particular environmental medium—for instance, air pollution or marine pollution. This limitation is due to administrative design in most countries. Epistemic-community members' expertise is highly specialized, so their influence and involvement in agency choices are limited to a specific regime and environmental issue area. Moreover, environmental responsibilities within governments vary by environmental issue because different agencies are responsible for coordinating national activities in different areas. Thus, epistemic-community members' influence in the short term is highly restricted. Few governments have the domestic potential for these lessons to spread within the government because the functional agencies have little influence over other agencies.

The prospects for improving national sensitivity to epistemic-community influence rest on recruitment and retention patterns within national administrations. Environmental agencies often lack the ability to spread their influence within government administrations because most of them are so specialized and their staffs hold their positions for only limited periods, so there is limited potential for individuals to educate each other.

Scientific consensus is often poorly represented to decision makers. Eloquent and articulate specialists from within scientific disciplines are needed to communicate across technical and cultural divides to policy analysts and politicians (Koehn and Rosenau 2002). CLRTAP shows that the gradual development of scientific consensus can allow policy consensus to emerge. In contrast, the IPCC has made it clear that it is much harder to mobilize knowledge to address politicized and high stakes issues when knowledge is being developed at the same time as policy debates are occurring.

Although constructivists are in general reluctant to rely on the rationalist assumptions of theories such as the principal–agent theory, in the particular issue of global warming the material incentive to the principals are so clear that the principal–agent theory's rationality assumptions are warranted. It is only in cases with greater uncertainty about material national interests and national effects from environmental harm that more constructivist approaches are warranted, although some dimensions of the global-warming issue are open to fruitful constructivist analysis (Miller and Edwards 2001; Martello and Jasanoff 2004). In conclusion, institutionalized knowledge has contributed to more effective multilateral environmental management. Further work on this topic can advance the understanding of the interplay between knowledge and power by studying when states are willing to create autonomous institutions and defer to their authority; of the interplay between different ideational factors in shaping collective action; and of the interplay between various nonstate actors at and across multiple scales of multilevel global governance (Annan 1998; Camilleri, Malhotra et al. 2000; Kanie and Haas 2004).

Acknowledgments

We are very indebted to Rolf Lidskog and Göran Sundquist for offering insightful and helpful editorial comments on earlier drafts of this chapter. The chapter expands and updates the analysis in Peter M. Haas, "When Does Power Listen to Truth? A Constructivist Approach to the Policy Process," *European Journal of Public Policy* 11(4) (2004): 569–592.

APPENDIX 16.1 Collection of information from tables

	Regime name	*Abbreviation*	*Committee organization*	*Expert selection*	*Type of science committee*	*Degree of social learning*
1	Vienna Convention for the Protection of the Ozone Layer (and protocols)	Stratospheric Ozone	Standing; sets agenda	Secretariat	Group of experts	Social learning
2	UN Framework Convention on Climate Change	Climate Change	Standing	COP	Group of experts	Little/none
3	Convention on Long-Range Transboundary Air Pollution	CLRTAP	Standing; sets agenda		Group of experts	Social learning
4	Convention on Biological Diversity	CBD	Standing; sets agenda	COP	Open	Little/none
5	Convention for the Conservation on Marine Living Resources	CCAMLR	Standing; sets agenda	COP	Open	Social learning
6	Ramsar Convention on Wetlands	Ramsar	Standing	COP	Group of experts	Some
7	Bonn Convention on Migratory Species	Bonn	Standing	COP	Open	Some
8	Convention on International Trade of Endangered Species	CITES	Standing; sets agenda	COP	Group of experts	Some
9	International Whaling Convention	IWC	Standing	COP	Group of experts	Some
10	International Maritime Organization	IMO	Standing	Other intergovernmental organizations	Open	Little/none
11	Convention on the Conservation and Management of the Pollock Resources in the Central Bering Sea	CCMPRCBS	Ad hoc			Little/none

12	Convention for the Conservation of Southern Bluefin Tuna	CCSBT	Standing		Open	Little/none
13	Inter-American Tropical Tuna Commission	IATTC	Ad hoc		Open	Little/none
14	International Commission for the Conservation of Atlantic Tuna	ICCAT	Ad hoc		Open	Little/none
15	Indian Ocean Tuna Commission	IOTC	Standing	COP	Open	Little/none
16	International Pacific Halibut Commission	IPHC	Ad hoc		Group of experts	Little/none
17	Northwest Atlantic Fishery Organization	NAFO	Standing		Open	Little/none
18	North Atlantic Salmon Conservation Organization	NASCO	Ad hoc		Open	Little/none
19	Southeast Atlantic Fishery Organization	SEAFO	Standing		Open	Little/none
20	Program for the Assessment and Control of Pollution in the Mediterranean Region	MEDPOL	Standing; sets agenda	Secretariat	Group of experts	Social learning
21	Convention for the Protection of the Marine Environment of the North-East Atlantic	North Sea	Standing; sets agenda	Secretariat		Some
22	Convention on the Protection of the Marine Environment of the Baltic Sea Area	Baltic Sea	Standing	Secretariat	Group of experts	Some
23	Convention for the Protection of the Natural Resources and Environment of the South Pacific Region (and protocols)	South Pacific	Ad hoc	Secretariat	Group of experts	Some

(Continued)

APPENDIX 16.1 (Continued)

	Regime name	*Abbreviation*	*Committee organization*	*Expert selection*	*Type of science committee*	*Degree of social learning*
24	Convention for the Protection of the Marine Environment and Coastal Zones of the South-East Pacific (Lima Convention)	Southeast Pacific	Standing	Secretariat	Group of experts	Some
25	Regional Organization for the Protection of the Marine Environment (Persian Gulf, pre-1990)	ROPME	Ad hoc	COP	Group of experts	Some
26	Program for the Environment of the Red Sea and Gulf of Aden	PERSGA	Ad hoc	COP	Open	Little/none
27	Basel Convention on the Control of Transboundary Movements of Hazardous Wastes and Their Disposal	Basel Convention	Standing	COP	Open	Little/none
28	Stockholm Convention on Persistent Organic Pollutants	Stockholm POPs	Standing; sets agenda	COP	Group of experts	Social learning
29	Rotterdam Convention on Prior Informed Consent	Rotterdam PIC	Standing	COP	Group of experts	Some
30	UN Convention concerning Desertification	UNCCD	Standing	COP	Open	Little/none
31	Agreement on Conservation of Polar Bears	Polar Bears	Standing; sets agenda	Other intergovernmental organizations	Group of experts	Social learning

Note

References

Adler, E. (2002). *Changing Identities: The Road to Peace*. Toronto, University of Toronto.

Agrawala, S. (1998). "Context and Early Origins of the Intergovernmental Panel on Climate Change." *Climatic Change* 39: 605–620.

Agrawala, S. (1998). "Structure and Process History of the Intergovernmental Panel on Climate Change." *Climatic Change* 39: 621–642.

Andresen, S. (2000). *Science and Politics in International Environmental Regimes: Between Integrity and Involvement*. New York, Manchester University Press.

Andresen, S. (2002). "The International Whaling Commission (IWC): More Failure than Success?". *Environmental Regime Effectiveness: Confronting Theory with Evidence*. E. L. Miles, A. Underdal, S. Andresen, et al. Cambridge, MA, MIT Press.

Andresen, S., T. Skodvin, et al. (2000). *Science and Politics in International Environmental Regimes*. Manchester, Manchester University Press.

Andresen, S., L. Walloe, et al. (2005). "The Precautionary Principle: Knowledge Counts but Power Decides?". *Biodiversity and Precautionary Principle: Risk and Uncertainty in Conservation and Sustainable Use*. B. Dieson and R. Conney. London, Earthscan: 39–54.

Annan, K. (1998). "The Quiet Revolution." *Global Governance* 4: 123–138.

Asher, W. (1986). "The Evolution of the Policy Sciences." *Journal of Public Analysis and Management* 5(2): 365–389.

Biermann, F. (2002). "Institutions for Scientific Advice." *Global Governance* 8(2): 195–219.

Botcheva, L. (2001). "Expertise and International Governance." *Global Governance* 7(2): 197.

Brown, G. (1997). "Environmental Science Under Siege in the US Congress." *Environment* 39(2): 12–20, 29–31.

Camilleri, J. A., K. Malhotra, et al. (2000). *Reimagining the Future*. Bandoora, Victoria, Australia, Department of Politics, La Trobe University.

Checkel, J. (1998). "The Constructivist Turn in International Relations Theory." *World Politics* 50(2): 324–348.

Christiansen, T., K. E. Jorgensen, et al. (1999). "The Social Construction of Europe." *Journal of European Public Policy* 6(4): 528–544.

Clark, W. C., Ed. (1990). *Usable Knowledge for Managing Global Climate Change*. Stockholm, Stockholm Environmental Institute.

Clark, W. C. (2001). "Knowledge Systems for Sustainable Development." *Environment* 43(8): 1.

Clark, W. C. (2007). "Sustainability Science: A Room of Its Own." *Proceedings of the National Academy of Sciences of the United States of America* 104(6): 1737–1738.

Clark, W. C. and G. Majone (1985). "The Critical Appraisal of Scientific Inquiries with Policy Implications." *Science, Technology and Human Values* 10(3): 6–19.

Dimitrov, R. (2010). "Inside Copenhagen: The State of Climate Governance." *Global Environmental Politics* 10(2): 18–24.

Dimitrov, R. S. (2003). "Knowledge, Power and Interests in Environmental Regime Formation." *International Studies Quarterly* 47(1): 123–151.

Drori, G. and R. Meyer (2006). "Global Scientization: An Environment for Expanded Organization". *Globalization and Organization: World Society and Organization* G. S. Drori, J. W. Meyer and H. Hwang. Stanford, CA, Stanford University Press.

Drori, G. S., J.W. Meyer, et al. (2003). *Science in the Modern World Polity: Institutionalization and Globalization*. Stanford, Stanford University Press.

Dryzek, J. S. (1997). *The Politics of the Earth: Environmental Discourses*. Oxford, Oxford University Press.

Environmental Business Journal (2006). "Global Market Overview." *Environmental Business Journal*, 19(5-6): 1–3.

European Commission (2001). Report of Working Group on Broadening and Enriching the Public Debate on European Matters: Democratising Expertise and Establishing Scientific Reference Systems. Brussels, European Commission.

European Commission (2002). Improving the Knowledge Base for Better Policies. Luxembourg, Communication from the Commission on the Collection and Use of Expertise by the Commission, Expertise and Guidelines.

Farrell, A. E. and J. Jager, Eds. (2005). *Assessments of Regional and Global Environmental Risks: Designing Processes for the Effective Use of Science in Decisionmaking*. Washington, D.C., Resources for the Future.

Fischer, F. and J. Forester, Eds. (1993). *The Argumentative Turn in Policy Analysis and Planning*. Durham, NC, Duke University Press.

Fritz, J.-S. (2000). The Second Report on International Scientific Advisory Processes on the Environment and Sustainable Development, United Nations Environmental Program.

Funtowicz, S. O. and J. R. Ravetz (1991). "A New Scientific Methodology for Global Environmental Issues". *Ecological Economics*. R. Costanza. New York, Columbia University Press: 137–152.

Funtowicz, S. O. and J. R. Ravetz (2001). "Global Risk, Uncertainty, and Ignorance". *Global Environmental Risk*. J. X. Kasperson and R. Kasperson. London, Earthscan.

Garcia-Johnson, R. (2000). *Exporting Environmentalism: U.S. Multinational Chemical Corporations in Brazil and Mexico*. Cambridge, MA, MIT Press.

Gore, A. (1996). *The Technology Challenge: What is the Role of Science in American Society*. American Association for the Advancement of Science, Baltimore, MD.

Gornitzka, A. and U. Sverdrup (2008). "Who Consults? The Configuration of Expert Groups in the European Union." *Western European Politics* 31(4): 725–750.

Guston, D. H. (1998). "Principal-Agent theory and the Structure of Science Policy." *Science and Public Policy* 23(4): 229–240.

Guston, D. H. (2003). "Principal-Agent Theory and the Structure of Science Policy, Revisited." *Science and Public Policy* 30(5): 347–357.

Guzzini, S. (2000). "A Reconstruction of Constructivism in International Relations." *European Journal of International Relations* 6(2): 147–182.

Haas, P. (2001). "Epistemic Communities and Policy Knowledge". *The International Encyclopedia of the Social and Behavioral Sciences*. N. J. Smelser, J. Wright and P. B. Baltes. New York, Elsevier.

Haas, P. (2004). "Science Policy for Multilateral Environmental Governance". *Emerging Forces in Environmental Governance*. N. Kanie and P. Haas. Tokyo, UNU Press: 115–136.

Haas, P. (2004). "When does Power Listen to Truth? A Constructivist Approach to the Policy Process." *Journal of European Public Policy* 11(4): 569–592.

Haas, P. M. (1990). *Saving the Mediterranean: The Politics of International Environmental Cooperation*. New York, Columbia University Press

Haas, P. M. (2000). "Letter to the Editor." *Marine Policy* 24(6): 499–500.

Haas, P. M. (2000). "Prospects for Effective Marine Governance in the NW Pacific Region." *Marine Policy* 24(4): 341–348.

Haas, P. M. (2001). "Environment: Pollution". *Managing Global Issues*. P. J. Simmons and C. de Jonge Oudraat. Washington, D.C., Carnegie Endowment for International Peace: 310.

Haas, P. M. (2004). "Science Policy for Multilateral Environmental Governance". *Emerging Forces in Environmental Governance.* N. Kanie and P. M. Haas. Tokyo, United Nations University Press: 115–136.

Haas, P. M. (2007). "Epistemic Communities and International Environmental Law". *Oxford Handbook of International Environmental Law.* D. Bodansky, E. Hey and J. Brunnee. Oxford, Oxford University Press.

Haas, P. M. (2008). "Climate Change Governance after Bali." *Global Environmental Politics* 8(3): 1–7.

Haas, P. M. and E. B. Haas (1995). "Learning to Learn: Improving International Governance." *Global Governance* 1: 255–285.

Haas, P. M. and E. B. Haas (2002). "Pragmatic Constructivism and the Study of International Institutions." *Millenium: Journal of International Studies* 31(3): 573–602.

Hajer, M. A. and H. Wagenaar, Eds. (2003). *Deliberative Policy Analysis: Understanding Governance in the Network Society.* Cambridge, Cambridge University Press.

Hasselman, K. and T. Barker (2008). "The Stern Review and the IPCC Fourth Assessment Report: Implications for Interaction Between Policymakers and Climate Experts. An Editorial Essay." *Climatic Change* 89(219): 229.

Haufler, V. (2001). *A Public Role for the Private Sector: Industry Self-Regulation in a Global Economy.* Washington, D.C., Carnegie Endowment for International Peace.

Hill, R. and C. Sendashonga (2006). "Conservation Biology, Genetically Modified Organisms, and Biosafety Protocol." *Conservation Biology* 20(6): 1620–1625.

Hopf, T. (1998). "The Promise of Constructivism in International Relations Theory." *International Security* 23(1): 171–200.

Jasanoff, S. (1990). *The Fifth Branch: Science Advisers as Policymakers.* Cambridge, MA, Harvard University Press.

Jasanoff, S. (2004). *States of Knowledge: The Co-production of Science and Social Order.* New York, Routledge.

Jasanoff, S., G. E. Markle, et al. (1995). *Handbook of Science and Technology Studies.* Thousand Oaks, CA, Sage Publications.

Jasanoff, S. and B. Wynne (1998). "Science and Decisionmaking". *Human Choice and Climate Change.* S. Rayner and E. L. Malone. Columbus, OH, Batelle Press. 1: 1–87.

Kanie, N. and P. M. Haas, Eds. (2004). *Emerging Forces in Environmental Governance.* Tokyo, UNU Press.

Kates, R. W., W. C. Clark, et al. (2001). "Sustainability Science." *Science* 292(5517): 641–642.

Koehn, P. and J. N. Rosenau (2002). "Transnational Competence in an Emergent Epoch." *International Studies Perspectives* 3(2): 105–127.

Koremenos, B., C. Lipson, et al. (2001). "Rational Designs: Explaining the Form of International Institutions." *International Organization* 55(4): 1–32.

Kubalkova, V., N. G. Onuf, et al., Eds. (1998). *International Relations in a Constructed World.* Armonk, NY, M.E. Sharpe.

Lee, K. N. (1993). *Compass and Gyroscope.* Washington, D.C., Island Press.

Levy, M. A., J. Cavender-Bares, et al. (2001). "Goal and Strategy Formulation in the Management of Global Environmental Risks". *Social Learning and the Management of Global Environmental Risks.* S. L. Group. Cambridge, MA, MIT Press: 87–114.

Lidskog, R. and G. Sundqvist (2002). "The Role of Science in Environmental Regimes: The Case of LRTAP." *European Journal of International Relations* 8(1): 77–101.

Lindner, J. (2003). "Institutional Stability and Change: Two Sides of the Same Coin." *Journal of European Public Policy* 10(6): 912–935.

Litfin, K. (1994). *Ozone Discourses: Science and Politics in Global Environmental Cooperation.* New York, Columbia University Press.

Martello, M. L. and S. Jasanoff (2004). "Introduction: Globalization and Environmental Governance". *Earthly Politics: Local and Global in Environmental Governance.* M. L. Martello and S. Jasanoff. Cambridge, MA, MIT Press: 1–29.

Miles, E. L. (1998). "Personal Reflections on an Unfinished Journey through Global Environmental Problems of Long Timescale." *Policy Sciences* 31(1): 1–33.

Miles, E. L., A. Underdal, et al., Eds. (2002). *Environmental Regime Effectiveness: Confronting Theory with Evidence.* Cambridge, MA, MIT Press.

Miller, C. A. and P. N. Edwards (2001). *Changing the Atmosphere: Expert Knowledge and Environmental Governance.* Cambridge, MA, MIT Press.

Milton, G. (1999). *Nathaniel's Nutmeg.* New York, Farrar, Straus and Giroux.

Mitchell, R. (2002). "International Environment". *Handbook of International Relations.* T. Risse, B. Simmons and W. Carlsnaes. Sage Publications: 500–516.

Mitchell, R. B., W. C. Clark, et al., Eds. (2006). *Global Environmental Assessments: Information and Influence.* Cambridge, MA, MIT Press.

Moe, T. M. (1984). "The New Economics of Organization." *American Journal of Political Science* 28(4): 739–777.

National Research Council (1999). *Our Common Journey: A Transition Toward Sustainability.* Washington, D.C., National Academy Press.

Nelkin, D. (1979). *Controversy Politics of Technical Decisions.* Beverely Hills, Sage.

Oberthur, S. and T. Gehring (2006). *Institutional Interaction in Global Environmental Governance.* Cambridge, MA, MIT Press.

Ortiz, E. F. and G. Tang (2005). *Review of the Management, Administration, and Activities of the Secretariat of the United Nations Convention to Combat Desertification (UNCDD).* Geneva, United Nations Joint Inspection Unit.

Ravetz, J. R. (1986). "Usable Knowledge, Usable Ignorance". *Sustainable Development of the Biosphere.* W. C. Clark and T. Munn. Cambridge, Cambridge University Press: 415–434.

Reinicke, W. H., F. Deng, et al. (2000). *Critical Choices: The United Nations, Networks, and the Future of Global Governance.* Ottawa, International Development Research Centre.

Ruggie, J. G. (1998). "What makes the World Hang Together? Neo-Utilitarianism and the Social Constructivist Challenge." *International Organization* 52(4): 855–885.

Schroeder, H., L. A. King, et al. (2008). "Contributing to the Science Policy Interface: Policy Relevance of Findings on the Institutional Dimensions of Global Environmental Change". *Institutions and Environmental Change: Principal Findings, Applications, and Research Frontiers.* O. Young, L. King and H. Schroeder. Cambridge, MA, MIT Press.

Siebenhuner, B. (2002). "How do Scientific Assessments Learn?" *Environmental Science & Policy* 5(5): 411–427.

Siebenhuner, B. (2003). "The Changing Role of Nation States in International Environmental Assessments." *Global Environmental Change* 13: 113–123.

Skodvin, T. (2000). "The Intergovernmental Panel on Climate Change". *Science and Politics in International Environmental Regimes.* S. Andresen, T. Skodvin, A. Underdal and J. Wettestad. Manchester, Manchester University Press: 146–180.

Skodvin, T. (2000). *Structures and Agents in the Scientific Diplomacy of Climate Change.* Dordrecht, Kluwer.

Skoie, H. (2001). *The Research Councils in the Nordic Countires – Developments and Some Challenges,* Norsk Institute for Studier av Forskning og Utdanning.

Social Learning Group (2001). *Social Learning and the Management of Global Environmental Risks.* Cambridge, MA, MIT Press.

Swart, R., J. Mitchell, et al. (2002). "Stabilisation Scenerioes for Climate Impact Assessment." *Global Environmental Change* 12(3): 155–165.

Tolba, M. and I. Rummel-Bulska (1998). *Global Environmental Diplomacy: Negotiating Environmental Agreements for the World, 1973–1992.* Cambridge, MA, MIT Press.

Tuinstra, W. (2008). "European Air Pollution Assessments: Co-Production of Science and Policy." *International Environmental Agreement: Politics, Law, and Economics* 8(1): 35–49.

Tuinstra, W., L. Hordijk, et al. (1999). "Using Computer Models in International Negotiations." *Environment* 41(9): 33–42.

UNEP (1998). *Report on International Scientific Advisory Processes on the Environment and Sustainable Development.* Paris, UNEP.

UNEP and UNSECO (2003). *Global Marine Assessments: A Survey of Global and Regional Marine Environmental Assessments and Related Scientific Activities.* Paris, UNEP.

Victor, D. G., K. Raustiala, et al., Eds. (1988). *The Implementation and Effectiveness of International Environmental Commitments.* Cambridge, MA, MIT Press.

Weiss, E. B. and H. K. Jacobson (1998). *Engaging Countries: Strengthening Compliance with International Accords.* Cambridge, MA, MIT Press.

Wildavsky, A. B. (1979). *Speaking Truth to Power: The Art and Craft of Policy Analysis.* Boston, Little, Brown.

Young, O. R., Ed. (1999). *The Effectiveness of International Environmental Regimes.* Cambridge, MA, MIT Press.

Ziman, J. (2000). *Real Science: What It is, and What It Means.* Cambridge, Cambridge University Press.

PART VI
Conclusion

These two brief pieces point towards the effects of the broad patterns and dynamics laid out in this volume. Social learning occurs as ideas are institutionalized through the political involvement of epistemic communities in world politics.

The mechanism behind institutionalization is essentially linkage politics. Substantively, shared conceptions of linkages become increasingly broader, by virtue of the spreading influence of ecological epistemic communities and their comprehensive notions of connections between coupled issues. Constitutive forces are also at play, as scientists and experts increasingly become legitimate participants in international decision making, thus consolidating their ideas in the political landscape, as they privilege specific international institutions and the application of their ideas.

"Promoting knowledge-based international governance" (Chapter 17) provides a roadmap for networked knowledge-based governance that can lead to more sustainable global governance. Networked knowledge may lead to a wider reliance on experts and epistemic communities. As decision makers grow more confident and experienced in their delegation to experts, the prospects improve for a broader and more coupled agenda between a diverse array of issues: between environmental issues and between environmental issues and other issues, such as trade, foreign aid, development policy, and security.

"The global spreading of ideas" (Chapter 18) looks at emergent patterns of environment governance resulting from institutionalized epistemic involvement in global environmental governance over the last 40 years. As the number of regimes informed by epistemic influence grows, and an increasing number of states ratify them and enforce them domestically, a broader social recognition of reliance on nature occurs. Causally such recognitions lead to changes in the governance of coupled issue areas. Broader norms of environmental protection and sustainability are also possible, as the new practices get codified and acquire normative consensus.

Direct and indirect forces guide future trajectories of governance. States and elites are directly influenced by epistemic communities and learn about new state systemic properties and adopt new state responsibilities. Indirectly, states become caught up in a broader network of institutionalized incentives and behavioral expectations for environmental protection. Political economy forces are created from these political foundations, leading to new environmental domestic corporate constituencies and the emergence of markets which favor environmental protection rather than its degradation. States become more likely to subscribe to normative commitments for the environment. Ecological ideas about linkages within and between systems become more widely taken for granted and applied to global governance.

Future work on epistemic communities and governance can help document and trace these new mechanisms of political influence and change.

17

PROMOTING KNOWLEDGE-BASED INTERNATIONAL GOVERNANCE FOR SUSTAINABLE DEVELOPMENT*

Peter M. Haas

Globalization comprises the dominant context for contemporary international relations. The challenge for decision makers is how to navigate the interstices of an increasingly complex global agenda. Sustainable development has become the new mantra for a coherent multilateral approach to addressing globalization. Sustainable development entails a widespread and comprehensive approach to dealing with such interlinked issues as environmental protection, economic development, democratization, the pursuit of human justice, and peace. Such an emergent paradigm or global vision appears to underlie the United Nations Millennium Development Goals.

Promoting sustainable development on this scale entails global governance that reflects guidance from an interdisciplinary knowledge base that is capable of generating accurate assessments of the world, and that relies for its legitimacy on a deliberative process that engages states and multiple non-state actors, including civil society, expert networks, the private sector, and possibly local communities as well (Reinicke, Deng et al. 2000; Ruggie 2004; Held and Koenig-Archibugi 2005).

Adequate global governance is unlikely to appear solely from national leadership, or "hegemony." The US is reluctant to address a broad agenda of issues, and lacks the legitimacy to command sustained voluntary support from its allies and the rest of the world (Ikenberry 2004; Rues-Smit 2004; Tucker and Hendrickson 2004). Europe is reluctant, and probably incapable of unilaterally providing global governance. Thus governance has to rest on formal and informal institutions. This means an increased role for international organizations, transnational scientific groups, civil society organizations and the private sector for articulating consensual views about governance.

Knowledge-based governance and usable knowledge

Global environmental threats embody many of the defining characteristics of globalization: ecosystems are complex so that management decisions must address multiple interacting subsystems, decision makers are often uncertain about the choices available to them, and accurate policy advice and guidance is at a premium. States and individual policy makers cannot clearly anticipate or address environmental threats without better mechanisms for environmental policy advice. The nature of and magnitude of environmental threats and the associated policy responses are often unclear to decision makers who require expert advice about the technical details of such challenges, including the connections to other issue areas, such as balancing economic development and environmental protection. Governance principles for managing complexity are controversial and contested. NGOs and some European governments urge application of the precautionary principle in such cases, whereas others urge a use of scientifically informed policies

A major informal institution for governance is usable knowledge. While there is broad deference to science in policy-making circles, widespread research has demonstrated that decision makers only selectively defer to science. Recently, social scientists have focused on "usable knowledge" as a better defined subset of "scientific knowledge" that is most likely to be converted to effective governance. Constructivist approaches to policy analysis suggest that to be usable, science must be developed authoritatively, and then delivered by responsible carriers to politicians. "Authoritative" and "responsible" in this regard refer to procedures that are viewed as socially legitimate by mass publics who will ultimately be affected by decisions resting on these expert judgments. In most instances this means that the procedure by which usable knowledge is developed will be by transparent international panels of experts who enjoy reputations for professional and impartial expertise. Of course all experts have their own preferences and norms, but if they are widely seen to be willing to subordinate their personal beliefs to their socialized professional training—i.e. the scientific method and consensus approaches to truth—then their arguments are more likely to reflect broader technical consensus as well as being accepted as being politically acceptable by various policy consumers (Haas and Haas 2002).

Epistemic communities are the actors involved in formulating and disseminating usable knowledge. The more autonomous and independent science is from policy, the greater its potential influence. Consensus in isolation builds value and integrity, and then its consequences should be discussed publicly. Measures of autonomy and integrity include the selection and funding of scientists by international organizations rather than by governments, their recruitment by merit on important panels, and reliance on individuals whose reputation and authority rest on their role as active researchers rather than policy advocates or science administrators. Accuracy can be achieved via peer review, interdisciplinary research teams, and independence from sponsoring sources.

Substantively, usable knowledge is accurate information that is politically tractable for politicians and policy makers. Such knowledge frequently exceeds the mastery of any individual disciplinary approach. Recent studies of global environmental assessments apply the criteria of credibility, legitimacy, and saliency (Siebenhunner 2002; Siebenhunner 2003; Haas 2004). Credibility means that the key knowledge producers and their consumers believe their product is true. Legitimacy means that the claims are believed to be legitimate, that is, developed through a process that minimizes the potential for bias and is more equitable in terms of participation by those who are dependent upon the information. Current research from comparative politics, international relations, policy studies, and democratic theory suggests that science remains influential if its expertise and claims are developed behind a politically insulated wall. Finally, saliency means that such information is provided in a timely manner and contains information that is useful for making public policy by decision makers: that is, in practice, it arrives in conjuncture with the policy process and provides advice which can be converted into laws or decisions by decision makers. In practice, credibility and legitimacy are mutually reinforcing, as a procedural approach to developing consensual knowledge is likely to generate both accurate and acceptable knowledge. Yet saliency and credibility may be at odds, as the long time often necessary for developing credible knowledge may interfere with the short-term needs for applying the knowledge to making policy. Existing knowledge that precedes the policy process is more likely to be usable than is knowledge that is being developed concurrently with the policy process because it will be less likely to be seen as politically tainted by its proximity to immediate policy decisions.

From usable knowledge to networked reflexive governance

This process of developing and circulating usable knowledge provides a provisional and rudimentary roadmap for a new model of reflexive governance, based on processes of social consensus. Participants create shared tangible understandings about how the world works and how national interests are affected. A new collective pattern of voluntary governance can emerge based on new appreciations of self-interest. Such an approach to governance is part of a broader system of governance without government that involves reasoned discourse and usable knowledge by many actors in international society (Clark and Munn 1986; Costanza 1991; Rosenau and Czempiel 1991; Hooghe and Marks 2001; Kates 2001; Haas and Haas 2002; Schimmelfennig and Sedelmeier 2004; Schimmelfennig and Wagner 2004; Martimort-Asso and Tubiana 2005). Reflexive governance replaces the traditional dichotomous concepts of global governance organized hierarchically or anarchically with a network model of complex decentralized global governance. The political dynamics are those of social learning.

Reflexive governance requires the participation of multiple actors. Private firms, NGOs, civil society and scientists all contribute elements of usable knowledge for decision makers. The broader process of reflexive governance parallels that of the process by which usable knowledge is formulated. Groups of relevant experts are brought together to formulate a view of the world. After reaching provisional group consensus, their views are refined and presented to decision makers in a structured discursive setting. The ideas are then applied as policy experiments.

Social scientists *cum* policy analysts then appraise the effectiveness of such policies, and engage in similar discussions with policy makers as occurred initially with epistemic communities. In turn, following focused discussions among epistemic communities, relevant experts, policy makers and social scientists, broader policy doctrines are formulated and ultimately institutionalized as they are converted to international regimes, national policies, and administrative procedures. Eventually, policy evolution and innovations through reflexive governance give rise to incentives for the emergence of new markets. Such a process provides a way to triangulate between the normative poles of globalization, regulation, and democratic sovereignty that leads slowly to robust institutionalized orders of governance through processes of social learning that come to embody regulatory approaches that may not reflect the initial interests of the major parties, as actors learn about the nature of the issues they seek to govern and about their own interest through the structured interactions with others (Frankel 2005).

In the environmental realm incremental social learning about the behavior of global ecosystems, their influence on national environmental and broader social conditions, and the need for meaningful collective action led states to endorse increasingly more comprehensive environmental treaties, stronger national environmental regulations, and ultimately to send sufficient market signals to firms to increase attention to environmentally friendly technology, leading to the creation of a new environmental technology market that didn't previously exist, and may contribute to more effective environmental protection for the future. Reflexive environmental governance proceeded through a number of institutional channels. International conferences from UNCHE (United Nation Conference on the Human Environment, Stockholm 1972) through UNCED (United Nation Conference on Environment and Development, Rio 1992) to the World Summit on Sustainable Development (Johannesburg, 2002) expanded the number of actors engaged in environmental governance and helped consolidate beliefs around a broad vision of sustainable development that had procedural and substantive components calling for more comprehensive environmental policies linked to other issue areas (including trade) as well as invoking broader participation in a discursive process of sustainability. International organizations helped to develop and popularize doctrines of ecological management, as well as enhancing the political profile of individuals (epistemic community members) involved in developing such usable knowledge. Furthermore, states helped create a more comprehensive and sweeping body of international law for international environmental governance (Caldwell 1996; Haas 1996; Haas 1997; Bernstein 2001; Haas 2001; Hoffman 2001; Haas 2002; Haas 2004).

Gradually the social goals of environmental protection have become expanded, through the application of usable knowledge over a 30-year period to include a broader international agenda that focuses on the linkages between security, international economic relations, and environmental protection, most recently expressed in the acclamation of the Millennium Declaration and subsequent development of the Millennium Development Goals. The aspirations and presumptive means of collective action been reframed toward broader governance. New principles of sustainability have been forged in which individual issues can be nested, based on a socially learned set of causal connections.

Conclusion

This volume, and the broader SUSTRA network from which it builds, marks a step in a new direction in sustainable development and reflexive governance. It seeks to advance multilevel governance through a conscious effort at pragmatic construction: bringing together the relevant expert and policy communities in an ongoing setting that promotes interchange and learning. Through the various stages the interplay between trade and sustainable development becomes better understood, and provides a stronger foundation for more effective policy to promote sustainability in trade.

Note

* Copyright © 2007 Ashgate Publishing. This chapter was first published as the Foreword in Sophie Thoyer and Benoit Martimort-Asso (eds) *Participation for Sustainability in Trade*. Brookfield, VT: Ashgate Publishing, 2007. Reprinted with permission of Ashgate Publishing.

References

Bernstein, S. (2001). *The Compromise of Liberal Environmentalism*. New York, Columbia University Press.

Caldwell, L. K. (1996). *International Environmental Policy*. Durham, NC, Duke University Press.

Clark, W. C. and R. E. Munn, Eds. (1986). *Sustainable Development of the Biosphere*. New York, Cambridge University Press.

Costanza, R. (1991). *Ecological Economics*. New York, Columbia University Press.

Frankel, J. (2005). "The Environment and Economic Globalization". *Globalizations What's New*. New York, Columbia University Press: 147.

Haas, P. (1996). "Is Sustainable Development Politically Sustainable?" *Brown Journal of World Affairs* 3(2): 239–248.

Haas, P. (2001). "Environment: Pollution". *Managing a Globalized World*. Washington, D.C., Carnegie Foundation.

Haas, P. (2002). "UN Conferences and the Constructivist Governance of the Environment." *Global Governance* 8(1): 73–91.

Haas, P. (2004). "Addressing the Global Governance Deficit." *Global Environmental Politics* 4(4): 1–15.

Haas, P. (2004). "When Does Power Listen to Truth? A Constructivist Approach to the Policy Process." *Journal of European Public Policy* 11(4): 569–592.

Haas, P. M. (1997). "Environmental Issues and Evolving International Environmental Governance". *Global Governance*. J.-Y. Chung. Seoul, Korea, The Sejong Institute.

Haas, P. M. and E. B. Haas (2002). "Pragmatic Constructivism and the Study of International Institutions." *Millenium: Journal of International Studies* 31(3): 573–602.

Held, D. and M. Koenig-Archibugi, Eds. (2005). *Global Governance and Public Accountability*. Malden, MA, Blackwell.

Hoffman, A. J. (2001). *From Heresy to Dogma*. Stanford, CA, Stanford University Press.

Hooghe, L. and G. Marks (2001). *Multilevel Governance and European Integration*. Boulder, CO, Rowman & Littlefield Publishers.

Ikenberry, J. (2004). "Illusions of Empire: Defining the New American Order." *Foreign Affairs* March/April.

Kates, R. (2001). "Sustainability Science." *Science* 292: 641–642.

Martimort-Asso, B. and L. Tubiana (2005). "International Environmental Governance: The Next Steps." *Les Syntheses de l'Iddri* 7.

Reinicke, W. H., F. Deng, et al. (2000). *Critical Choices: The United Nations, Networks, and the Future of Global Governance*. Ottawa, International Development Research Centre.

Rosenau, J. N. and E.-O. Czempiel, Eds. (1991). *Governance Without Government: Change and Order in World Politics*. Cambridge Studies in International Relations. New York, Cambridge University Press.

Rues-Smit, C. (2004). *American Power and World Order*. Cambridge, Polity Press.

Ruggie, J. G. (2004). "Reconstituting the Global Public Domain." *European Journal of International Relations* 10(4): 499–531.

Schimmelfennig, F. and U. Sedelmeier (2004). "Governance by Conditionality." *Journal of European Public Policy* 11(4): 661–679.

Schimmelfennig, F. and W. Wagner (2004). "External Governance in the European Union." *Journal of European Public Policy* 11(4): 657–660.

Siebenhunner, B. (2002). "How do Scientific Assessments Learn?" *Environmental Science & Policy* 5(5): 411–427.

Siebenhunner, B. (2003). "The Changing Role of Nation States in International Environmental Assessments." *Global Environmental Change* 13: 113–123.

Tucker, R. and D. Hendrickson (2004). "The Sources of American Legitimacy." *Foreign Affairs* November/December.

18

THE GLOBAL SPREADING OF IDEAS*

Social learning and the evolution of multilateral environmental governance

Peter M. Haas

Political scientist Karl W. Deutsch, a major figure in the WZB's early history, was a seminal influence in work on political communication, social learning, and collective community formation. Among the major themes that continue to resonate, is the question: how do societies understand their settings, and how do they adapt? In the environmental domain we can ask: how do states and societies recognize and respond to impending environmental risks, and how do they organize their collective responses? What constitutes an effective response? Deutsch did not fall prey to the autistic theorizing of rational choice about decisions largely independent of social context, or to heroic assumptions about individual and collective knowledge and decision-making skills. Rather these questions of understanding and adaptation were historicized and treated as empirical questions about which meaningful social analysis could be conducted, leading to insights about social change, emancipation, transformation, and maybe even, in that not passé term, progress.

This can be demonstrated by looking at the empirical domain of multilateral environmental governance, or international environmental regimes. Thirty years ago this area of governance didn't really exist. It has been constructed. There has been an enormous proliferation of environmental treaties post-1970s: both in numbers, but also in novel areas as pollution replaced species protection and oil pollution of the open oceans as the primary focus. Markets for environmental goods have grown from non-existent to roughly 300 billion dollars a year during this period. Still, green markets aren't that big, and follow from governance rather than precede it. Changes in linguistic usages reflect some degree of underlying social change in collective understanding and meaning. "Swamps" have turned into "wetlands," what used to be "jungles" are now "rain forests," and "varmints" today are "endangered species."

What has been going on is the emergence of a new understanding of how nature affects humans, the gradual application of policies, and the improved management of the environment, at least selectively. In short, social learning (Haas 2001; Haas and Haas 2002; Haas 2004; Bodansky, Juttaa et al. 2007).

How did this occur? The historical explanation is that a new body of ecological understanding percolated up from the academy. Experts (including natural scientists, engineers and some lawyers) formed an "epistemic community," spreading their ideas to states and international organizations. The ideas didn't flow on their own, or without friction. Formal organizations were needed, and their influence was deployed through persuasion, administrative usurping, and by deploying material capabilities on behalf of their ideas to induce others to accept them. But the core point is that without the ideas new practices wouldn't have been possible, and part of the causal mechanism by which the ideas diffused was through persuasion.

A new environmental management doctrine based on ecological principles emerged in the 1960s. Paul Sears called this approach "subversive" because it challenged the reductionist understanding of social and physical systems by illuminating some of their unanticipated consequences. The ideas were developed and transmitted by an epistemic community that shared common normative goals, causal understandings, and truth tests, and were actively engaged in a common policy project.

Since 1972, this ecological epistemic community has increasingly institutionalized this new management doctrine in state policies and practices, in the programmatic activities of international institutions, and in international regimes. As a consequence of the international institutionalization of ideas held by the ecological epistemic community, states have undertaken more comprehensive styles of environmental management for transboundary and global environmental threats, leading to selective improvements in environmental quality. Members of the ecological epistemic community subscribed to holistic ecological beliefs about the need for policy coordination subject to ecosystemic laws. Their ideas about ecological management were based on a systems perspective of environmental and social systems. Ecological management proposals favor setting comprehensive environmental standards based on conservative estimates of the ability of ecosystems to sustain stress, subject to the epistemic community's technical understanding of the behavior of particular ecosystems. They promoted international environmental regimes that are grounded on policies that offered coherent plans for the management of entire ecosystems, sensitive to interactions between environmental media (such as air and water), sources of pollution, and contending uses of common property resources—rather than limiting themselves to more traditional policies for managing discrete activities or physical resources spaces within fairly short-term time horizons. They proposed treaties, in which bans and emissions limits were set for multiple contaminants, with environmental standards for each contaminant set according to scientific understanding about its environmental impact and its interactive effects with other contaminants.

What is striking is how these ideas prevailed over prevailing liberal and neoliberal policy ideas applied to the management of transboundary economic flows. In the early 1970s, when international institutions were founded to address global environmental threats under the influence of the United Nations Conference on the Human Environment (UNCHE), there were few contending epistemic communities in the environmental realm. In fact, there were only three major intellectual approaches to international environmental management. Traditional Resource managerial approaches had been widely discredited by the Limits to Growth study and the 1973 Oil Crisis. Market-based approaches to environmental management were weakly advanced by economists, the World Bank and the GATT at UNCHE, but they and their institutional backers generally paid little heed to the conference. What is analytically striking is the relative infrequency with which the negotiated outcomes that correspond to the dominant expertise paradigm—economics—have been adopted, and that, even in the Climate Change regime, market mechanisms are instrumentally applied to achieve the substantive goals identified by scientists of ensuring that anthropogenic greenhouse gas emissions do not exceed the carrying capacity of the atmosphere. Even with the popularity of market-based discourses in the 1990s, negotiated outcomes generally reflect the command and control approach of the ecological epistemic community.

International organizations were created as well to deal with new environmental threats. The United Nations Environment Program (UNEP) was established in 1973 and was staffed principally by young epistemic community members eager to put their professional knowledge to work. The UN Economic Commission for Europe played a strong role in managing European air pollution. The UNECE's environmental unit was led by a former UNEP official who carried the ecological management ideas from UNEP to UNECE. After 1987 the World Bank also became active in international environmental matters. As part of the environmental reforms introduced at the World Bank was the recruitment of ecological epistemic community members and their assignment to key posts in evaluating the environmental consequences of development projects.

Diffusion occurred principally through efforts of these major environmental international organizations (IOs) and through environmental regimes. The IOs encouraged other IOs to internalize environmental concerns into their missions through joint projects, and encouraged governments to pursue more comprehensive environmental policies through public education campaigns, publicizing environmental monitoring findings, resource transfers, training national officials, and through demonstration effects. Over time an increasing number of negotiated treaties and regimes came to reflect this new technical consensus.

Ecological epistemic communities, often working with UNEP, helped draft comprehensive international environmental regimes governing marine pollution, acid rain, stratospheric ozone protection, wetlands protection, protecting migratory species, polar bears protection, and the preservation of Antarctica. Such comprehensive efforts entail differentiated national obligations, substantive commitments, based on experts' consensus on the causes of environmental degradation. An increasing proportion of all environmental regimes are now based on the comprehensive ecological approach promoted by the ecological epistemic community as the epistemic community has increased in vigor and influence, and countries have institutionalized their ideas. In 1973, 3 out of 11 international environmental regimes were based on ecological management styles, in 1985, 7 out of 22 and in 1995, 10 out of 25. The application of ecological management ideas to environmental regimes spanning a number of geographic areas and functional activities means that most states have accepted ecological obligations for governing a wide variety of human activities.

Conversely, treaties and regimes concluded without epistemic communities yielded political compromises that were based on across the board reductions, or least common denominator-type negotiated outcomes. Instances of these non-social learning regimes include fisheries management, whaling, marine dumping, and the Baltic and North Seas. For instance, North Sea arrangements are based on a political formula of 30–50 percent reductions of a list of substances that had already been controlled by member governments. Thus the goal is devoid of any scientific justification, and is unlikely to be effective in a meaningful sense.

Ecological practices based on ecological management ideas have become locked in through a variety of mechanisms. Following ratification of international regimes, governments enforce these obligations domestically. Ecological practices get institutionalized domestically through legal precedents, bureaucratic standard operating procedures, and policy enforcement. In many countries they have acquired domestic constituencies—composed of lawyers and civil engineers who subcontract services, firms selling pollution control technologies, and environmental NGOs—that contribute political pressure for continued state enforcement of policies grounded on these ideas.

Exogenous forces have also reinforced these state commitments. Some degree of change is due to the emergence of modern values and green political parties, but it is not clear how much. Technological change facilitated making hard political choices, but by and large the regimes creating technological change—such as with the development of CFC substitutes—rather than the reverse.

Social learning is now better understood as a path-dependent process of collective social change. New ideas are solicited by uncertain policy makers in the aftermath of well-publicized disasters or shocks, which encourage decision makers to seek guidance for responding to new and uncertain issues. Such conditions are more frequent under conditions of globalization and complexity. Through the steps described above, new ideas imparted to decision makers by epistemic community members have the effect of changing state understandings of the policy environment and of their own national interests and eventually to changes in state practices as well. These beliefs become institutionalized through the redeployment of state resources and through the consolidation of administrative practices and laws.

Where does this leave us? The implications for environmental governance are not bad. With the notable exception of climate change and biodiversity, and those involving only the poorest countries of the world, most environmental regimes are fairly effective. The ones that have generated the strongest improvements in environmental quality are those that involve epistemic communities and UNEP.

More fundamentally, we have seen a profound change in how states recognize their own interests. Before the 1970s states were essentially parochial, focusing primarily on conditions within their own territories, and asserting sovereignty over decisions that would affect them at home. But following this era of social learning, states have come to recognize that domestic conditions are contingent on conditions elsewhere in the world. Thus economic planning has been altered to reflect this new awareness of how state interests are changed. Whereas in the past states essentially pursued the goals of wealth and power, they now qualify those goals in light of their environmental effects, and also pursue, to some extent, ecological integrity as a core national interest.

The future of effective environmental governance seems to revolve around two things. One is the future of UNEP itself. The second has to do with the enduring social legitimacy of science, which is being questioned by civil society, and at least the ideal type of 19th century glorified impartial science is being disentangled by the social studies of science.

For international relations theory I would be remiss if I didn't end with the argument that we cannot explain the dramatic changes in the last 30 years of environmental governance without invoking the changes of perceptions that have led to new practices and more vigorous multilateralism. Change occurs through social learning. Karl Deutsch would not have been overly surprised.

Note

References

Bodansky, D., B. Juttaa, et al. (2007). *The Oxford Handbook of International Environmental Law.* Oxford, Oxford University Press.

Haas, P. M. (2001). "Epistemic Communities and Policy Knowledge." *International Encyclopedia of Social and Behavioral Sciences*: 11578–11586.

Haas, P. M. (2004). "When Does Power Listen to Truth? A Constructivist Approach to the Policy Process." *Journal of European Public Policy* 11(4): 569–592.

Haas, P. M. and E. B. Haas (2002). "Pragmatic Constructivism and the Study of International Relations." *Millennium: Journal of International Studies* 31(3): 573–601.

INDEX

Note: *Italic* page numbers indicate tables; **bold** indicate figures.